Optical Wireless Communications for Broadband Global Internet Connectivity

Optical Wireless Communications for Broadband Global Internet Connectivity

Fundamentals and Potential Applications

Arun K. Majumdar

ELSEVIER

Elsevier
Radarweg 29, PO Box 211, 1000 AE Amsterdam, Netherlands
The Boulevard, Langford Lane, Kidlington, Oxford OX5 1GB, United Kingdom
50 Hampshire Street, 5th Floor, Cambridge, MA 02139, United States

Notices

Knowledge and best practice in this field are constantly changing. As new research and experience broaden our understanding, changes in research methods, professional practices, or medical treatment may become necessary.

Practitioners and researchers must always rely on their own experience and knowledge in evaluating and using any information, methods, compounds, or experiments described herein. In using such information or methods they should be mindful of their own safety and the safety of others, including parties for whom they have a professional responsibility.

To the fullest extent of the law, neither the Publisher nor the authors, contributors, or editors, assume any liability for any injury and/or damage to persons or property as a matter of products liability, negligence or otherwise, or from any use or operation of any methods, products, instructions, or ideas contained in the material herein.

Library of Congress Cataloging-in-Publication Data
A catalog record for this book is available from the Library of Congress

British Library Cataloguing-in-Publication Data
A catalogue record for this book is available from the British Library

ISBN: 978-0-12-813365-1

Publisher: Susan Dennis
Acquisition Editor: Anita A Koch
Editorial Project Manager: Michelle Kubilis
Production Project Manager: Denny Mansingh
Cover Designer: Christian J.Bilbow

Typeset by TNQ Technologies

This book is dedicated to the students, researchers, and entrepreneurs all over the world who might be inspired by, and "true believers" of, the unlimited possibilities created by global access to Internet connectivity that provides access for all.

Contents

Biography

Arun K. Majumdar, PhD

Dr. Majumdar has demonstrated leadership skills in Research and Development for the last 29 years from industry, university, and national laboratory settings in the areas of atmospheric turbulence effects on free-space laser communications, propagation, and imaging. His current research areas are free-space laser communications and advanced FSO communications, image corrections, and communications through atmospheric turbulence. His previous significant professional experience includes:

U.S. Department of the Navy (Ret.) (senior scientist/principal investigator), University of Colorado (full professor, Electrical Engineering and Computer Science Dept.), senior research scientist at Lockheed-California Company, NIST (visiting professor), Caltech's Jet Propulsion Laboratory, MIT Lincoln Laboratory (staff member), Brno University of Technology, Brno, Czech Republic (visiting professor), editor-in-chief of JOFCR (Journal of Optical and Fiber Communications Research, Springer, New York (2009–11)), and editor of the SPIE Proc. (2006–13). He has published over 60 refereed journal and conference papers and technical reports, and serves as reviewer for IEEE, OSA, SPIE, Springer, and Elsevier journals. He was the co-chair of the SPIE's conferences of Free-Space Laser Communications until 2013. He has been invited to provide seminars/short courses in various organizations, some of which include SPIE; The Boeing Company; BAE Systems; UCLA; Colorado State University at Pueblo, Colorado; University of California at Riverside, California; Gujrat University, NITK, Kolkata, India; Tripura University, India; Indian Institute of Science (IISc) Bangalore, India; University of Dayton, Ohio; and Oxford University, UK.

Published books:

- *Advanced Free Space Optics (FSO): A Systems Approach*, Springer: NY, 2014. http://www.springer.com/physics/optics+%26+lasers/book/978-1-4939-0917-9

- *Free-Space Laser Communications: Principles and Advances,* with co-author Jennifer C. Ricklin Springer-Verlag: NY, 2008. http://www.springer.com/west/home/default?SGWID=4-40356-22-70197108-0

Books under contract:

- *Optical Wireless Communications for Broadband Global Internet Connectivity,* to be published 2018.
- *Principles and Applications of Free Space Optical Communication,* Ed., Institution of Engineering and Technology, UK, to be published 2018.

ABOUT THE AUTHOR

Professor Arun K. Majumdar is a research scientist and a former university professor. In the past, he has worked in the optical communications and research industries for over 30 years, and was the Director of Research at his own company, LCResearch, Inc. in California. He served as a full professor at the University of Colorado for 5 years, and served as a visiting professor at NIST (Colorado), Brno University of Technology in the Czech Republic, University of Dayton (Ohio), and Tripura University (India). Professor Majumdar has published over 60 articles in international scientific peer-reviewed journals, has written over 25 technical reports, and is the author/editor of two other published books in Free -Space Laser Communications and Free Space Optics (FSO). He has been a member of IEEE, OSA, and SPIE.

He received his PhD in Electrical Engineering from the University of California, Irvine.

Affiliations and Expertise

Previously with U.S. Department of the Navy (Ret.), China Lake, CA, USA.

Preface

The author's recent books have paved the way to address the immediate next challenge and today's need for providing practical solutions and systems design for global Internet connectivity. The book provides a solid foundation for optical wireless communication (OWC), which is an emerging area where interests and research activities are growing. This book has been designed to provide a comprehensive, unified tutorial to further understand the fundamental issues of OWC through the Earth's atmosphere, terrestrial, and space. The driving force behind this rapid growth is the advanced development of free-space optical (FSO) technology to provide high-bandwidth information with high capacity compared to existing radio frequency (RF) communication technology. FSO will be one of the most unique and powerful tools to address connectivity bottlenecks that have been created in high-speed networks during the past decade due to the tremendous success and continued acceptance of the Internet. The next generation of Internet connectivity will push the limits of existing infrastructure with high-bandwidth applications such as videoconferencing, streaming multimedia content, and network-enabled portable devices. Clearing these bottlenecks is crucial for the future growth and success of the contemporary Internet society. The bandwidth of optical communications access and edge networks will absolutely be needed to satisfy these demands.

Simultaneously, the global Internet demand specifically for global broadband connectivity is extremely timely for combining FSO with other wireless technologies. For example, modular advanced communication systems are considered, which will allow worldwide access to the Internet by combining satellite communications, airborne/balloon platform, FSO wireless local area network (LAN), local multipoint distribution system (LMDS), and terrestrial digital video broadcast (DVB-T). The world's communication trend now is toward mobile users where wireless LAN offers uninterrupted connectivity in a network cell, while FSO allows established broadband wireless links through air and satellite communications in order to provide access to fixed and mobile services, and is a key element of the Internet backbone between various locations worldwide to close the Internet connectivity loop between any two as well as among various distant locations.

This book has been designed to provide a comprehensive, unified tutorial for further understanding of the fundamental techniques to achieve this goal. The motivation for writing this book started when the author saw how rapidly Internet use was growing, and that no comprehensive book currently addressed the fundamental technology needed to satisfy today's demand. The author's recently published books, *Free-Space Laser Communications: Principles and Advances*, and *Advanced Free-Space Optics (FSO): A Systems Approach*, both published by Springer, NY, provide an excellent background in both fundamentals and advanced knowledge of the state-of-the-art in the technologies available today. The books not only discuss in detail the subject matter, but also clearly point out the challenges. This book starts with these challenges and discusses the technologies for OWC as a practical solution for creating a three-dimensional global broadband communication grid with Internet access *anytime, anywhere* offering bandwidths far beyond possible RF range and capacity.

The writing of this book is particularly timely because of the following current facts:

1. The explosion of growth of demand for global Internet access requires a thorough understanding of the subject, from fundamentals to the advanced state-of-the-art applications, including those that are faster and compact, which are available today.
2. A recent success in 2014 demonstrating the accomplishment of optical communications for transferring data at the rate of 50Mb/s from a low-earth orbit (LEO) to a ground station brings confidence in developing potential space-to-ground optical communication architectures that will be extremely important to offer multi-FSO–based technology for multigigabits data transfer potential for future fast global Internet.
3. An emerging topic, the Internet of Things (IoT) for a *more connected world* has captured the world's imagination, raising billions of dollars needed to understand deeply and consider how all these *Things* should connect to the *Internet*.

There is thus a pressing need for a textbook on this topic. Some excellent books are already published in the area on indoor or outdoor optical wireless systems. There is no one single book that covers and combines the fundamentals and the most recent advanced technology of achieving global Internet access and connectivity in a concise but comprehensive manner as explained in this book. The FSO techniques and technologies are growing at much higher rate than covered by any book. This book is intended as an *all-inclusive source* to serve the needs of those who require information about the fundamentals of OWC as well as up-to-date advanced knowledge of the state-of-the-art of technologies available today. All these issues are addressed as a systems approach in this book to learn about the exciting potential of FSO: hence the book writing!

The reader will find all the important topics of an OWC system in one place with in-depth coverage. Each chapter will have a section on fundamentals (with necessary derivations of the equations), theory of operation, and recent developments with examples. The reader will benefit from the systems approach taken in this book since many of the conference proceedings and journal articles focus mainly on the author's research work without spending space on preliminaries and derivations. There is continuity between chapters so that the information and learning from any one chapter is connected to that of the other chapters.

This is the first book, which will be one of the most coherent and comprehensive books available to date, in the area covering advanced OWC/FSO technology with state-of-the-art applications. Many numerical examples for different applications are provided to help the readers design their own systems. This book, combined with *Free-Space Laser Communications,* will provide a complete in-depth understanding of this subject with up-to-date knowledge of the most recent applications, and could be used as a textbook, a reference, and research text.

This the first book that describes a proposed OWC/FSO communication-based system that combines FSO, satellite-based optical communications (providing a high-speed backbone between distant locations worldwide), and wireless LAN that operates under realistic atmospheric conditions, explaining the unique advantages of each of the technologies in order to develop the most flexible broadband communication system offering broadband access to the rest of the world without any geographical and terrestrial infrastructure.

The specific unique features are:

1. Contains a complete comprehensive treatment of a proposed combined system of FSO, satellite-based optical communications (providing a high-speed backbone between distant locations worldwide), and wireless LAN in order to develop the most flexible broadband communication system offering broadband access to the rest of the world *anytime, anywhere* without any geographical and terrestrial infrastructure.
2. Provides solutions and architectures to solve a secure global Internet connectivity problem with the integration of a number of different techniques to be able to handle indoor, outdoor, and terrestrial and space links for successful secure global Internet connectivity.
3. Provides two chapters discussing the recent developments of first satellite-based laser communication: Demonstration of successful LEO-to-ground optical communication experiments and optical communication relay

for LEO to geostationary equatorial orbit (GEO) and GEO-to-ground (input from JPL/NASA: discussions with Dr. Malcolm Wright on their research results in this area).

4. Discusses small-cells approach for bidirectional communications to solve backhaul communication, LED for Wi-Fi to increase 100 times speed, and Li-Fi, which refers to visible light communications (VLC) technology to enhance any existing Wi-Fi as an element of potential broadband solutions.
5. Discusses IoT relevant to OWC/FSO.

The book analyzes basic elements of future airborne communication networks to address future high-bandwidth challenges. A proposed communication network topology includes FSO links to enable large-bandwidth access to the terrestrial Internet via LEO satellites or any suitable high-altitude airborne platform system to ground station.

Some of the topics such as wireless network protocols (LTE, Wi-Fi, Bluetooth, WiGig, Z-Wave/Zigbee), WiMAX, Wi-Fi wireless networks, and TCP/IP (transmission control protocol/internet protocol) are not covered in detail in this book since they are beyond its scope.

The various chapters within the book provide a coherent source of knowledge in this subject. The book can serve as a textbook for a graduate-level course, a reference book for professional engineers/scientists involved in the design of OWC systems, and useful material for interested researchers who intend to learn the subject.

This book provides a comprehensive overview for readers who require information about the basics of the OWCs, as well as up-to-date advanced knowledge of the state-of-the-art technologies available today. The author has tried to make this book as useful as possible as a resource for engineers, scientists, researchers, and students interested in understanding and designing OWCs system for global channels. The reader will further benefit from the knowledge of how related technologies of OWCs can be integrated into achieving worldwide Internet connectivity. There is continuity between chapters so that the information and learning from any one chapter is connected to that of the others. This book will contain a collection of information not found in any single source.

Intended for research scientists, engineers, and students with an interest in the field of OWC for solving global Internet connectivity problems, the book can be taught at a graduate level. The proposed book could be an advanced graduate text for specialized courses for students with a background in optical and microwave communications.

The book is organized into eleven chapters, each having complete entities in and of themselves.

Chapter 1 is an introduction to broadband global Internet connectivity, the need, the existing technology, and the need for FSO communications in the proposed technological plans for achieving this goal. Wireless Internet access can be delivered through mobile phone towers to computers, cell phones, and other digital devices using portable modems that can be built into laptop computers, tablets, cell phones, and other devices. This chapter discusses statistical projections for the number of Internet users worldwide by 2018. Specifically, it is important to note the following: globally, mobile data traffic will grow 11-fold from 2013 to 2018, a compound annual growth rate of 61% reaching 15.9 exabytes/month by 2018 (1 exabyte $\sim$ 1 billion gigabytes), and the total Internet video traffic will be 75% of all Internet traffic in 2018, up from 57% in 2013 and finally, broadband speed will triple by 2018. This chapter explains why OWC is a viable technology to solve these huge demands for Internet access as well as the recent development of the relevant technology developments such as LED-based Wi-Fi and Li-Fi using VLC. Providing Internet access to more people in the world will allow taking advantage of all the opportunities available over the Internet. The UN reports and declares *Internet access a human right.*

Chapter 2 discusses how FSO communication/OWC is one of the key technologies for realizing very high speed multi-Gb/s large-capacity terrestrial and space communications. The author's recent books on *Free-Space Laser Communications and Free-Space Optics (FSO)* published by Springer have already paved the way to address the immediate next challenge and today's need for providing practical solutions and systems design for global Internet connectivity. The chapter provides and explains the basics of the modular advanced communication systems, which will allow worldwide access to the Internet by combining satellite communications, airborne/balloon platform, FSO wireless LAN, LMDS, and DVB-T. The chapter describes in detail the architectures for all of these scenarios, and will show how to establish bidirectional (duplex) mobile global Internet connectivity. The world's communication trend now is undoubtedly toward mobile users where wireless LAN offers uninterrupted connectivity in a network cell, FSO allows established broadband wireless links through air, and satellite communications provide access to fixed and mobile services and is a key element of the Internet backbone between various locations worldwide to close the Internet connectivity loop between any two or among various distant locations. The evolving technology is emerging to provide tens to hundreds of gigabits/sec (Gb/s) or more OWC-based Internet connectivity for global broadband Internet access to *anytime, anywhere.*

This chapter provides an in-depth treatment of the critical issue of limitations imposed by atmospheric effects and analyzes the basic physics of terrestrial, aerial, and space connectivity, the basic understanding of existing connectivity using FSO technology with various platforms including satellites for global coverage as well as aerial platforms such as drones, high-altitude balloons, with proper optical networks. LEO, microsatellites, or fixed high-altitude platform (HAP) stations can provide high-bandwidth relay services or across to the terrestrial Internet. An extension of this communication technology will include unmanned air vehicles (UAVs), which will be a major part of the communications-on-the- move part of the FSO broadband communication architecture.

Chapter 3 points out that the so-called last-mile problem (or last-mile bottleneck), which is to connect the high bandwidth from the fiber optic backbone to all of the business with high bandwidth networks, is not quite solved as of today, as promised over the last few years even if the backbone is realized with fiber. Less than 5% of all buildings in the United States have a direct connection to the very high speed (2.5–10 Gbps) fiber optic backbone, yet more than 75% of businesses are within 1 ... mile of the fiber backbone. Most of these businesses are running some high-speed data network within their building, such as fast Ethernet (100 Mbps) or Gigabit Ethernet (1.0 Gbps). The chapter provides an architecture showing the *small-cells* concept to develop basic technology blocks to achieve *all-weather 99.999% ("five-nines") availability* based on an analysis of link budgets and visibility-limiting weather conditions (even in the worst case of dense fog!) indicating to meet carrier-class availability (99.999%). Examples show FSO links to be normally be less than 140 m (there are cities like Phoenix and Las Vegas where this 99.999% distance limitation increases significantly). The design architecture describes a hybrid concept combining small FSO cells with high-speed Wi-Fi to achieve the broadband connectivity for the last-mile problem. Finally, with the small-cells approach for bidirectional communications to solve backhaul communication, LED for Wi-Fi to increase 100 times speed and Li-Fi with VLC visible light communications have been explained for potential broadband solutions.

Chapter 4 discusses the basics of the OWC system and subsystem structure, which includes optoelectronic components (transmitter/receiver), OWC channel model, modulation for optical wireless channel, communication parameters for evaluating performance, link budget and OWC performance: bit error rate (BER), and signal-to-noise ratio (SNR). The chapter then presents OWC network and access technology, providing an up-to-date introduction to optical networks, and introduces optical wireless technology specifically addressing the network aspect into the global communications infrastructure. The chapter explains how OWC systems play an important role as broadband access technology where directly propagating an optical beam from a fiber termination points through free-space using a full-OWC transceiver. The technology explains how at the receiving end a corresponding transceiver can directly couple the free-space propagated optical beam either into an optical fiber connection port or into another free-space optical port similar to a relay system. The global connectivity can thus be developed. The chapter presents a basic design concept of full-OWC link with a potential of offering stable, error-free transmission at multiple Gb/s data rates globally. The chapter describes the basics of indoor OWC using multiple sources including the VLC-based multiple-input-multiple-output technique to provide reliability improvement and bandwidth efficiency increase. Communication blocks in an indoor OWC system include intensity modulation/direct detection, optical transmitters and receivers (VLC and LED arrays and fly eye receivers), optical wireless channel propagation characteristics and channel modeling, channel impulse response, system performance (BER), and outage for indoor OWC systems, orthogonal frequency division multiplexing for Indoor OWC and aircraft (commercial airplane) on-board wireless communication. The chapter analyzes overall FSO link performance, and theory and analysis are provided based on the fact that the physics of optical wave propagation are entirely different for different OWC scenarios and environments. Providing a comprehensive theory of optical propagation, this chapter shows some typical examples that are useful for OWC engineers who want to develop end-to-end system design for achieving Internet connectivity. Mitigation techniques for designing atmospheric OWC systems are also discussed. Eye-safety issues for limiting power of a source (transmitter) for indoor and outdoor environments are clearly pointed out.

Chapter 5 will introduce a big picture for establishing global Internet connectivity using *all-optical* connectivity, which includes both FSO and some portion with broadband fiber optics. An all-optical network concept for establishing global Internet connectivity including remote locations is presented and discussed relating to the trends in FSO communication technology development. Based on the recent successful demonstrations of achieving optical links between a satellite terminal and the ground station using laser satellite this chapter establishes the potential of future high data rate communication *anytime, anywhere*. This chapter explains communication based on integrating the constellation of satellites, small and nanosatellites, HAPs, UAVs, and balloons, which show a clear pathway for the real possibility of establishing multigigabit-level Internet connectivity all over the world in almost any totally remote location. On the home-based and even street levels the newest Li-Fi technology will play an extremely important role for providing Internet hotspots for both fixed and mobile terminals, which will be all connected eventually to the global Internet gateways and optical nodes. This global Internet connectivity will be only possible if airborne, satellite, and fixed terrestrial terminals can all be integrated with optical devices and laser transceivers.

Chapter 6 discusses how free-space laser communication systems have the potential to provide flexible, high-speed connectivity suitable for terrestrial long-haul satellite-to-ground and intersatellite links. Recent advances in fast, compact, optical, and electrooptical technologies can be leveraged to design and develop global wideband fiber networks with multiple Tbit/sec capabilities which can eventually pave the pathway to establish all-optical broadband global Internet connectivity *anytime, anywhere*. This chapter describes the successful demonstrations of space-to-ground and airborne links for achieving high data rate FSO communications. Satellite (GEO and LEO), small satellites, UAV/airplane, and HAPs are considered discussing different communication channel propagation characteristics for each of the scenarios. Space-to-ground FSO communications links have been successfully demonstrated from several platforms over the last two decades (Geolite, GOLD, NFire, Russian ISS, OPALS, OICETS, Spot, LLCD, etc.), even culminating in an operational system deployed by ESA (AlphaSat). The challenges of link acquisition from fast slewing spacecraft in LEO configurations and beam breakup from atmospheric scintillation effects have been overcome in a variety of experiments that not only validate the hardware but also the concept of operations for moderate data rates of tens of Mbps up to several Gbps. The combined optical links of satellites, UAVs, and HAPs operating simultaneously can truly have the potential for establishing Internet connectivity worldwide.

Chapter 7 explains and discusses various advanced techniques used to achieve the critical security required in communication for FSO links on ground and space to satisfy the increasing demand in data throughput. In order to achieve secure communication, the first step is to develop encryption, which prevents unauthorized access to transmitted information. Starting with conventional computationally secure encryption, information-theoretic secrecy, chaos-based, and then quantum cryptography offer progressively higher levels of security. Chaos key distribution and quantum key distribution (QKD) are both powerful, depending on the extent of applications, and QKD is the most powerful adversary allowed by physics. This chapter provides different system architectures, approaches, and most experimental results on (1) acousto-optics chaos-based secure FSO communication links, (2) chaotic technique implementation of secure atmospheric optical wireless OWC/FSO communication, (3) quantum Internet for global use using free-space and atmospheric quantum communications to teleport quantum information, and (4) N-slit interferometry-based secure satellite-to-satellite optical communication.

Chapter 8 describes the concept of IoT, wherein machines and everyday objectives can be connected via the Internet. In order to satisfy the requirements of connecting using IoT to more than 75 billion networked devices by 2020, the chapter discusses the role of OWC/FSO in the goal of IoT to enable things to be connected *anytime, anywhere,* with anything and anyone. The chapter provides potential architecture of OWC/FSO integration as an integral part of IoT and machine-to-machine communication to meet the need for fast, high-capacity networks and achieving global IoT. Finally, a hybrid terrestrial/satellite as well as terrestrial/constellation of satellites into the IoT for seamless and ubiquitous communications for IoT are explained.

Chapter 9 discusses how the Internet is now really a part of our social fabric and is essential to how we connect, communicate, share, and collaborate. Digital technologies, information and communication technologies, and broadband global connectivity when combined offer various opportunities to improve people's lives in worldwide societies by exchanging and sharing information. Potential applications of broadband global connectivity include social connections; health, well-being, and telemedicine; education; business and financial transactions; emergency response; and search and rescue. These all are possible with Internet access and are discussed.

In Chapter 10, technology based on free-space and fiber-based large bandwidth optical communications in advancing all-optical concepts technology for creating global Internet connectivity and access to provide access for all is discussed. This chapter discusses some of the recent research efforts and points out some future directions. All-optical networks and advanced photonics materials and components that will be useful in establishing future global Internet connectivity and access are discussed.

Chapter 11 provides conclusions and discussions on how all-optical concepts to establish global Internet connectivity and access at high-speed described in this book, covering all aspects of fundamentals of FSO communications and high bandwidth fiber-optics backbones to design optical links suitable for various scenarios in a number of applications. The concepts of the technology proposed in this book are the first step toward accomplishing the ultimate goal of providing Internet access for all, and the next step in our development is to apply the concepts and the proposed optical link schemes to build true all-optical global Internet connectivity systems to have Internet access for all, *anytime, anywhere.*

I want to thank Dr. Anita Koch, Acquisition Editor, Physics & Astronomy Books, Elsevier, Amsterdam, The Netherlands for initiating this book project and providing advice, comments, and inspiration. Also I wanted to thank Peter Jardim, Editorial Project Manager, Science & Technology Books, Elsevier, Inc., Cambridge, MA, USA, for many helpful business details in preparing this book. Also I wanted to thank Denny Mansingh, Project Manager, Elsevier Global Book Production for efficiently organizing my book production process. I would like to express my sincere thanks to Dr. Rajarshi Roy (University of Maryland), Dr. Monish Chatterjee (University of Dayton), Dr. William Brown (Colorado State University—Pueblo), Dr. Malcolm Wright (California Institute of Technology, JPL), Dr. Timothy Brothers (Georgia Institute of Technology), and Steve Lundin for valuable technical discussions and suggestions throughout the preparation of this book.

Over the years, I have benefitted from technical discussions and informal conversations with many colleagues in the academic, research, and applied areas of FSO communication, for which I am gratefully indebted.

Finally, I wish to thank my wife, Gargi, for her support, and both my daughter Sharmistha and her newly married husband Nehemiah Benjamin for their constant inspiration throughout the preparation of the book, helping to create an optimum environment with encouragement in completing my book under extreme adverse conditions during my critical writing.

Arun K. Majumdar
San Diego, California

CHAPTER 1

Introduction

1.1 INTRODUCTION

Chapter 1 is an introduction to broadband global Internet connectivity: the need, the existing technology, and the need for free-space optical (FSO) communications in the proposed technological plans for achieving this goal. Wireless Internet access can be delivered through mobile phone towers to computers, cell phones, and other digital devices using portable modems that can be built into laptop computers, tablets, cell phones, and other devices. This chapter discusses statistical projections for the number of Internet users worldwide by 2018. Specifically, it is important to note the following: globally, mobile data traffic will grow 11-fold from 2013 to 2018, a compound annual growth rate of 61% reaching 15.9 exabytes/month by 2018 (1 exabyte ~1 billion gigabytes); the total Internet video traffic will be 75% of all Internet traffic in 2018, up from 57% in 2013; and finally, broadband speed will triple by 2018. This chapter explains why optical wireless communication (OWC) is a viable technology to solve these huge demands for Internet access as well as the recent development of the relevant technology developments such as LED-based hotspots to replace Wi-Fi and Li-Fi using visible light communication. Providing Internet access to more people in the world will allow taking advantage of all the opportunities available over the Internet. The UN reports and declares *Internet access a human right.*

It is therefore obvious that the demand for bandwidth in communications, either fixed platform-based or mobile, is growing as the numbers of users worldwide is increasing significantly. Future increasing demand will continuously grow higher in the next decade for high-quality multimedia services and for developing more efficient access networks addressed by a number of technologies to meet the end users' communication needs. Some of the applications where higher capacity to support various broadband wireless services are necessary include high-definition TV, broadband computer network applications, mobile videophones, video conferencing, and high-speed Internet access. Optical wireless communication can offer possible solutions to provide high-speed Internet access for broadband services to locations everywhere, including remote places. This chapter gives an overview of the book, which

Optical Wireless Communications for Broadband Global Internet Connectivity.
https://doi.org/10.1016/B978-0-12-813365-1.00001-1

addresses the problem of all-optical technology for establishing global Internet connectivity and access using FSO communication and high-bandwidth, fiber-based solutions. FSO must be included in the set of solutions for meeting the bandwidth requirements of the modern Internet. FSO communication is the most practical alternative to solve the bottleneck broadband connectivity problem, and as a supplement to conventional radio frequency/microwave links. FSO is recently very active with many exciting fundamental and technological challenges to improve its performance in a range of scenarios. The data rate provided by FSO links continues to increase in both long- and short-range applications. FSO will be one of the most unique and powerful tools to address connectivity bottlenecks that have been created in high-speed networks during the past decade due to the tremendous success and continued acceptance of the Internet. The next generation of Internet connectivity will push the limits of existing infrastructure with high-bandwidth applications such as videoconferencing, streaming multimedia content, and network-enabled portable devices. Clearing these bottlenecks is crucial for the future growth and success of the contemporary Internet society. The bandwidth of optical communications access and edge networks will be needed to satisfy these demands. A scheme of FSO communication technology with optical wireless and different user connectivity involves encoder, optical source, and atmospheric channel (or fiber-optics channel), optical receiver (photodetector), and decoder [1,2].

Continued global demand for increased bandwidth and security issues, free-space optical, and fiber-based access networks are the only viable solutions to offer high-capacity Internet services to the end users. OWC is a promising technology for indoor and outdoor applications for voice and data, video and entertainment, business connectivity, disaster recovery, and data communications, and will play a significant role in future information superhighways.

A number of chapters mainly discuss the free-space optical connectivity in today's network. FSO communications refers to a line-of-sight technology transmitting modulated visible or infrared beams through the atmosphere to establish optical communications. FSO has been considered one of the most viable technologies for next-generation broadband communicating and wireless networks. The greatest disadvantage of FSO is the weather. When the communication link includes part of the atmosphere, clear-air turbulence and a possible boundary-layer turbulence (e.g., one end of the communication link is on an aircraft) cause serious phase distortions and fading to the link. FSO networks may also suffer because of insufficient availability and low reliability due to weather turbulence. Weather conditions such as fog, smog, snow, rain, and dust particles affect FSO communication performance.

This chapter gives an overview of the optical communications technologies to develop schemes for indoor, outdoor, terrestrial, and space applications and to address the immediate next challenge and today's need for providing practical solutions and systems design for global Internet connectivity. The chapter provides and explains the basics of the modular advanced communication systems that will allow worldwide access to the Internet by combining satellite communications, airborne/balloon platform, FSO wireless local area network (LAN), local multipoint distribution system, and terrestrial digital video broadcast. The chapter describes in detail the scenarios, and mentions the need for establishing bidirectional (duplex) mobile global Internet connectivity. Whereas wireless LAN offers uninterrupted connectivity in a network cell, FSO allows established broadband wireless links through air, and satellite communication provides access to fixed and mobile services and is a key element of the Internet backbone between various locations worldwide to close the Internet connectivity loop between any two or among various distant locations. The evolving technology is emerging to provide 10s to 100s of gigabits/sec (Gb/s) or more OWC-based Internet connectivity for global broadband Internet access *anytime, anywhere.* The ultimate goal of developing various communications technologies is to develop innovative Internet for remote areas for establishing broadband connectivity *anytime, anywhere,* which eventually should be cost effective and affordable for everybody. In order to achieve this goal to meet the growing demands of Internet access everywhere a number of most recent advanced technological developments worldwide are described, which are independent but closely related to Internet connectivity. Global *all-optical* broadband network around the world has the potential solution. Various broadband optical communication technologies and scenarios are necessary to determine conceptual designs and architectures. Some of them can be summarized as follows, and each one of them is described in detail in different chapters:

1. Terrestrial optical communication from one user to another user, including horizontal and slant paths
2. Longer distance terrestrial and connecting to the Internet (access points to optical tower via LAN, connection to fiber backbone: FSO communication as a solution for last-mile problem
3. Maintaining Internet connectivity on the fly
   a. Mobile ad-hop platforms and communications, concept for optical antenna, retroreflector for automatic alignment
   b. Small cell solution: All-optically
4. Optical tower to optical communication satellite
5. Optical ground station to optical communication satellite
6. Optical ground station to various optical access points/optical towers

7. Remote locations via optical communication satellite
    a. Low earth orbit (LEO) to other LEOs, or between small satellites
    b. LEO to airplane or geosynchronous earth orbit (GEO) to airplane platform
    c. Airplane platform to ground optical tower/cells
    d. GEO to LEO
    e. GEO to another GEO
    f. Constellation of satellites, small satellites, and CubeSat
8. Remote locations via drones, balloon, high altitude platforms (HAP) for optical communication
9. Internet access for Internet of Things devices to integrate with 5G communication networks
10. Additional innovations in design
    a. Li-Fi—based hotspots to replace Wi-Fi in future
    b. Modulated retroreflector based on my patent for achieving two-way (duplex) communications
    c. Chip-based compact optical devices such as optical router, optical switching to be integrated with the optical networks systems for Internet accesses by smartphones
11. Quantum entanglement and communications for achieving the most secure Internet communication links over a distance

The Internet is now really a part of our social fabric and is essential to how we connect, communicate, share, and collaborate. Digital technologies, information and communication technologies, and broadband global connectivity when combined offer various opportunities to improve people's lives in worldwide societies by exchanging and sharing information. Some of the potential applications of broadband global connectivity include social connections, health, well-being and telemedicine, education, business and financial transaction, emergency response, and search and rescue. These are all possible with global Internet access and connectivity. Some of the current advanced research relevant to establishing all-optical connectivity for future optical communications includes the following: fast luminescent detector, nonlinear silicon photonic, high-speed photonics components, plasmonic device, graphene-based technologies, on-chip-twisted light emitter, and quantum Internet.

References

[1] A.K. Majumdar, Advanced Free Space Optics (FSO): A Systems Approach, Springer, New York, 2015.

[2] A.K. Majumdar, J.C. Ricklin, Free-space Laser Communications; Principles and Advances, Springer, 2008.

CHAPTER 2

Basics of Worldwide Broadband Wireless Access Independent of Terrestrial Limitations

2.1 INTRODUCTION

The free-space optical (FSO) communications/optical wireless communication (OWC) is one of the key technologies for realizing very high speed multi-gigabits/sec (Gb/s) large-capacity terrestrial and space communications. The author's recent books on free-space laser communications and advanced FSO published by Springer have already paved the way to address the immediate next challenge and today's need for providing practical solutions and systems design for global Internet connectivity. The chapter provides and explains the basics of the modular advanced communication systems that will allow world-wide access to the Internet by combining satellite communications, airborne/balloon platform, FSO wireless local area network (LAN), local multipoint distribution system (LMDS), and terrestrial digital video broadcast. The chapter describes the scenarios in detail, and mentions the need for establishing bidirectional (duplex) mobile global Internet connectivity. The world's communication trend now is undoubtedly toward mobile users where wireless LAN offers uninterrupted connectivity in a network cell, FSO allows established broadband wireless links through air, and satellite communications provide access to fixed and mobile services, and is a key element of the Internet back-bone between various locations worldwide to close the Internet connectivity loop between any two or among various distant locations. The evolving technology is emerging to provide tens to hundreds of Gb/s or more OWC-based Internet connectivity for global broadband Internet access to *anytime, anywhere.*

This chapter discusses the fundamentals of engineering aspects of free-space laser/FSO communications protocols and describes technology concepts to establish worldwide broadband access to the Internet or other networks independent of terrestrial infrastructure. The technology will involve a combined satellite communications, FSO, and wireless and LAN in the following way: (1) wireless LAN offers connectivity in a network cell to either mobile, fixed, or both users, and LMDS, (2) FSO can be used for installation

Optical Wireless Communications for Broadband Global Internet Connectivity.
https://doi.org/10.1016/B978-0-12-813365-1.00002-3

of broadband fixed wireless links instead of a cable (e.g., to establish connectivity between the end point of a broadband fiber optic backbone and the nearest optical cells, thus addressing the so-called last-mile problem, to be discussed later in this book), and (3) satellite communications can bridge the gap of providing a backbone between remote locations on the world. In order to maintain the overall global system operability, all these systems must work together. The overall technology to achieve this task obviously involves different technologies since communications scenarios are totally different for each of these technologies, and involve different signal propagation phenomena (depending on terrestrial atmosphere including turbulence and scattering media, free space propagation, techniques of maintaining connectivity without loss of signals, availability of access points in remote locations, etc., to name just a few).

We live in a constantly changing world with continuous technological advances, progress and development of new telecommunications resulting in online environment from Internet technologies, telecommunications, and the media and entertainment industries. The next generation of access networks will have the potential to contribute to the improvement of all aspects of broadband technology and broadband services with the necessary speed and capacity to deliver the content with high resolution, maintaining the speed and demanding different applications for broadband connections. One of the challenges in achieving the goal to provide global *broadband* Internet connectivity access is still to develop basic network infrastructure from urban to some of the extremely remote rural locations. Therefore a combination of FSO and satellite-based optical connection will be the only viable solutions for providing the technologies that can eventually be affordable, economical, and thus cost-effective, keeping in mind the different requirements for developing technological infrastructures may be different for different population densities.

2.1.1 Need and Geographic and Socioeconomic Challenges to Achieve Global Access to the Internet

Access to the Internet has become both faster and relatively cheaper, offering novel applications and services that results in a critical infrastructure enabling employment, education, health care, digital economy, social networks, and much more. The world is becoming smaller and smaller in this respect, and therefore Internet access should be *universal* both in terms of availability and ability to be a part of and to participate in the evolving wider community, which enables true digital inclusion. However, there is increasing geographic and socioeconomic challenges in providing Internet access to those who cannot afford access to its services. Current technologies exist and are available today to achieve global Internet access, but the question remains at what cost is it affordable for everybody in very remote locations? This chapter will attempt to develop innovative Internet for remote areas and provide concepts for establishing broadband Internet connectivity with current technologies *anywhere, anytime* that hopefully are eventually cost-effective and affordable for everybody.

In the United States, lawmakers want to advance high-speed Internet funding and stakeholders from the federal government to the municipal level are urging the Trump administration to include broadband in its $1 trillion infrastructure plan. This includes a robust infrastructure bill for broadband. "Broadband Internet access is necessary infrastructure, and key to prosperity," the mayors and municipal leaders from various parts of the country and part of Next Century Cities wrote. "It empowers entrepreneurship and economic growth, arms our teachers and students for success in the classroom, and gives our citizens a voice in the national dialog on our future." People don't currently have broadband (and computers are really slow) in rural areas where it hasn't been profitable to build there. Now that we have 5G on the horizon, it will be more possible for each those areas [1]. The trend for developing robust infrastructure in rural and remote locations is clearly evident. This chapter will show the readers a systematic way to build the broadband global Internet connectivity in rural and extremely remote areas with all-optical concepts.

2.2 OPTICAL WIRELESS COMMUNICATION TECHNOLOGIES

FSO communication [2] is considered to be one of the key technologies for realizing very-high speed multi-Gb/s large-capacity aerospace communications. FSO lasercom uses lasers as signal carriers and can provide a line-of-sight, wireless, high-bandwidth, communication link between remote sites. The wireless aspect of FSO lasercom has an advantage particularly in metropolitan area networks (MANs) where in cities, laying broadband fiber optic cables is expensive. FSO lasercom offers substantial advantages over conventional radio frequency (RF) wireless communications technology such as much higher bandwidth (data rates), low probability of intercept for higher security, low power requirements, and much smaller packaging to make them portable. As will be discussed later in this book, even high-speed optical/photonic devices, which can include transmitter, modulator, necessary optics to transmit and receive, as well as fast signal processing chips, Microelectromechanical systems (MEMS), and nanotechnology, can all be integrated in a very small device that eventually will be incorporated with existing smart phones, laptops, or tablets. This approach gives tremendous freedom to operate all the modern gadgets needed for Internet connectivity among various users on the fly to transmit and receive digital information *anytime, anywhere* without signal drop.

FSO is a practical solution for creating a three-dimensional global broadband communication grid. The main drivers behind the emergence of advanced FSO [3] are recent high demands in bandwidth, thanks to the recent growth of Internet usage for Internet protocol television (IPTV), voice over Internet protocol (VoIP), and YouTube and Twitter, and the immediate future and coming applications in Internet of Things (IoT). With the rapid growth of data-centric devices and the general deployment of broadband access networks, high data rates from 10 Gbps to more spectrally efficient 40 Gbps or 100 Gbps/channel and beyond is a recent trend of development. To provide the most flexibility, communication channels must be *wireless*, so there numerous challenges exist today to accomplish higher transmission rates due to degradation in the signal quality as a result of atmospheric channel impairments. Atmospheric effects such as turbulence and scattering that are important for optical communication channels from the ground to an altitude of a few km are the most dominant effects relevant to optical communications.

2.3 OPTICAL COMMUNICATIONS SYSTEM FOR GLOBAL BROADBAND ACCESS TO THE INTERNET: VARIOUS SCENARIOS

Unlike wired or fiber-optic communication channels, the concept architecture for establishing all-optical broadband Internet access developed and described in this book now includes both terrestrial atmosphere (effects of turbulence and/or scattering media) for short-range optical data links and/or free space through air or vacuum. The various scenarios are described next.

Communication between two terminals on the ground, say a few hundred meters to about 1–5 km apart, or communication terminals between a ground station and an airplane or a fixed terminal on a mountain creating slant path of a few km is one scenario. The next scenario can be an example of communication links between a ground station and drones, balloons, unmanned air vehicles (UAVs), or high altitude platforms (HAP) (initially Facebook started at an altitude of 2150 feet (0.66 km) and eventually planned for 60,000 to 90,000 feet (18–27 km)). The other scenarios for establishing communication links are between a ground station and a low earth orbit (LEO) satellite at an altitude of 2000 km (1200 mi), an international space station (ISS) in an LEO about 400 km (250 mi), and between a ground station and a geostationary equatorial orbit (GEO) (~35,786 km (22,236 mi)).

This chapter points out the critical issue of limitations imposed by atmospheric effects and analyzes the basic physics of terrestrial, aerial, and space connectivity; and the basic understanding of existing connectivity using FSO technology with various platforms including satellites for global coverage as well as aerial platforms such as drones and high-altitude balloons, with proper optical networks. LEO, microsatellites, or fixed HAP stations can provide high-bandwidth relay services across the terrestrial Internet. An extension of this communication technology will include UAVs, which will be a major part of the communications-on-the-move (COTM) part of FSO broadband communication architecture. It is therefore important to analyze the basic physics of terrestrial, aerial, and space connectivity using FSO technology with various platforms including satellites for global coverage as well as aerial platforms such as drones, high-altitude balloons, and the efficient mesh networks.

The chapter analyzes basic elements of future airborne communication networks (ACN) to address future high-bandwidth challenges. A proposed potential communication network topology includes FSO links to enable large-bandwidth access to the terrestrial Internet via LEO satellites or any suitable HAP system to ground station. Basically, the proposed technology would be used between aircraft and would connect to the global Internet via ground-based terrestrial Internet access points. LEO, microsatellites, or fixed HAP stations can provide high-bandwidth relay services across the terrestrial Internet. Free-space optical capability to aircraft can thus be realized by installing each aircraft with optical terminals that would then function as a network node, with signal regeneration and add-drop capability. In essence, the aircrafts will form a high-capacity backbone network. An extension of this communication technology will include UAVs which will be a major part of the COTM part of the FSO broadband communication architecture. Broadband FSO communication is just now reaching wide market penetration in many commercial applications because of the advanced research development in mitigating atmospheric effects and outage probabilities in improving terrestrial FSO link availability.

Geographical challenges start with the physical remote locations where installations of the required infrastructures such as communication cell towers and the associated hardwires will not be very cost-effective for Internet service providers (ISPs). In many cases, where achieving broadband connectivity with wired technologies or by extending high-speed fiber lines is either impractical or impossible, the only solution for addressing the last-mile problem are appropriate solutions between the high-speed fiber-optic backbone network and the closet Internet access point. The distances between the nearest access points and the fiber-optic backbone can be very large or remote, which does not justify the cost. Because of poor connectivity ISPs in many developing countries, routing local traffic over very expensive international links does not solve the cost problem. The next challenge are socioeconomic issues; for example, the United Nation reveals that just to have access to fixed broadband costs almost 40 to 100 times the national average. In developed countries, appropriate credit approval is also needed, as well as an often long-time commitment. These are some of the challenges for accomplishing future Internet capability to offer appropriate Internet access to all parts of society. *There is thus need for global access to the Internet for all.* The aim of this chapter is to outline and describe innovative concept architectures for broadband global Internet connectivity.

2.3.1 Establishing a Global All-Optical Broadband Network Around the World

A proposed communication network topology includes free-space optical links to enable large bandwidth access to the terrestrial Internet via LEO satellites or any suitable HAP system to ground station. Basically, the proposed technology would involve an airborne platform (e.g., balloon, UAV, drone, etc.) that would connect to the global Internet via ground-based terrestrial Internet access points, which can also be cell towers equipped with optical antennas and transceiver systems. FSO capability to airborne platform can be realized by installing each platform with optical terminals, which would then function as a network node, with signal regenerating and add-drop capability. In essence, the airborne platforms will form a high-capacity *backbone network* and will be a major part of the COTM part of FSO broadband communication architecture. The all-optical wireless communication (AOWC) described here operates by directly transmitting/propagating an optical/laser beam from a termination point of a broadband fiber optic backbone through free space using an AOWC transceiver. At the receiving end a corresponding/matching transceiver couples the free-space propagated optical/laser beam into an optical fiber connection point and demodulates the signal to recover the information. As mentioned earlier, the optical transceiver can be located at any location—ground, airplane, or satellite. The goal is to develop the concept architecture and necessary technology to establish a seamless coupling and hand off a free-space propagated optical beam to a fiber at every possible transmitting and receiving communication end. To accomplish successful Internet connectivity, the communication system must be able to seamlessly hand off from the previous optical beam whether received from different optical towers or moving platforms, or from different drones, balloons, or satellites as the case might be. This, in theory, will mean no or fewer dropouts of the signals.

There are various broadband optical communication links, including the following scenarios, which will determine conceptual designs and architectures.

1. Terrestrial optical communication ($\sim$400 m–10 km) from one user to another including horizontal and slant paths
2. Longer distance terrestrial connecting to the Internet (access points to optical tower via LAN, connection to fiber backbone; also, the last mile problem)

3. Maintaining Internet connectivity on the fly
   a. Mobile ad-hop platforms and communications, concept for optical antenna, retroreflector for automatic alignment
   b. Small cell solution: all-optically
4. Optical tower to optical communication satellite
5. Optical ground station to optical communication satellite
6. Optical ground station to various optical access points/optical towers
7. Remote locations via optical communication satellite
   a. LEO to other LEOs or between small satellites
   b. LEO to airplane or GEO to airplane platform
   c. Airplane platform to ground optical tower/cells
   d. GEO to LEO
   e. GEO to another GEO (120 degrees apart?)
8. Remote locations via drones, balloon, HAP platforms for optical communication
9. Reverse the communication process to the Internet user
10. Additional innovations in the design
    a. Li-Fi—based Wi-Fi
    b. Modulated retroreflector based on my patent for achieving two-way (duplex) communications

2.4 THE OPTICAL WIRELESS CHANNEL FOR PROVIDING BROADBAND ACCESS TECHNOLOGY: ALL-OPTICAL NETWORK ARCHITECTURE

Starting from a basic terrestrial OWC system this section will describe step-by-step a systematic procedure for establishing optical communications for both short-range (~a few to hundreds of meters) to very long range (~few thousands of km). The section will start from very basic indoor optical communications, then will discuss terrestrial horizontal FSO links—Uplink (from an optical ground station (OGS) to an airplane, balloon, UAV, or satellite) and Downlink (from satellite, UAV, balloon, airplane to ground station)—and satellites to remote locations (say in a different country). Technology and architecture concepts will also explain reverse communication links, from remote locations (in a different country back to satellite, satellite to UAV, balloons, and airplane) back to the optical ground station where the FSO link started. This is a very complex architecture not only because complex technological issues such as FSO systems and subsystems will have to work together and be compatible, but also because of logistics for maintaining the networks from end to end. One of the most important points to remember is that the network should be capable of switching from one satellite or cell tower to another where the second receiver looks on to the new beam and *seamlessly hands off*. This will ensure fewer signal dropouts.

The section will build up the complete all-optical architecture starting from very basic indoor optical wire communication technology and then move to the next and so on, ultimately establishing broadband networks around the world.

2.4.1 Some Basics of Network Architecture

In order to understand various network communication technologies that will address optical global Internet connectivity, it is important to understand some basics of network architecture: a framework specifying not only network topology, network type, and components and their functionality but also possible data communication protocols, data formats, and the services supported. Network *topologies* define how the network devices are interconnected by links to accomplish a functional communication network. The details of all of these subjects are beyond the scope of this book, and the reader can find detailed descriptions in any recent book or paper in the area of communication networks. In the discussion note that a *router* will be referred to as a network device that interconnects and relays packets from one network to another according to their destination address and thus operates at network layer and interconnects and enables data exchange between different networks. The routers thus communicate with each other and determine the best path between any two hosts and forward the data packets on those paths. A *gateway* is a network device that extends the functionality of a router according to an application layer.

2.4.2 Some Examples of Network Types

Network types discussed in this section depend on communication technology used for a specific application. For any network architecture, the major network components are their nodes and corresponding interconnecting links. For simplicity, the types of network can be classified as *broadcast* or *point-to-point*. In a broadcast network, a message sent by a node is received by all other nodes connected to the network and share the same communication medium. A typical example is a TV network where the same TV channel is delivered to all devices attached to the network, as shown in Fig. 2.1. In point-to-point networks, a source node may be connected to a destination node by multiple routes and as shown in Fig. 2.2, the same content (TV channels) is delivered to all network devices attached to the network. In a point-to-point network, a source node may be connected to a destination via multiple routes traversing multiple interconnect devices forming a communication route, also shown in Fig. 2.2.

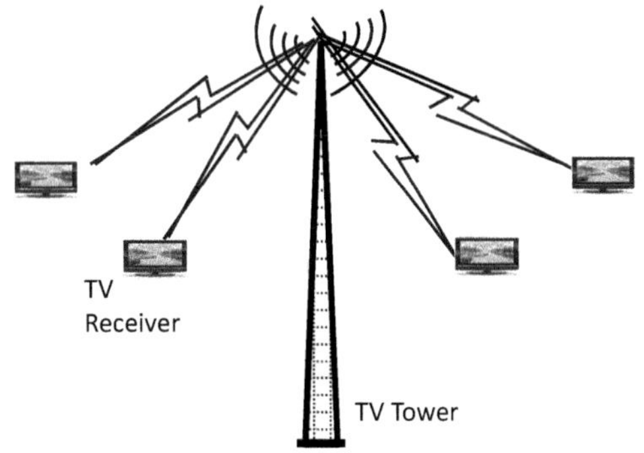

FIGURE 2.1
An example of broadcast network (TV).

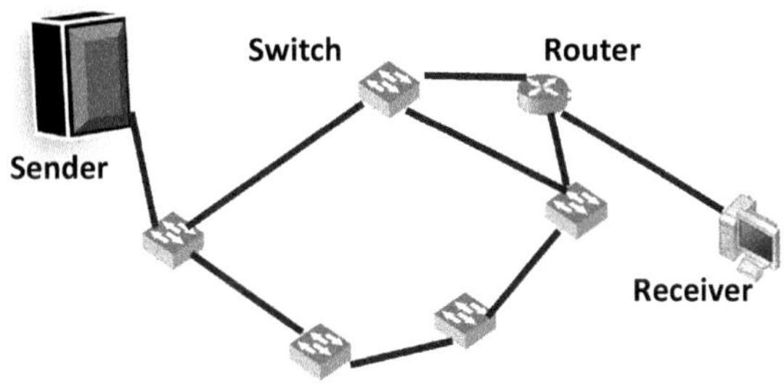

FIGURE 2.2
Message transportation from the sender to the receiver via multiple routes.

2.4.3 Network Categories

Some of the network categories include the following.

2.4.3.1 Personal Area Network

Personal area networks (PANs) are networks serving one person, covering short range (~ 1 m) transmission to link various peripherals such as printers, computer keyboard, and mouse. Wireless transmission for operating smart phones, video cameras, remote control units for TVs, stereos, and such will also fall under this PAN. In fact, most electronic entertainment items are operated with remote controls and smart phones and Bluetooth wireless devices are part of wireless personal area networks (WPANs). This is shown in Fig. 2.3.

2.4.3.2 Local Area Network

LANs consist of a main functional unit located in a university, office building, factory, for example, such that multiple computers, printers, and data storage units are all connected from the central functional unit. The data rate for transmission can be few Mbps to hundreds of Mbps and even up to 10 Gbps

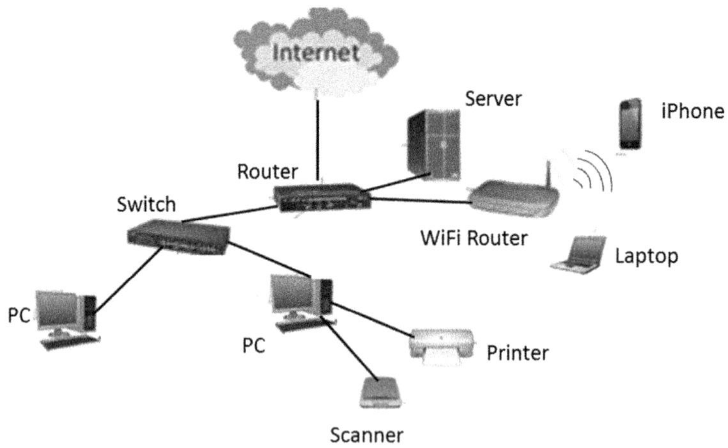

FIGURE 2.3
Personal area network.

with faster technologies, Ethernet being one popular LAN. A wireless local area network (WLAN) has the advantages of flexibility and mobility while still enjoying speed up to 10 Gbps. See Fig. 2.4 for some typical LAN architecture.

2.4.3.3 Metropolitan Area Network

MANs are larger than LANs and generally are not owned by a single organization; instead their communications links and equipment are generally owned by

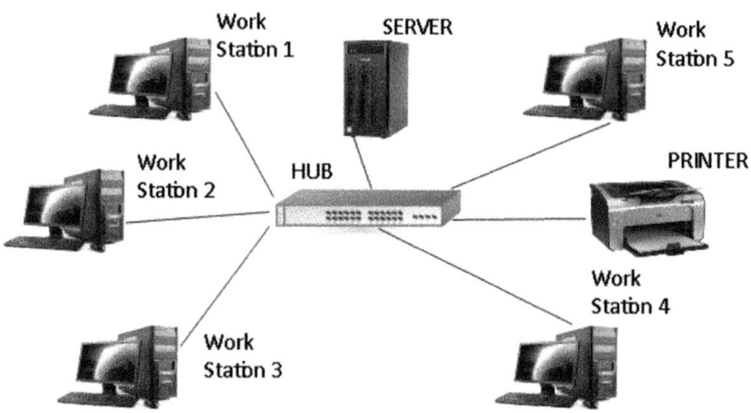

FIGURE 2.4
Some typical local area network architecture.

either a consortium of users or by a single network provider who sells the service to the users. MANs typically cover transmission ranges between 5 and 10 km in diameter and may cover an area the size of a city, or even a group of small buildings. MANs often act as a high-speed network that allows sharing of regional resources and also provide a shared connection to other networks. MANs can also mean the interconnection of several LANs by bridging them with backbone lines. Wireless metropolitan area networks (WMANs) are used to establish wireless connections between multiple locations within a metropolitan area, such as between multiple office buildings in a city or on a university campus, without the high cost of laying fiber or leasing lines. For optical wires communications, infrared light can be used to transmit data. Broadband wireless access networks can provide high-speed access to the Internet. For example, point-to-point high data rates (100 Mb/s to tens of Gbp/s) can achieved over short distances using FSO communication with infrared lasers, unlicensed requiring line-of-sight (LOS), and is affected by weather (Fig. 2.5).

2.4.3.4 Wide Area Network

Wide area networks (WANs) are communication networks that span a wide geographical area such as across cities, states, or countries (global). The Internet can be interpreted as the largest WAN because it connects lots of smaller LANs or MANs, providing the backbone infrastructure to transport data between the interconnected networks. WANs may use several technologies

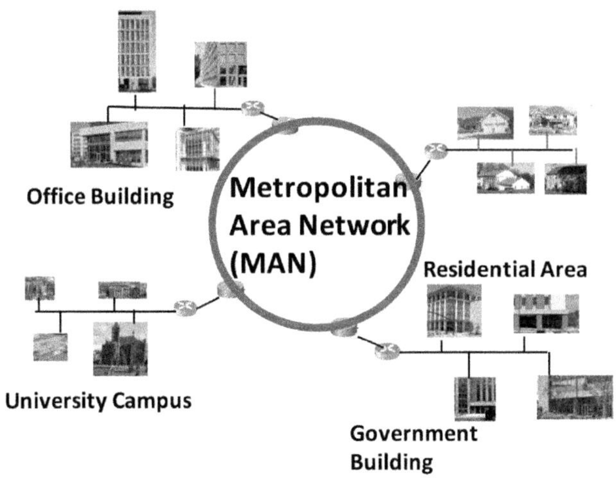

FIGURE 2.5
Metropolitan area network.

for the communication subsystem. A data transfer within a WAN can include broadband fiber optics and optical wireless technologies including terrestrial or space-based (such as satellite, UAV, drone, HAP, and airplane). Basic communication elements are communication channel (e.g., fiber optics, atmospheric, and FSO links) and switching elements (i.e., routers) connecting two or more communication channels. WAN technology can be used to form networks that can connect arbitrarily many computers covering a long distance. A WAN consists of various electronic devices called *packet switches* interconnected by communication lines. A packet switch contains a processor, memory, and I/O interfaces that either connect to a local computer or another packet switch. An arriving packet is placed in the memory of a packet switch until the processor can forward the packet to its destination.

2.4.4 The Internet

The Internet is a network of networks and thus is a collection of diversified networks in terms of topologies and communication technologies using a common set of protocols to offer certain devices. Internet seems to be a single, seamless communication system. Thus, the ability of networks to be joined together to form larger networks has resulted in what is called the Internet. Internet therefore enables worldwide collection of connected computers able to communicate with each other. The Internet, which is a large seamless network, uses a uniform addressing scheme where each computer is assigned a unique Internet protocol (IP) address where all Internet applications use the address when communicating with each other. The Internet can be thought of as a global system of interconnected computer networks that use the Internet protocol suite (transmission control protocol [TCP]/IP) to link devices worldwide. The Internet protocol defines an IP *datagram*, a basic unit of transfer across a TCP/IP Internet. Internet routing and routing protocols are important because they allow propagation of routing information in the Internet. IP routing can be partitioned into two broad categories: static and dynamic routing. The static routing creates a forwarding table when the system does not change entries, whereas in dynamic routing the propagation software runs on the system continuously updating the forwarding table to follow an optimum route. For routing in the global Internet, routers and networks in the Internet are divided into groups, and all routers within a group exchange routing information using a routing hierarchy. Internet routing protocols have two categories, interior gateway protocols (IGPs) and exterior gateway protocols (EGPs). Protocols used to pass routes inside an autonomous system are called IGPs, and EGPs are protocols used to pass routes between autonomous systems. Network performance can be measured by *delay*, the time required to send a bit from one computer to another, as well as by *throughput* (a measure of network capacity), which is the number of bits per second that can be transmitted across the network. In data networks a measure of variance in delay is termed *jitter*, and the Internet uses the protocol approach to handle the transmission of real-time audio and video. Other important areas of the Internet networks, including *multimedia and IP telephony* (VoIP), *network security*, and *network management*, are not discussed in this section because they are beyond the scope of this book. The subjects are discussed in many excellent books such as Ref. [4].

In the Internet, the end user usually does not get direct access to the backbone whereas distribution networks are connected to the backbones. Basic networks are frequently classified by the geographical boundaries the network spans; for example, LAN and WAN or MAN as mentioned earlier. A *gateway* plays an important role in the Internet network, and can be simply defined as the link between two computers to connect to the Internet or another network and works like a portal among two programs by means of communications between protocol, permitting them to share data on the same computers or among different computers. A gateway is a network point that can be an access point to another work. In the Internet, a node can be either a gateway node or sometimes a node joining the network. Gateways can be implemented either by software, hardware, or a combination of both. In the following subsection fundamentals of Internet networks describing how an Internet works with an overview schematic and data communication of networking concepts that define IP in the Internet are discussed. The discussions are helpful to understand exactly how Internet connectivity is accomplished from local and metro regions and eventually to global remote locations.

2.4.4.1 Internet Architecture

Internets are connected by routers that need to know about the topology of the Internet beyond the networks to which they are connected. Using the destination network, not the destination, host routers route a packet.

2.4.4.2 Basic Networking Concepts in Internet

In order to learn the Internet global connectivity and how that can be accomplished optically, the reader should first have very clear knowledge about how the Internet works (independent of RF or optical) to send a message from the sender's location that reaches their ultimate destination, which can be local, between cities of a country, or even to another country. A review of the networking concepts that define TCP/IP (transmission control protocol/ Internet protocol), the suites of protocols that allow conversations to happen on the global Internet, and the role of Internet routing is discussed later. To start, a message is split up into many individual *packets* labeled with their source and destination. The packets are then sent by the computer to a *router* to decide where to send them; for example, how to get to the local network, the best route to a few other local networks, and one route to a gateway to the rest of the Internet (using a routing table). The router checks the destination addresses, sends the packet to the closest match in the routing table, and allows the nest router to continue the same process until the packet eventually arrives at the destination. Using a common addressing scheme and protocol for forwarding packets the packets are thus flown through the global Internet, which makes it possible to exchange information (emails, documents, pictures, videos) on a global scale. Fig. 2.6 shows a schematic diagram of an Internet networking architecture. In order to perform this, note that a common set of communication protocols to regulate when and how computers can speak to each other needs to be regulated first, the most well-known being the *OSI (international standard for open systems interconnection) model*. The network traffic is divided by the OSI model into a number of independent *layers*, building services from the lower layer and providing services to the next higher layer. Both hardware and software can implement each layer following the interface standard between layers. However, the TCP/IP model of networking is not an international standard and is the protocol stacking most commonly used on the global Internet. *IP addressing* is typically a 32-bit number, written as 8-bit numbers expressed in decimal form and separated by periods with every possible IP addresses ranging from 0.0.0.0 to 255.255.255.255 with a total of more than four billion possible IP addresses ($255 \times 255 \times 255 \times 255 = 4,228,250,625$). Each IP address must be unique and generally cannot be used in different places on the Internet at the same time so that routers would be able to send best packets to them, and duplicate addresses cannot be used by different networks. Global IP addresses are assigned and distributed by regional Internet registries (RIRs) to ISPs, which allocate smaller IP blocks to their clients as required. IP addresses are

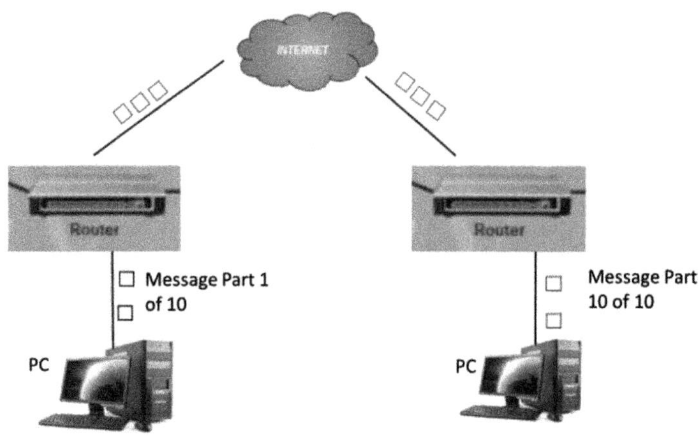

FIGURE 2.6
Internet networking where packets are forwarded between routers until they reach their desired destination.

provided by ISPs to all Internet users and the available 4 billion available IP addresses are administered by the *Internet Assigned Numbers Authority*. Packets between networks by routing are now possible once the IP addresses are agreed upon to participate in the global Internet. *In summary, in Internet networking, each network segment has a router with two IP addresses to make it link local to two different networks, and the packets are forwarded between routers until they reach their ultimate destination.*

2.4.4.3 Configurations of Wireless Networks

There are three basic topologies for a link:

Point-to-point link: Allows a remote site to share a central Internet connection (which already has an Internet connection). In excess of 30 km is possible for a LOS link.

Point-to-multiple link: This type of network layout is a way to connect several nodes (capable of sending and receiving data on a network) to a central point of access. Connection can be shared among all the clients if one of the nodes happens to be in an Internet gateway.

Multiple-to-multipoint clouds: This type of network layout can also be ad hoc or *mesh network* where there is no central authority with all nodes communicating with each other directly. In a mesh type of arrangement, every point can reach another at a very high speed or use the central connection to reach the Internet.

Factors affecting TCP/IP performance over a satellite connection: Satellite connection can involve high bandwidth but also high latency due to the long distance to the satellite and the constant speed of light. Satellite 35,000 km from ground antennas separated by thousands of km can add for example, about 520 ms to a packet's round-trip time (RTT). This impacts TCP/IP performance in long RTT, large bandwidth delay product, and transition errors.

2.4.4.4 Communications, Networks, and the Internet: Some Basics

Communication is all about sharing information, local or remote. Specifically, *data communication* is the exchange of data between two devices (part of the communication system made up of hardware/physical equipment and software/program) via a transmission medium such as a wire cable or air/water (wireless). Data communications use physical phenomena to transfer information; for example, optical fibers used for high-speed data transfer rely on the properties of light propagation and transmission through fiber medium, and optical wireless data communications will depend on the attenuation, absorption, and scattering of air/free space. The modulation technique also affects the network design and influences the design of many protocol layers; for example, the specific choice of modulation depends on the bandwidth directly determining a network throughput. Techniques like multiplexing are used to combine information from multiple sources to be transmitted across a shared medium, and later to be separated for delivery to multiple destinations. In addition, encryption concept can be introduced in data communications for network security.

The essential components of data communication are:

Message: Information (data) to be communicated (e.g., text, numbers, pictures, video)
Sender: The device that sends the data message (e.g., computer, video camera)
Receiver: The device that receives the message (e.g., a computer, workstation, television)
Transmission Medium: The physical path the message travels, from sender to receiver, such as a wire cable like twisted-pair wire, coaxial cable, fiber-optic cable and medium like air/water through which electromagnetic waves like radio or optical waves can propagate
Protocol: A set of rules that govern data communications, an agreement between the communicating devices

Information data can be represented in different forms such as text, numbers, images, audio, and video.

A *network*, as discussed throughout this chapter, is a set of devices (often referred to as *nodes*) that are connected by communication links. A node is a computer, printer, or any device that can send and/or receive data generated by other nodes on the network. Data passes among protocol layers when computers communicate across a network where each computer has a set of layered protocols and data passes through each layer.

The *Internet* is a collaboration of hundreds of thousands of interconnected networks. The advent of powerful, affordable, personal computers and communication technologies provided the computational power needed for complex computation and graphical displays. This has obviously shifted the focus of the Internet from resource sharing to general-purpose communication; for example, from text to multimedia.

2.4.4.5 How Do Computer Networks and the Internet Operate Together to Communicate?

The emphasis given in this book is for readers and students to learn both bottom-up (dealing the lowest-level details and then going to the next higher levels) and top-down (starting with a high-level application and going to the next lower level with enough learning about how the application can operate) approaches in the subject. It is therefore essential for readers and students to understand clearly the different network applications and the communication paradigms such as all OWCs for global Internet connectivity. This section therefore explains the discipline of networking and the Internet, without putting too much emphasis on the plethora of acronyms where readers and students can easily get lost and lose interest in actually learning.

The Internet provides a general-purpose communication mechanism with different services such as web browsing, email, text messaging, and video teleconferences supplied by application programs that run on computers attached to the Internet. A unique address, and for each service such as email or web access, a unique identifier (a protocol port number) are assigned to each computer. Two basic Internet communication mechanisms are *stream and message transports*. The stream transport transfers a sequence of bytes and a sending application may generate 1 byte at a time or blocks of bytes. A protocol is specified when a server starts and the address of the computer on which the server runs is specified with the protocol port number. However, the transfer of the corresponding bytes by the sending applications to the sending applications is not guaranteed. In the message transport the network accepts and delivers messages transmitted by a sender so that the receiver will find exactly the same number of bytes in the incoming message originated from the sender. In this case a message can also be sent from an application on one computer directly to an application on another, or the message can be broadcast to all computers on a given network, or even can be sent to some of the computers on a network. Note that the Internet stream service is *connection-oriented*, meaning a connection is created at the request of two applications and then allows the applications to send data in either direction. The *client–server model* of interaction allows a pair of applications running on two independent computers to coordinate a request for a connection at the same time. A client runs in a standard computer, makes the connection through the Internet to the server, which runs in a server-class computer. When using a single server, there can sometimes be a bottleneck for the network connection between the server and the Internet. One possible method to avoid this is to use a scheme called peer-to-peer architecture, where data is distributed equally among a set of servers.

Standard Internet Applications: Some Basics and Reviews. A programmer specifies details of communication defining an *application-layer protocol* in order for two applications to communicate over a network. An application-layer protocol standard is created independent of any implementation. For the World Wide Web, the most widely used service in the Internet, many protocol standards exist to specify various aspects and details. Some of the standards the World Web service use are hyper text markup language (HTML), uniform resource locator (URL), and hyper text transfer protocol (HTTP). A browser architecture contains client code for each of the protocols used; for example, how to interact with a server and how to interpret responses.

Another widely used application is email, which allows a user on one computer to send a message directly to a user on another computer. The email architecture involves the original email configuration with direct transfer from a sender's computer to a recipient's computer. A browser uses HTTP as the primary transfer protocol to interact with a web server and a client extracts a server name from a URL and contacts the server. A usual protocol reference for URLs is *http://*. In this context, it is relevant to know what a file transfer process is: a service that allows a copy of a *file* (a fundamental storage abstraction) to be sent from one computer to another. A file can represent a document, spreadsheet, computer program, graphic image, or data. For example, the extension *.jpg* is used for a *Joint Picture Encoding Group* (*JPEG*) image; another extension, *.jpeg*, a file transfer protocol (FTP), is the most widely used file transfer service in the Internet.

Coming back to email, the most widely used Internet application, simple mail transfer protocol (SMTP) is the standard protocol used to transfer a mail message across the Internet to a server. An ISP, when functioning as an email server, provides a mailbox for each subscriber (interface software), allowing users to access their mailbox. Another important Internet-related service for understanding ISPs, mail services, and mail access is the *domain name system (DNS)*, which maps human-readable symbolic names to computer addresses used by most browsers, mail software, and other Internet applications and thus provides automated mapping from human-readable names to computer addresses. Each domain name consists of a sequence of alphanumeric segments separated by periods, and is an identification string that defines a specific administrative autonomy or authority and control within the Internet. Each name registered in the DNS is a domain name. For example for a publishing company, Elsevier, the domain name might be *elsevier.com*. The domain names that begin with *www*, which many organizations assign, usually represent the service they provide; for example, www.journals.elsevier.com.

Data Communications: An Overview. This subsection discusses the actual transmission of information across physical media such as wires, optical fibers, and electromagnetic waves. These can include radio and optical waves in order to accomplish a complete communication system. Data communication in a physical sense lies at the intersection of three basic disciplines and foundations: physics, electrical engineering, and mathematics. Data communication is all about exchange and transfer of information and it is therefore important to understand the sources, forms, and the ways information can be transformed into another. Electricity and electromagnetic radiation are fundamental for transmitting information. An overall organization of a complete communication system is shown in Fig. 2.7 for a conceptual framework of a data communication system.

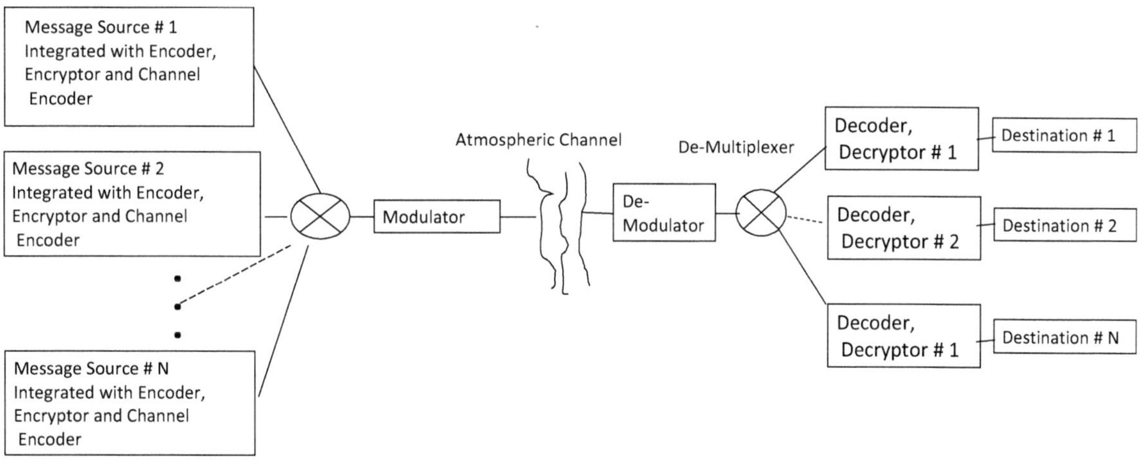

FIGURE 2.7
A general data communication system with information/messages sent from different sources 1...N to destinations 1...N.

Signals, Modulation Techniques, and Information Exchanges in Data Communications. This subsection discusses the physics and electrical engineering aspects of data communication to explain how the information from one or more *sources* are delivered to a specified *destination*. The aim of this book is to develop technology concepts to achieve all-optical global Internet connectivity and not to repeat the same topics of Internet networks, with details of network programming, data transfers with various protocols, and so on, which are found in many excellent text books on the subjects of networks and the Internet (e.g., see Ref. [4–8]). A communication system requires a signal at the source to generate and transmit data, manipulates the data (information) in appropriate formats to be send to a communication channel whether fiber-optics or free-space, and ultimately is received at the destination where the data (information) can be recovered. For a global Internet network, the source and destination of information are a pair of application programs. For devices other than computers the source of information can be arbitrary as long as they all generate some kind of signals, either analog or digital. The signal is then transmitted/propagated through a transmission medium, which can be guided (such as wired, fiber-optics) or unguided (free-space), and is ultimately detected by a receiver at the destination point (can be a computer at the end). Basics of communication can be found in any good book on Communications Engineering, and the specific topics are beyond the scope of the present book. For transmission medium, such as free-space, the readers are encouraged to review the author's books [2,3] on laser/optical communications. OWC can involve infrared communications technology (specifically suitable for indoor communications), point-to-point laser/optical communications (terrestrial, a few km to hundreds of km for slant paths such as between a ground station to a mountain or airplane), and ground-based station to satellites or between satellites. Communication satellites used for establishing global communications can be classified into three broad categories depending on the orbit type:

LEO: Geocentric orbits with altitudes from 160 to 2000 km (100–12,400 miles); have the advantage of low delay but the satellites move across the sky from the observer's point of view.

Medium earth orbit (MEO): Geocentric orbits ranging in altitudes from 2000 km (1240 miles) to just below 35,786 km, an elliptical, not circular, orbit that primarily can provide communications at the North and South poles.

GEO: Geosynchronous orbit at 35,786 km (22,236 miles) has the main advantage that the satellite remains at a fixed position with respect to the observer on earth but because of the distance can introduce longer delays for transmitting and receiving communication signals.

For a GEO communication satellite the orbital period is exactly the same as the rate at which the earth rotates. If the location on the earth is chosen to be above the equator, a GEO satellite remains exactly at the same position over the earth's surface at all times. For an alignment point of view, between a ground station and the satellite this has the advantage of not disturbing the ground station equipment once the initial alignment is completed. But because of this long distance it introduces additional delay to a communication signal.

The round-trip time for a laser between the GEO satellite and a ground station at the speed of light, 3×10^8 m/s, is $(2 \times 35{,}785 \times 10^3\ \text{m}/3 \times 10^8\ \text{m/s}) = 0.238$ s. From a communication point of view this delay of 0.238 s can be extremely important in many practical applications such as a telephone call, video teleconferencing, telemedicine, or financial stock exchanges. Therefore, mitigation techniques of the time delays in communication signals are absolutely necessary. At a minimum three GEO satellites are needed to cover the earth, positioned around the equator with 120 degrees separation. The maximum number of GEO satellites that can be placed on the orbit depend on a number of factors such as the transmitting power, frequency allocation to avoid interference, and so on, but may require an angular separation of between 4 and 8 degrees. To cover the entire 360 degrees circle above the equator, therefore, determines the number of GEO satellites that can be placed, which is between 45 and 90 satellites.

LEO satellites offer the advantage of shorter time delay (typically 1–4 ms) and can be placed up to an altitude of 2000 km. However, the LEO satellite must be above a certain altitude, such as 500 km, to have much less atmospheric effects due to initial terrestrial turbulence to encountering atmospheric gases to minimize absorption or scattering. An LEO satellite appears to move across the sky from an observer's point of view, which may require tracking the satellite signal. This can be difficult, especially if the ground station has a large antenna (RF or optical). Other nonmoving techniques are therefore desirable. One possible practical solution is to use a cluster of GEO satellites, which can work together to forward messages to the appropriate member of the cluster that is currently over a given area of the earth where the ground station is located.

Finally, the choice LEO or GEO communication satellite achieving Internet connectivity depends on a number of practical factors such as cost, data rate, delay, atmospheric effects (attenuation and scattering), susceptibility to interference, and electrical noise and security. Measures of performance of the communication system can be determined by propagation delay and channel capacity after evaluating the effects of noise (and signal-to-noise ratio) on communication.

Networks Throughout the Internet by Data Communications. There are two types of transmission modes, namely, serial and parallel. In the serial mode one bit at a time is sent from a sender to a receiver, whereas in parallel mode multiple bits are sent over separate media simultaneously. When multiple wires are used for a parallel transmission, it is important to make sure that all the wires are synchronized, operating simultaneously, and a bit travels across each wire at the same time. Note that data in a computer is divided into bytes and a byte is divided into bits (8 bits/byte). Sometimes the least-significant bit of the most-significant byte is sent first during transmission. There are basically three types of serial transmission of two data items: *asynchronous* (arbitrary delay between transmission), *synchronous* (continuous with no gap between the transmission), and *isochronous* (regular intervals with a fixed gap between the transmission). In the case of synchronous transmission, an interface called a *frame* starting with a special sequence of bits is added to a synchronous mechanism to accept and deliver a *block* of bytes. *Isochronous transmission* provides steady bit flow for multimedia applications containing voice or video, and the network is designed to accept and send data at a fixed rate with the interface also capable of handing data to the network at the same rate. There are three ways to transfer data from a sender to a receiver: *simplex* (sends data in one direction), *full-duplex* (transfer data in two directions simultaneously), and *half-duplex* (a two-way data transfer, but allows transfer in one direction at a given time).

Information From the Sender to Destination Receiver: Various Techniques for Mapping Information. Data communication starts with the use of high-frequency electromagnetic wave signals to carry information. Various techniques exist for analog and digital inputs. This section mainly discusses the techniques most applicable for optical communications including free-space as a transmitting medium since the aim of this book is to describe how to accomplish global OWC and establish broadband Internet connectivity with all-optical technologies. The process of modulation is the method that maps the information onto a magnetic medium (a carrier). Three parameters of the carrier can be varied to perform the modulation process for mapping the information into changes in either the amplitude, phase, or frequency. At the receiver end demodulation is the reverse process of removing the carrier signal to obtain the original signal waveform (and thus the message/information) back. The source of information can be analog or digital. In digital communication, the signal produced by this source is converted into a digital signal consisting of *1s* and *0s*. Using as few binary digits as possible the object is to represent the signal in such a way that an efficient representation of the source output results in little or no redundancy. A source encoding serves this purpose. A typical communication system configuration is shown in Fig. 2.8, where the communication channel can be fiber-optic and/or free-space because the section mainly concentrates on optical transmission/propagation. The different purposes and functions of the individual blocks in the communication system are the following.

The communication signal s(t) modulator must pass the transmitter, the communication channel (transmission medium), and the receiver (not shown) before it can reach the demodulator. The received signal at the input of the demodulator can be described as

$$r(t) = A(t)[s(t) \times h(t)] + n(t) \tag{2.1}$$

where h(t) is the impulse response function of the channel filter, and denotes convolution, A(t) is the multiplicative contribution due to a fading

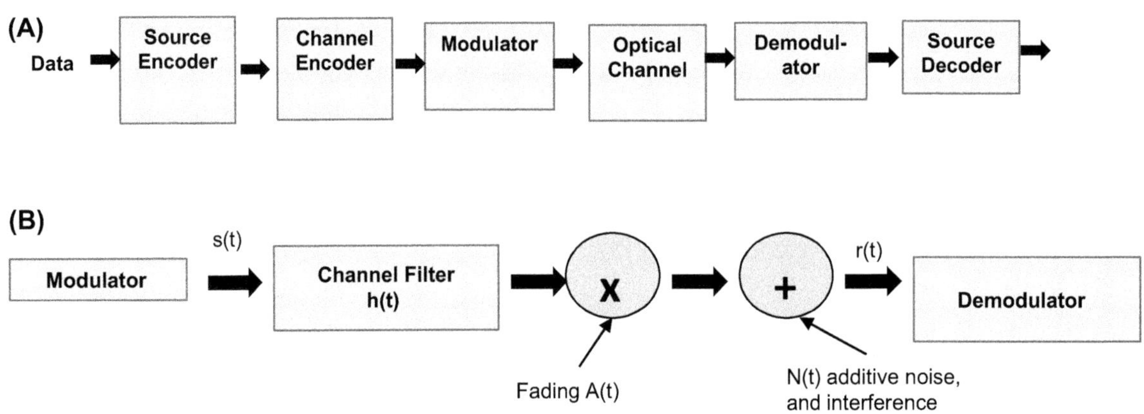

FIGURE 2.8
Simplified block diagram of a typical digital communication system (A) and a system model (B).

(temporally varying as in free-space, but not present in a fiber-optics medium) channel, and n(t) is the additive noise and interference. The channel transfer function can be lumped together with the transmitter, channel, and the receiver transfer function as follows:

$$H(f) = H_T(f)\, H_C(f)\, H_R(f) \tag{2.2}$$

where $H_T(f)$, $H_C(f)$, and $H_R(f)$ are the transfer functions of the transmitter, channel, and receiver, respectively.

Taking the inverse of Eq. (2.2), the impulse response of the channel filter is obtained:

$$h(t) = h_T(t) \times h_C(t) \times h_R(t) \tag{2.3}$$

where $h_T(t)$, $h_C(t)$, and $h_R(t)$ are the impulse responses of the transmitter, the channel, and the receiver, respectively. The term A(t) in Eq. (2.1) represents the fading from the atmospheric channel and is not present if we deal with a fiber-optics medium only. In real atmospheric channel, A(t) is a random signal characterized by the random intensity fluctuations of an optical wave when propagated through atmospheric turbulence or scattering media. The exact form of the random signal is not deterministic but the statistical distributions can be estimated from the recent propagation models. FSO communication channel models are described in Ref. [3] and are not repeated here.

Modulation Schemes in Optical Communications. The intensity of an optical source is modulated to transmit signals. For digital data transmission, digital modulation provides source coding (data compression), channel coding (error detection/correction), and easy multiplexing of multiple information streams. Either a bit-by-bit basis (binary encoding) or bit-word basis (block encoding) can be performed to transmit the digital data. The optical carrier can be modulated in its amplitude, frequency, phase, and polarization. The most viable modulation is intensity modulation, where the desired waveform is modulated onto the instantaneous power of the carrier. At the receiver side, the most practical method is direct detection, where a photon detector produces a current proportional to the received instantaneous power. Some of the modulation schemes commonly used for an optical communication system are *on-off keying, pulse position modulation,* and *binary phase-shift keying modulation,* discussed in detail in Ref. [3]. Communication parameters like *signal-to-noise ratio (SNR)* and *bit error rate* associated with communication system performance also must be evaluated and compared for each type of modulation scheme to select the optimum design.

Channel Capacity and Coding for Optical Communications. The two parameters, namely channel capacity and channel coding, are important in developing the overall communication system where the network is part of the system. Any real-world communication system has to operate in the presence of various noises like atmospheric turbulence and detectors and therefore is always susceptible to errors in the transmission of information. The three resources like signal power, time, and bandwidth can be traded against each other such that errors that occur in transmission can be reduced to an arbitrarily small probability. The object is often to achieve maximum data transfer, in a minimum bandwidth while maintaining an acceptable quality of transmission for reliable data communication. The capacity is defined as a maximum rate of information data stream within the communication channel, and the spectral efficiency is defined as a ratio of the capacity and bandwidth of the communication channel. For a binary alphabet, the maximum rate of 1 bit per channel use can be achieved. The Shannon–Hartley formula can be used to calculate the capacity of the channel with *additive white Gaussian noise* channel [3]:

$$C = B_w \log_2 [1 + \mathrm{SNR}] \text{ bits/s} \tag{2.4}$$

where B_w is the bandwidth. The spectral efficiency can then be expressed as

$$\hat{C} = (C/B_w) = \log_2[1 + \mathrm{SNR}] \tag{2.5}$$

There are various coding techniques available for communication system design to reduce the probability of error to an arbitrary small probability. Coding usually involves adding extra bits of information that, when processed at the receiver, aid in correcting errors that were made during transmission. There are different types of channel coding techniques [3] such as *convolution codes (CC), linear block codes, Viterbi decoding of CC, Reed-Solomon codes, low density parity-check,* and *turbo codes.*

Multiplexing and Demultiplexing for Data Communications. Instead of sending information from one single sender to another single receiver, it may be necessary to send and receive by multiple senders and receivers. *Multiplexing* is a mechanism that allows pairs of senders and receivers to communicate over a shared medium. Information streams from a multiple source are combined in multiplexing for transmission over a shared medium by a *multiplexer* that actually implements the combination. The reverse process, *demultiplexing,* separates the combination back into separate information streams implemented by a *demultiplexer.* Multiplexing and demultiplexing are not restricted to hardware or to individual bit streams but are fundamental in communication used in many parts of computer networking. Fig. 2.9 shows the concepts of multiplexing and demultiplexing, where independent pairs of senders and receivers share a transmission medium.

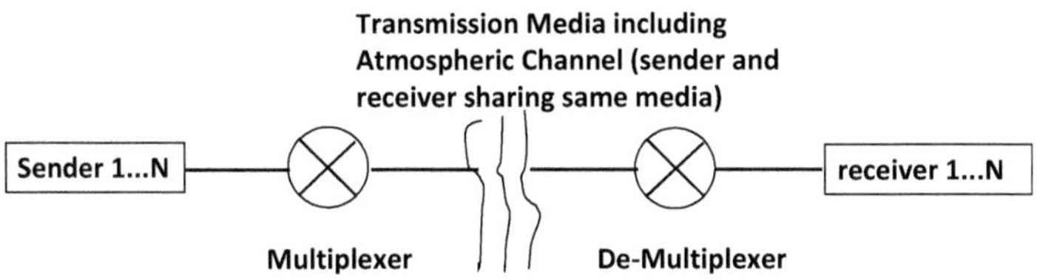

FIGURE 2.9

The concept of combining and separating pairs of senders and receivers for simultaneous transmitting and receiving over a common medium: multiplexing and demultiplexing mechanism.

In addition, there exists practical multiple–input multiple output (MIMO) systems that can be used in achieving a high data rate indoor optical communication system. MIMO configurations can be realized by various schemes that include MIMO using parallel single-input single-output links, MIMO using spatial multiplexing, and MIMO using spatial modulation.

There are basically four types of multiplexing mechanism: *frequency division multiplexing (FDM)*, *wavelength division multiplexing (WDM)*, *time division multiplexing (TDM)*, and *code division multiplexing (CDM)*. Depending on the nature of the carrier signal, physics of the transmission medium, desirable data rate minimizing noise and interference, complexity and cost, optimum multiplexing/demultiplexing methods can be selected. A simple example for FDM is where a set of radio stations, for example, can transmit electromagnetic signals using a separate channel assigned with different frequency simultaneously without any interference. At the receiving end a demultiplexer uses a set of filters to extract a small range of frequencies near one of the carrier frequencies of the transmitting signal. WDM refers generally to the multiplexing process dealing with optical communications through transmission medium of fiber-optics and/or free-space. WDM is a multiplexing technique working in the wavelength domain that multiplexes a number of optical carrier signals onto either a single optical fiber or over the same FOS medium by using different wavelengths (i.e., colors) of laser light. The advantage of this technique is that it enables bidirectional communications over the transmission medium and also has the multiplication capacity. The multiplexer at the transmitter end joins the several signals and a demultiplexer at the receiver end splits them apart. The system offers to generate very high data rate signals; for example, a system using 160 signals can expand to a basic 100 Gbit/s system over a single fiber pair to over 1 Tbit/s. When WDM is combined with an optical amplifier the system can accommodate several generations of technology development without having to overhaul the backbone network. By upgrading the multiplexers and demultiplexers the overall capacity of a given link can be expanded.

Recently dense WDM has been developed, which uses the optical wavelength of 1530–1565 nm transmission window with denser channel spacing. TDM is the digital multiplexing technique where the channel/link is not divided on the basis of frequency but on the basis of time, where total time available in the channel is divided between several users, each allocated a particular time slot. All the signals to be transmitted are not transmitted simultaneously but one by one. When applied to synchronous networks where no gap occurs between items, TDM is known as synchronous TDM. In some situations, multiplexing related to the Internet is used; for example, an ISP requiring higher bit rates than are available an inverse multiplexing—that is, multiplexing is used in reverse. In that case, high-speed digital input is spread over multiple lower-speed circuits for transmission and then the results are combined at the receiving end. CDM is the multiplexing technique where multiple data signals are combined for simultaneous transmission over a common frequency band. *Code division multi-access,* a specific version of CDM, is used in cell phones This type of multiplexing technique uses a combination of analog-to-digital conversion and spread spectrum technology where the transmitter encodes the signal using a pseudo-random sequence, thus involving the original signal with a spreading code. Using the code generated and transmitted by the transmitter, the receiver now can decode the received signal. Different random sequences correspond to a different communication channel from a multiple station.

Internet Access Technologies for Data Communications. Internet is a global system of interconnected servers and computers to link thousands of millions of devices worldwide using standard protocols and different technologies. As technology grows, the need for bigger, better, and faster Internet connections also grows. Also, the way content is presented via the Web has also changed drastically. The devices in the Internet are linked and connected by various networking technologies such as arrays of electronics, wired lines, wireless, and optical fiber. Today, Flash, animation, online gaming, streaming video, database-driven websites, and mobile applications are becoming more standard. It is therefore important to understand the Internet access technology that refers to a data communication system connecting an Internet subscriber to an ISP. Internet connection speeds also vary between ISPs. The Internet subscriber can be a residence or business who is the customer for the ISP, the Internet connection provider. There are a number of different names for selling Internet access such as broadband, high speed, wireless, fiber, and so on, but without any details of how the subscriber is connecting to the Internet. Most Internet users follow an asymmetric pattern; that is, a typical residential subscriber receives more data from the Internet than they send. When viewing a web page, a browser sends a URL that comprises a few bytes, but a web server sends content that may have thousands of bytes of text or an image comprising tens of thousands of bytes. On the other side the business as web server sends more data than it receives. Internet access technologies need to be designed to transfer more data in one direction than the other. In access technologies, *downstream* refers to data traveling from ISP to a subscriber, and *upstream* refers to data traveling from a subscriber to ISP.

Typical types of Internet connections are *dial-up, cable DSL/ADSL, wireless broadband (Wi-Fi, Wi—MAX,* etc.*), mobile broadband, fiber optic,* and *satellite broadband.* Different technologies are used to provide the Internet connection to the devices that are reliable, fast, and cheap. Wired technologies for the Internet access include dial-up connection to PCs through a modem on telephone lines, broadband connection on coaxial cable, and fiber-optic cable. Wireless technologies include Wi-Fi, satellite, and cellular mobile technologies (2G, 3G, 4G, and future 5G). In a dial-up connection, the Internet is connected to the user's computer through a telephone network and modem. Since telephone lines are analog in nature, while information over the Internet is in digital form, a modem is used between the user's computer and telephone line to encode and decode voice signals suitable for digital communication in the Internet. Dial-up is very low speed Internet access technology, typically up to 56 kbps.

Wired broadband Internet access technologies include integrated services digital network transporting voice and digital data for voice, video conferencing, and broadband data services at data rates up to 128 kbps, and also leased lines (either optical fiber or RF) to provide an ISP to connect a LAN or campus network to provide data rates from 56 kbps to 1.5 Mbps (T1) or 2.0 Mbps (E1). Another Internet access technology includes cable Internet access, which provides Internet access through hybrid fiber coaxial cables (originally used for television channels and with existing cable TV connection customers) with a download data rate of 400 Mbps for business and corporate sector, and 250 Mbps for residential. The upload data rates can be from 350 kbps to 2 Mbps. Digital subscriber line (DSL) service provides Internet access through existing telephone networks and uses high frequencies for Internet data and low frequencies for voice communication over telephone networks. Asymmetric DSL is another form of DSL that provides a download data rate from 256 kbps to 20 Mbps, and lower upload rates . Symmetric DSL provides equal download and upload data rates. Another type of DSL is very high bit rate DSL (VDSL), which provides download data rates of 52 and 16 Mbps for upload over copper cables. It supports application line high-definition television and VoIP. Another version of VDSL is VDSL2, which provides data rates for download and upload up to 100 Mbps.

Fiber to home is another broadband using fiber cable, which provides much higher data rates for long distances since fiber cable is connected to most of the country now. Fiber-optic cable can provide very high data rates in the range of terabits/sec for longer distance in the range of km. Internet access technologies for wireless connections include Wi-Fie, Wi-MAX, satellite broadband, and mobile broadband. Wi-Fi is a technology for wireless networking with devises based on IEEE 802.11 standards and uses devices like PC, video game consoles, smart phones, digital cameras, tablet computers, smart TVs, digital audio players, and modern printers. The devices can be connected to the Internet via a WLAN and wireless access point (hot spots) with a range of about 20 m (66 feet) indoors and a greater range outdoors. Wi-Fi uses the 2.4 GHz UHF and 5 GHz SHM ISM radio bands. Wi-Fi ad-hoc mode refers to devices talking directly to each other without the need to first talk to an access point (base station). Wi-Fi technology may also be used to provide Internet access to devices that are within the range of wireless network that is connected to the Internet. Interconnected access points (hot spots) can cover an area of a few rooms to a large area of many square kilometers using a group of access points with overlapping coverage. In homes and other buildings, routers (with a digital subscriber line modem or a cable modem) are often used to provide Internet access and internetworking to all devices connected to them, wirelessly or via cable. Recently city-wide Wi-Fi is being constructed in many cities of the world to become Wi-Fi—enabled cities. Wi—Max (worldwide interoperability for microwave access) enables wireless broadband Internet access as an alternative to cable or DSL networks and provides data rates up to 1 Gbit/s offering MAN with a signal radius of around 50 km. Internet connections via satellite provide fixed, portable, or mobile Internet access and uses a parabolic or dish antenna aligned toward a satellite requiring line-of-sight path. Data rates from 2 kbps to 1 Gbps for downstream and 2 kbps to 10 Mbps for upstream are provided. Satellite scenarios with different locations such as LEO and GEO will be discussed in other sections.

2.4.4.6 Technical Trends for Broadband Access Technologies: Broadband Wireline Access Networks and Broadband Wireless Access Networks

Broadband access paradigm with tremendous growth for demands all over the world required to satisfy the technological demands for the Internet access *anywhere, anytime* and different means of connectivity are continuously improving as a result of significant technological evolution of both broadband wired and wireless access networks. Technological developments of various passive and fast optical network architectures employing different rates including 155 Mb/s, 622 Mb/s, 1.25 Gb/s, and 2.5 Gb/s, make it now possible to connect the broadband networks via fiber, which exists in many cities in the country and is the most popular access network worldwide. Wireline access network technologies thus include xDSL (ADSL and VDSL), twisted pair, coaxial cable, and hybrid fiber-copper. Global broadband strategies relate to economic growth and priorities given by different countries, for example between fiber technologies or upgrading xDSL. Broadband wireless access networks can serve users over much wider geographical areas. Short-range wireless links covering a home or building need to be connected to the nearest wired (fiber-optics) backhaul access technologies for broadband access to distant users. WPAN (IEEE 802.15) can cover about 30 feet to connect the personal devices using Bluetooth, infrared, wireless USB, and wireless HD. WMANs (IEEE 802.16) can serve as backups for copper or fiber cabling.

Two other types of broadband access technologies are the IEEE 802.11-based LAN (WLAN) standard known as Wi-Fi, and 3G, 4G, and future 5G cellular technology. Future Wi-Fi technology (IEEE 802.11ad) will have the potential to target a frequency band of 60 GHz, maximum data rate of 7 Gbp/s using the technology multiple-user—MIMO. The *mobile network technology* provides the connectivity over a wide geographical area supporting hand-over mobility and cell roaming, two very important and desirable features. Finally, *Fiber-Wireless* networks offer the future of broadband access technology, which combine the advantages of both the huge bandwidth of optical fiber and the flexibility of wireless, and holds great promise for future evolved broadband access networks. In order to achieve broadband wireless access technology, fiber-based distributed antenna systems using radio-over-fiber (RoF) links have been developed and demonstrated to improve wireless coverage in buildings (indoors or outdoors) with small-size remote antenna units and centralized management. Cellular, Wi-Fi, Wi—MAX, and emergency services can be included. The generated RoF signal would be available for both optical baseband and wireless communications services applications. Recently 100 Gb/s mm-wave wireless links have been demonstrated using advanced modulation schemes such as QPSK/16QAM/64QAM and MIMO antenna arrays. Light emitting diodes as visible light source for communication are being used and are very practical for indoor short-range communications. Currently, wireless systems will require data rates of 1 Gb/s and 10 s of Gb/s, or even more for a number of applications such as high-definition video and high-speed Internet.

Optical communications using wavelength is a viable solution, which is the main emphasis of this book: developing innovative technology concepts to accomplish optical wireless Internet connectivity, worldwide. The FSO system can be very fast, uses subcompact optical devices, is portable, and does not require licensing. One of the disadvantages of the FOS communication systems is the atmospheric effects (attenuations and scattering of optical wavelength over the atmosphere including turbulence). The communication link usually has to be LOS with a very narrow transmitting beam, and can thus be very secure. Whereas fiber-optics can offer Tb/s data rate, tens and hundreds of Gb/s rates can be achieved for free-space links of a few km to hundreds of km. Communication links between satellites do not suffer atmospheric attenuation and many Gb/s links can be easily achieved, which also can offer global coverage in remote locations. However, in satellite-based systems, accurate pointing and tracking is required. Satellite-based optical Internet connectivity is the most promising future solution for high bandwidth and worldwide coverage. This book will describe this potential technology in detail. An excellent review describing the evolution and trends of broadband access technologies can be found in Ref. [9].

Networks of low capacity may be connected together via a *backbone* network, which is a network of high capacity such as a WAN network, while LANs and WANs can also be interconnected via digital lines. Note that *protocols* used in understanding Internet network architectures define the rules that govern the communications between two computers connected to the network. The following devices are normally used to achieve networks according to the protocols involved:

Bridge: Links two similar LANs based on the same protocol
Router: Different types of networks with different protocols are connected by the router, which is basically a communication computer
Bridge/Router: A single device that combines both functions of a bridge and a router
Gateway: A network device that connects two different systems, using direct and systematic translation between protocols

The *networks interconnection/Internet* concept was introduced and implemented by the Defense Advanced Research Project Agency Network (Arpanet) in 1966 in the United States, and consists of connecting several computer networks based on different protocols. Obviously, it requires the definition of a common interconnection protocol on top of the local protocols. This is exactly what the IP accomplishes by defining unique addresses for a network and a host machine. The chief function of the IP is to decompose the initial information flow into packets of standardized size reassemble at the destination, and route a packet through successive networks from the source machine to the destination identified by its IP address. The initial data to be sent can be decomposed into datagrams with a header including the IP address and the destination port number, sent to selected gateways such as IP routers, connected to the same local network and to an IP service provider network. Ultimately datagrams are transferred from gateway to gateway until they arrive at their final destination.

2.4.4.7 A Broadband Home Access Architecture: Basic Principles

Independent of RF or optical spectrum, it is important to understand the basics of a broadband home access architecture, which represents an ultimate technical solution for the Internet-based vendors to the customers located in the comfort of their own homes. In general, the architecture normally is an integrated and interoperable system in order to successfully penetrate the mainstream marketplace of the broadband local loop, the residential gateway, the home access network, and the network devices, which are a set of intelligent processing tools used in home environments. Understanding the role of each of these components will be helpful in later optical technology concept development; for example, to extend the basic network access principles of RF to modify to optical domain using the recently developed optical wave technology. The broadband local loop is widely used to describe the connection from the local provider to the customers: bringing high-speed information to homes and offices. A complete system will obviously have to include both stationary (fixed) and mobile users and applications.

Fig. 2.10 shows a general network architecture of establishing Internet and data communication including WPANs, WLANs, wireless local loops, cellular WANs, and satellite networks. The residential gateway is the interface device interconnecting the broadband local loop to the in–home network offering bidirectional communication channels to every home-networked device. Simply, this is the access point between home and the outside world that will bridge the different broadband and LAN technologies and PANs. The home area network provides the interconnectivity to all processing tools used in home environments such as computers, appliances, and electronic devices in order to communicate with each other as well as to the outside world (e.g., Internet) via broadband local loop. The last-mile broadband wireless

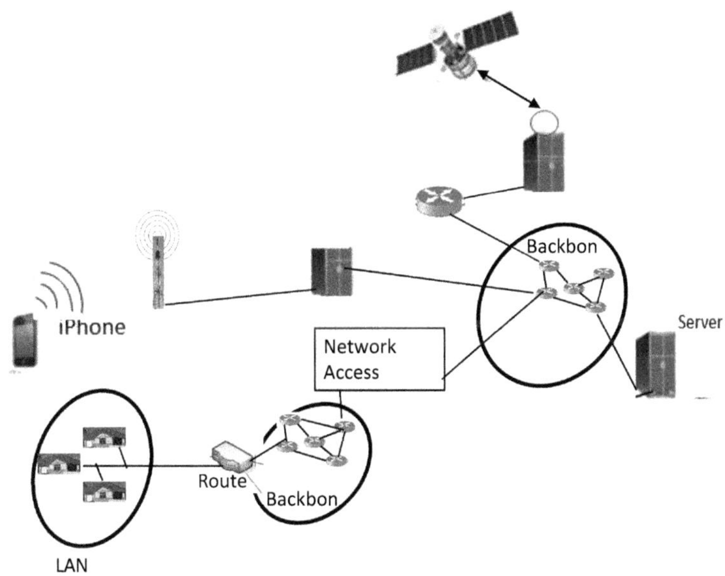

FIGURE 2.10
Overview of the schematic of the Internet and data communication in the Internet.

access provides networking capabilities between the end users, which include point-to-point, point-to-multipoint, and multipoint-to multipoint to bring broadband communications information. The architectures of broadband wireless access services include Internet access, audio, voice, and streaming video, and such; and the deployment scenarios (for service area coverage), which can be *supercells* (large service areas of up to 30 mi radius), *macrocells* (a cell radius of about 5 mi with non-LOS propagation links), and *microcells* (typical cell radius of 1 mi using omnidirectional antennas). There should be a sufficient communication link margin to provide indoor coverage.

2.4.4.8 A Broadband Home Access Architecture

A broadband home access architecture must address the next generation of Internet-based vendors with consumers located in the comfort of their own homes. Some of the technological components for home-based integrated and interoperable systems consist of broadband local loop (connection from the local provider to the consumer at homes or offices), the residential gateway (the interface device that interconnects the broadband local loop to the in—home network and serves to bridge the different broadband and LAN technologies), the home area network (the high-speed in—home network that distributes the information), and electrodomestic network devices (a set of intelligent processing tools used in home environments such as computers, appliances, and electronics with the capability of communicating with each other, and also connecting to the outside world; e.g., the Internet).

2.5 SUMMARY

This chapter introduces the fundamentals of engineering aspects of free-space laser/optical communications protocols and describes technology concepts to establish worldwide broadband access to the Internet or other networks independent of terrestrial infrastructure. Advanced FSO development are the main drivers for satisfying recent high demands in bandwidth thanks to the recent growth of Internet usage, IPTV, VoIP, and You Tube/Twitter and the immediate future and coming applications in IoT. With the rapid growth of data-centric devices and the general deployment of broadband access networks, high data rates from 10 Gbps to more spectrally efficient 40 Gbps or 100 Gbps/channel are feasible. Some basics of network topology, network type, and components with their functionality as well as examples of networks types and categories are also discussed in this chapter. How the optical networks eventually connected to establish interaccess and connectivity are explained. The next chapter will discuss the last-mile problem (or last-mile bottleneck), describing the issues and potential solutions for connecting high bandwidth from the fiber optic backbone to all businesses with high bandwidth networks.

References

[1] Google News as seen on June 19, 2017 Washington Examiner, Politics Policy News Opinion: Magazine Beltway The Race To 2020, 2017.

[2] A.K. Majumdar, J.C. Ricklin, Free-space Laser Communications: Principles and Advances, Springer, New York, 2008.

[3] A.K. Majumdar, Advanced Free Space Optics (FSO): A Systems Approach, Springer, New York, 2015.

[4] D.E. Comer, Computer Networks and Internet, Pearson, Prentice Hall, Pearson Education, Inc., Upper Saddle River, New Jersey 07458, 2009.

[5] W. Stallings, Computer Networking with Internet Protocols and Technology, Prentice Hall, 2003.

[6] D.E. Comer, Internetworking with TCP/IP, Volume One, sixth ed., Pearson, May 5, 2013.

[7] J.F. Kurose, K.W. Ross, Computer Networking: A Top-down Approach, fifth ed., Pearson, March 31, 2009.

[8] K.R. Fall, W. Richard Stevens, TCP/IP Illustrated, Volume 1: The Protocols, second ed., November 25, 2011. Addison-Wesley Professional.

[9] Y. Ma, Z. Jia, in: M. Tornatore, et al. (Eds.), Chapter in Book, Fiber-wireless Convergence in Next-generation Communication Networks, Optical Networks, Springer International Publishing Switzerland, January 2017, https://doi.org/10.1007/978-3-319-42822-2_2.

Further Reading

[1] H.L. Minh, D. O'Brien, G. Faulkner, L. Zeng, K. Lee, D. Jung, Y.J. Oh, E.T. Won, 100-Mb/s NRZ visible light communications using a postequalized white LED, IEEE Photonics Technology Letters 21 (15) (August 2009).

[2] J. Vučić, C. Kottke, S. Nerreter, K.-D. Langer, W. Waleski, 513 Mbit/s visible light communication link based on DMT-modulation of a white LED, Journal of Light Technology 28 (24) (2010).

[3] J. Vučić, C. Kottke, K. Habel, K.-D. Langer, 803 Mbit/s visible light WDM link based on DMT modulation of a single RGB LED luminary, in: Optical Fiber Communication Conference and Exposition (OFC/NFOEC), 2011 and the National Fiber Optic Engineers Conference, Conference Publications, March 2011, pp. 1–3.

[4] http://www.laserfocusworld.com/articles/2012/09/laser-light-communications-commercial-fso-2017.html.

[5] E. Leitgeb, M. Gebhart, U. Birnbacher, Optical networks, last mile access and applications, in: Chapter in Book, Free-space Laser Communications: Principles and Advances, Springer, New York, 2008.

[6] E. Leitgeb, M.S. Awan, P. Brandl, T. Plank, C. Capsoni, R. Nebuloni, T. Javornik, G. Kandus, S. Sheik Muhammad, F. Ghassemlooy, M. Loschnigg, F. Nadeem, Current optical technologies for wireless access, in: 10th International Conference on Telecommunications- ConTEL 2009, Crotia, Zagreb, June 8–10, 2009, ISBN 978-953-184-131-3, pp. 7–17.

[7] H. Elgala, R. Mesleh, H. Haas, Indoor optical wireless communication: potential and state-of-the-art, IEEE Communications Magazine 49 (2011) 56–62 (VLC Technology).

[8] A. Jovicic, J. Li, T. Richardson, Visible light communication: opportunities, challenges and the path to market, IEEE Communications Magazine 51 (2013) 26–32 (VLC Technology).

[9] R.E. Meyers, K.S. Deacon, A.D. Tunick, Free-space and atmospheric quantum communications (Chapter 10), in: A.K. Majumdar (Ed.), Advanced Free Space Optics (FSO): A Systems Approach, Springer, New York, 2015, pp. 343–387.

[10] https://www.hhi.fraunhofer.de/en/departments/pn/research-groups/free-space-optical-systems/research-topics/terrestrial-free-space-optics.html.

[11] "10 Gbps through the Air". Arto Link. New Artolink Wireless Communication System with the Highest Capacity: 10 Gbps, Full Duplex [..] Artolink M1–10GE Model, October 27, 2014. http://artolink.com/page/what/.

[12] D. Tsonev, S. Videv, H. Haas, Towards a 100 Gb/s visible light wireless access network, Optics Express 23 (2) (January 26, 2015).

[13] A. Gomez, K. Shi, C. Quintana, M. Sato, G. Faulkner, B.C. Thomsen, D. O'Brien, Beyond 100 Gb/s indoor wide field-of-view optical wireless communications, IEEE Photonics Technology Letters 27 (4) (February 15, 2015) 367–370.

[14] H. Kaushal, G. Kaddoum, Optical Communicationin Space: Challenges and Mitigation Techniques, IEEE Communications Surveys & Tutorials, Firstquarter 2017 19 (1) (2016) 57–96, https://doi.org/10.1109/COMST.2016.2603518.

[15] http://oneweb.world/.

[16] F. Khan, Mobile Internet from the Heavens" Computer Science/Networking and Internet Architecture (cs.NI), August 9, 2015 arXiv:1508.02383v1 [cs.NI].

[17] R.W. Kingsbury, Optical Communications for Small Satellites (Ph.D. thesis), Submitted to MIT, the Department of Aeronautics and Astronautics, September 2015. SSL # 16–15.

[18] F.A. d'Oliveira1, F.C.L. de Melo, T.C. Devezas, High-altitude platforms — present situation and technology trends, Journal of Aerospace Technology and Management 8 (3) (July–September 2016) 249–262. São José dos Campos.

CHAPTER 3

Last-Mile Problem Revisited With Potential Solution for Broadband Connectivity

3.1 INTRODUCTION

The so-called last-mile problem (or last-mile bottleneck), which is to connect the high bandwidth from the fiber optic backbone to all businesses with high bandwidth networks, is still not quite solved as promised even if the backbone is realized with fiber. Less than 5% of all buildings in the United States have a direct connection to the very high speed (2.5−10 Gbps) fiber optic backbone, yet more than 75% of businesses are within 1 mile of the fiber backbone. Most of these businesses are running a high-speed data network within their building, such as fast Ethernet (100 Mbps) or Gigabit Ethernet (1.0 Gbps). This chapter provides an architecture showing the *small-cells* concept to develop a basic technology block to achieve *all-weather 99.999% (five nines) availability* based on an analysis of link budgets and visibility-limiting weather conditions (even in the worst case of dense fog!) to meet carrier-class availability (five nines). Examples show free-space optical (FSO) links to be normally be less than 140 m (there are cities like Phoenix and Las Vegas where this 99.999% distance limitation increases significantly). The design architecture describes a hybrid concept combining small FSO cells with high-speed Wi-Fi to achieve the broadband connectivity for the last-mile problem. Finally, with the small-cells approach for bidirectional communications to solve backhaul communication, LED for Wi-Fi to increase 100 times speed and Li-Fi with visible light communication (VLC) have been explained for potential broadband solutions. In completing the last-mile to connect as a part of global connectivity network it will then be possible to establish actual global Internet connectivity.

3.2 THE LAST-MILE PROBLEM

For efficient telecom and Internet connectivity proper infrastructure is required to develop both the backbone network and last-mile access. The last-mile problem mentioned earlier for connecting the businesses with only broadband fiber cables is not practical because digging up the ground and laying down new

Optical Wireless Communications for Broadband Global Internet Connectivity.
https://doi.org/10.1016/B978-0-12-813365-1.00003-5

fiber to extend the fiber optic backbone is prohibitively expensive. Therefore, the connectivity gap between the backbone network and the last-mile access needs to be solved. FSO is one of the most promising new, viable access technologies and offers the most practical solution for the last-mile problem. In FSO communications optical transceivers transmit and receive laser beams through the atmosphere to form point-to-point high-speed communication links. In fact, more than 95% of all enterprises including businesses, Internet service providers (ISPs), university campuses, banks, and all types of industry are within a mile of the fiber optic backbone. Telecommunication carriers offering more and more data speed to the customers for various online applications including real-time videos and movies on demand require a cost-effective and innovative solution for the last-mile problem. High-speed communication systems provide very large bandwidth where a data rate of 100 Gbit/s can be achieved over a distance of 1–4 km.

FSO communications is attractive because of unlicensed spectrum and a very narrow beam width. The FSO communication technology is fundamentally based on transmitting an optical signal through free space and the air as a communication channel for signal transmission and can offer very high data rates in the gigabits/sec range for a few km propagations. Extremely narrow laser beam width offers spatial and temporal multiplexing and multiple link capabilities to mitigate the atmospheric effects as well as to accomplish extremely high data rate through adverse atmospheric conditions. Local area network–based FSO systems have the potential to solve the last-mile problem in the future as a bandwidth in excess of 2 THz or more. Data carrying light beams are confined to a very narrow area because the light beam is of small diameter, which therefore provides greater security to the data, and the FSO link consumes less power with no electromagnetic interference. However, this solution has its own set of problems due to the atmospheric effects such as turbulence and scattering on the FSO communication systems performance. The signal is attenuated when transmitting through rain, snow, and especially fog. The optical signal can suffer a loss of 20–300 dB/km when transmitted through various types of fog. Also, the laser beams need to be focused between transmitter and receiver and the two beams need to be within line-of-sight (LOS) of each other.

3.2.1 Atmospheric Effects on Free-Space Optical Communication Relevant to Last-Mile Problem: Analysis of All-Weather Broadband Free-Space Optical 99.999% Availability

Fog interference can be significant at FSO wavelengths because of the scattering process. FSO communications can still be achieved for low to medium fog conditions for a few km (~1–4 km) distance without appreciable signal loss as long as some light (photons) reach the detector receiver. For heavy (thick) fog the signal can be reduced to a point where connectivity will be lost. The link availability is the percentage of time over a year that an FSO link will be operational and five-nines is defined as:

5 Nines = 99.999% = Down 5 min/year

The atmospheric effects on FSO communications is discussed in detail in another chapter of this book. FSO parameters to analyze communication performance relevant to all-weather effects are:

Attenuation: Caused by two main factors: (1) absorption and (2) scattering. Atmospheric conditions responsible for attenuation relevant to FSO communications are fog, snow, rain, and turbulence.

Link Margin: The performance of a lasercom system is generally quantified by the power received and link margin (LM). The LM calculation is therefore essential to design an acceptable system and is defined as

$$\mathrm{LM} = \frac{\text{Available received power}}{\text{Receiver power required to achieve a specified Bit Error Rate (BER)at a given rate}}$$

Note that the link margin value shows how much margin an FSO system has at a given range to compensate for scattering, absorption (due to fog, snow, and rain), and scintillation (turbulence). Table 3.1 shows the all-weather broadband availability showing the visibility, attenuation in dB/km at 785 nm, effective link range for an FSO system under various weather conditions [1,2].

A typical value of attenuation caused by turbulence (not shown in the table) for a range of 850 m is 3.2 dB. Fog has the worst effect on FSO communication because of scattering effects and can be even 350 dB/km. Rain and snow have less adverse effects for optical links and can be in the ranges of 45 and 150 dB/km in heavy rain or snow [1]. For a transmitter power of 30 mW and using a detector of sensitivity 25 nW, the researchers showed that the link margin is 54 dB, and the range of 140 m is still possible even in the presence of worst-case 350 dB/km fog (corresponding to 99.999% link availability) [1]. Other results of availability and reliability of FSO links from the estimation from measured fog parameters [3] show from a visibility experiment on October 31, 2011 for 312 m range on a foggy day that the link was down only 90 times per day (0.104% of the day). Their calculated availability of the FSO link was 99.896% for a 312 m path [3]. For a 100 m path, it was estimated that the availability should reach 99.999%. The results from the visibility curves on November 15, 2011 showed that all values of visibility were higher than the thresholds for the FSO system and therefore the availability of the FSO system was 100%.

Table 3.1 All-Weather Broadband Availability [1]

Weather condition	Precipitation (mm/hr)	Visibility	dB/km Loss (785 nm)	FSO System Range
Dense fog		0 m		0 m
		50 m	-339.6	110 m
Thick fog		200 m	-84.9	330 m
Moderate fog	↑	500 m	-34.0	610 m
Light fog	Outburst (100)	770 m	-20.0	780 m
		1 km	-14.2	880 m
Thin fog	Snow Heavy rain (25)	1.9 km	-7.1	1.02 km
		2 km	-6.7	1.03 km
Haze	Medium rain (12.5)	2.8 km	-4.6	1.08 km
		4 km	-3.0	1.13 km
Light Haze	↓ Light rain (2.5)	5.9 km	-1.8	1.18 km
		10 km	-1.1	1.22 km
Clear	Drizzle (0.25)	18.1 km	-0.6	1.25 km
		20 km	-0.53	1.26 km
Very Clear		23 km	-0.46	1.26 km
		50 km	-0.21	1.28 km

3.2.1.1 Impulse Response Function in the Case of Multiple Scattering

In calculating the FSO communications system performance operating in the presence of multiple scattering media such as dense rain, light fog, fog and cloud using transmitting laser pulses, it is necessary to evaluate the impulse response function of the scattering medium. The theoretical model and analysis is discussed in [4], showing impulse response functions for various atmospheric conditions of fog and rains, and therefore is not repeated here. The analysis is important to determine the possibility of using FSO communication as a potential solution for the last-mile problem. In [4] bit-error-rate (BER) versus required transmitter power for multiple scattering media (fog and cloud) for a short input laser pulse is calculated, which is helpful in designing an FSO communication system for the last-mile.

Last-Mile Problem Fiber backbone already has the availability of multigigabit connectivity. Basically, the *last-mile problem* is the problem of utilizing the multigigabit (or even more) connectivity point to the buildings, campus, and industries within about a mile without any existing fiber links for establishing high speed Internet connection. This is still a challenging issue to address the network connectivity problems between the buildings using advanced transceivers and fast, compact optical devices that can handle multigigabits or higher data rates.

3.2.2 Possible Solutions for the Last-Mile Problem

3.2.2.1 Direct Fixed Point Terminals for Covering Long Range

FSO communications can offer solutions to the last-mile problem, affording flexible and rapid connectivity. Atmospheric effects limiting the extensive implementation of FSO are briefly addressed earlier with a special emphasis on fog, which affects most the optical propagation because of large loss of the signal. There are some alternate solutions for the last-mile problem. On a macro level, whether a link is operational or not depends on the received signal-to-noise ratio (SNR) level, whether or not it is too low for communicating at an acceptable BER. If a specific link is inadequate there may be alternate ways to transmit the data by avoiding the failed link as potential solutions that can be adjusted at the optical wireless networks that are part of the transceivers in the last-mile links. Some of the possible solutions include the following.

Buffering can be applied in some severe weather-related situations such as fog where the received signal is severely attenuated, resulting in considerable error in detection. The data can be stored and then transmitted when the weather is favorable using appropriate optical networking. However, huge data storage can be required for even brief outages. In this case a hybrid solution of part-optical and part-electronic buffer can be practical without any total loss of communication at any time. This solution also will depend on the needed quality of service.

Retransmission In some special cases such as communication links to operate in strong turbulence or in the presence of severe building sway, repeated transmission can be helpful.

Rerouting Optical wireless link paths are accomplished by networking between nodes to establish alternative routes from sender to recipient when the direct connection is disabled. The solution will require point-to-multipoint transceivers with dedicated transmitters for each link.

Multitransceiver FSO communication structures FSO communication has the potential of delivering wireless communication links at optical-level speeds.

Since maintenance of LOS between transceivers during an on-going transmission, even in the presence of turbulence or scattering as well as in cases like building sway, is very important since the transmitters are highly directional. In Ref. [5], a prototype implementation of a multitransceiver electronically steered-communication structure is described, which assigns logical data streams to appropriate physical links. Using multiple directional transceivers, it is demonstrated that optical wireless links can be maintained with minimal disruptions caused by the relative mobility of communicating nodes. This concept of using multitransceiver FSO structures between various buildings located within a mile from the ISP location serving a large number of end nodes can be a viable solution for the last-mile problem. The researchers developed the prototype of a spherical surface covered with FSO transceivers where the optical antenna was designed based on spatial reuse and angular diversity via directional transceivers tessellated on the surface of the spherical node. The design also established required alignment of two transceivers in the LOS of each other. Their concept for auto-alignment mechanisms and cross-layer buffering schemes to overcome the FSO directionality has the potential solution for the last-mile problem as well as mobile inner-city settings. Some solutions for the last-mile are described next.

3.2.2.1.1 Small Cells and Mesh Networks Solution

One possible solution would be to connect broadband fiber to Wi-Fi (capable of receiving and transmitting messages at very high speeds in the order of Gbit/s or more) with high bandwidth, secure exchange (using various coding and modulation techniques, especially suitable for optical regimes) for distances of 1−10 kms. A combination of small-cells approach with possible mesh networks with point-to-multipoint configuration can be an optimum solution for solving the last-mile problem. Some of the timing and technology development issues to handle this problem include:

- Recent developments in electrooptical components, fiber optical and high-speed devices due to the success of fiber optics, new semiconductor materials and devices, and processing technology. For example, high data rate optical data transmissions over long distances (e.g., for relaying) are becoming possible with the use of low-noise optical amplifier (EDFA) and low-noise single-mode fibers with extremely small coupling loss and very high-speed detectors. Designs of new network architectures for high bandwidth data communications are needed.
- Novel ways of repurposing fiber optical components and high-speed devices in FSO space technologies were developed to deal with atmospheric turbulence mitigations. For this reason, the new technology can now address and solve the so-called *last-mile problem*

where the signals from the broad telecommunication backbone (like fiber backbone) along the relatively short distance (hence, the last-mile) to and from the home or business via broadband FSO communications.

- Integrating with an optical/small-cells system will open up the *five nines* 99.999% *reliability* of using the established technology for data transmission under any atmospheric condition—a real viable solution for bad weather conditions.
- Proliferation of smart phones/tablets, video content, and future adoption of 4 K TV and such are increasing demand on broadband wireless. FSO links are the most potential viable solution.

Earlier analysis of link budgets and visibility-limiting weather conditions indicates that to meet carrier-class availability (99.999%), FSO links should normally be less than 140 m.

3.2.2.1.2 Wide Field-of-View Amplified Fiber-Retromodulator Unit

The niche area of free-space laser communications and data links using modulating retroreflectors, or retromodulators, is very active [6]. This is a growing area of interest since technology can now support "shutters" that can achieve usable communication rates. Retromodulators require very little power draw and offer extremely small form factors and mass, which is very practical to install on building roof corners. The system can be designed to minimize the challenge of beam alignment and deployment.

Some of the features of retromodulators are the following:

- A novel approach to an optical system for remote optical communications at a high data rate between a base station and a remote station that includes a new type of retro-reflector consists of two sets of lenslets coupled with a single-mode fiber array (fiber-retro) and can operate under atmospheric turbulence.
- The fiber-retro modulates the optical beam according to the input signal/data and redirects the modulated optical beam to the base station for detection by a receiver.
- The technology provides a means of extremely fast remote data transfer with additional advantage of maintaining secure data transfer (recovery) achieving a wide field-of-view.

A new concept in modulated retroreflector technology based on fiber-based amplified-retromodulators using single-mode fiber is recently described [6,7] in detail. The concept is based on a recent patent by the author and his colleague [7] (Patent No. US 8, 301, 032 B2, Oct. 30, 2012) describing a wide field-of-view amplified fiber-retro system. The effects of atmospheric turbulence on the amplified fiber-retro modulator (AFRM) are discussed [6] with a communication systems link analysis of BER in the presence of atmospheric turbulence and scattering medium. It is also possible to design and develop a modulated retroreflected-based FSO communication system that can be duplex (i.e., perform two-way optical communications). Fig. 3.1 shows the concept of a full-duplex FSO communication system based on a modulated retroreflector.

In the small-cells concept (described next), AFRM units can be installed on each building and can establish point-to-point and point-to-multipoint optical communication two-way links.

3.2.2.2 Small-Cells Networks

As one of the potential solutions for the last-mile problem, it is possible to develop network architecture using street-shaped small-cells networks. Small-cells networks can also be useful in large cities with tall office buildings where the AFRM-based optical devices can be installed and point-to-multipoint communications can be performed. Street-level small cells can offer a new coverage model and can be located close to the mobile customer with available local street light poles (for integrating Li-Fi, to be discussed later) to provide a complete solution for the last-mile problem. The small cells can be designed to cover a few tens of meters to a couple of hundred meters of fiber backhauled to the network via FSO links, and can deliver two-way optical link communications. Fig. 3.2 shows an artist's small-cells concept to develop a basic technology block to achieve all-weather 99.999% availability.

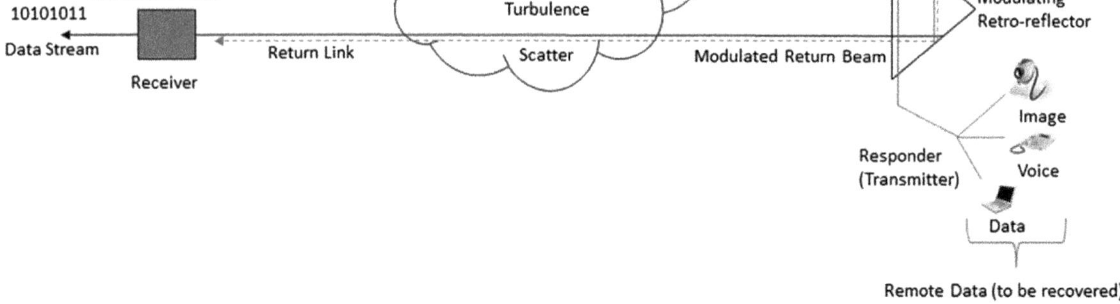

FIGURE 3.1
Modulated retroreflector-based FSO communication concept (full-duplex).

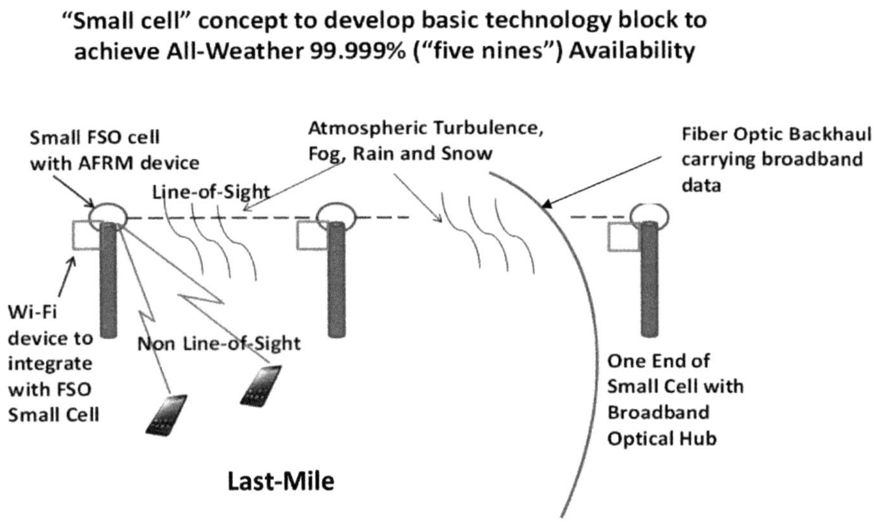

FIGURE 3.2
An artist's concept for small-cells deployment to solve the last-mile problem.

Since we are using FSO communications as a potential solution for solving the last-mile problem, the Wi-Fi's in Fig. 3.2 for developing the concept of small-cells deployment can be replaced by Li-Fi's, which use common household LED light bulbs to enable data transfer, boasting speeds up to 224 Gigabits per second [8]. Li-Fi, coined by University of Edinburgh Professor Harald Haas during a talk in 2011, has envisioned light bulbs that could act as wireless routers. Li-Fi and Wi-Fi are quite similar as both transmit data electromagnetically; however, Wi-Fi uses radio waves and Li-Fi runs on visible light. Li-Fi is therefore is a visible light communication (VLC) system and accommodates a photodetector to receive light signals and a signal processing element to convert the data into streamable content. When data is fed into an LED light bulb (with signal processing technology), it then can send data embedded in its beam at rapid speeds to the photodetector. The changes in the rapid dimming of LED bulbs is then converted by the receiver into an electrical signal, which is then converted back into a binary data stream that is recognizable as web, video, and audio applications that run on Internet-enabled devices [8]. Fig. 3.3 shows an artist's concept for Internet connectivity with street lights equipped with Li-Fi, which therefore can either add or replace the Wi-Fi–based solutions for small cells described earlier. Li-Fi is discussed in another chapter of this book about VLC with Li-Fi and Internet connectivity and therefore is not repeated in detail here.

3.2.2.3 Mesh Network

FSO communication can therefore provide a last-mile solution to broadband access. The signal power losses were discussed earlier for evaluating

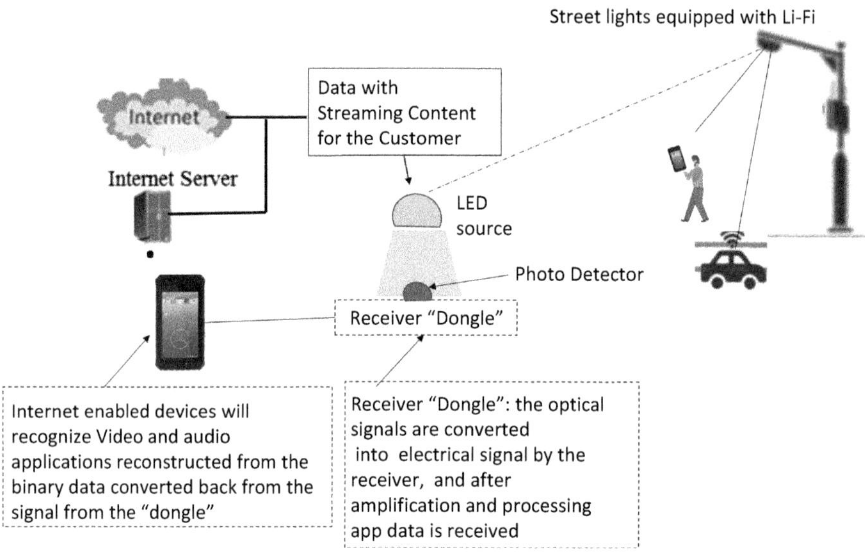

FIGURE 3.3
An artist's concept showing Internet connectivity with street lights equipped with Li-Fi.

atmospheric effects on optical propagation through rain, fog, and so on. For fog the signal can suffer an appreciable loss of up to 50–300 dB/km and it is also important to keep the beams carefully focused between transceivers within LOS of each other. A mesh topology can be a potential solution to keep the optical links sufficiently short to avoid any outage during fog. Fig. 3.4 shows an FSO network with different architectures. For high reliability, the optimum network architecture is a meshed network.

From the earlier analysis when distances are kept less than ~150 m, it is possible to have five-nines of availability: 99.999%. Using buildings located near the hub as regenerators of the signal the distances can be extended so that further buildings can be reached. Thus a single point of fiber can reach a wider set of end points connected with the small-cells concept described earlier. Network-related issues for solving the last-mile problem include optimal locations of base stations, architecture of optimal connectivity and routing, managing mobility, and supporting cell base stations. The wireless optical meshes can provide cost-effective, local broadband access at high capacity with flexibility over a wide service area carrying simultaneously voice, IP, video, and Automated Teller Machine (ATM) cell streams.

In conclusion, one of the most practical solutions for the last-mile is the combination of the small-cells approach with meshed network architectures, which will be capable of establishing high-speed optical communications even in the presence of atmospheric turbulence and scattering media like fog.

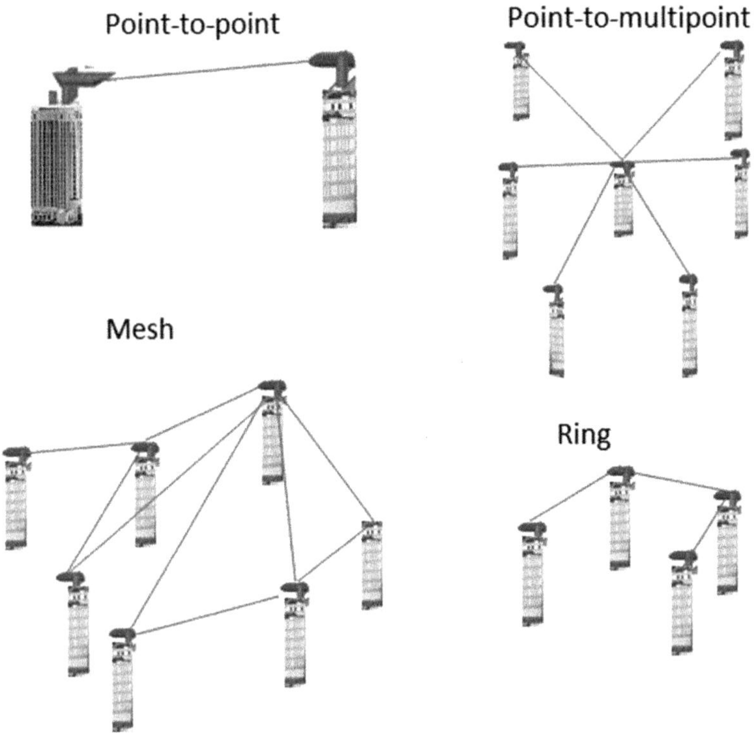

FIGURE 3.4
FSO network with various architectures.

3.2.2.3.1 Other Related Solutions for Last-Mile

3.2.2.3.1.1 Hybrid Optical Communication System. Researchers reported in Ref. [9] a design and simulation of a hybrid optical communication system that is a combination of fiber link and wireless optical diffused link (DL) to provide a possible solution for solving first- and last-mile problems. The proposed system consists of the front and back end of the system being a short-haul free-space DL setup and in between a long-haul fiber dense wavelength division multiplexing (DWDM) setup. The proposed system is composed of an 8-channel DWDM fiber optic link for long-haul, and wireless optical diffuses link for short-haul, multicasting applications. Optical DL links at either end can solve the first-mile and last-mile problem of an optical communication network. The wireless optical DL uses the optimized coherent optical quadrature phase shift keying modulation as the best choice for operating the system in the presence of atmospheric turbulence. The fiber-based DWDM system uses the symmetrical dispersion compensation technique. This optimized modified duo-binary return-to-zero modulation format is the most efficient in performing in presence of fiber nonlinearity and dispersion. The optimized coverage distance was found to be 126 m of wireless optical diffuse link and 450 m of fiber optic link. The researchers used the free-space diffused optical link setup for the possible solution of the last-mile and first-mile problem. The details of the system setup are described in Ref. [9]. The bit rates considered in their simulation were 160, 100, and 40 Gbit/s per channel. The optimized coverage distance of 126 m for wireless diffuse link for a bit rate of 160 Gbit/s per channel is very promising since five-nines availability in the presence of severe fog can be achieved, solving the last-mile problem.

3.2.2.3.1.2 Multichannel FSO Communication System in Backhaul for a Potential Solution for the Last-Mile. Multichannel FSO communication system using compact (maximum one square-foot), two-dimensional antennas with multiple communication links between them has been explored [10]. High bandwidths of hundreds of Gbit/s are considered [10]. Addressing the multichannel interference issues for such arrays to reduce per channel capacity by efficient array design and employing optical orthogonal codes (OOCs) for FSO communication multi-Gbit/s bandwidths have been achieved. This unique concept technology can be applied directly to multihop broadband access networks or mesh networks and in backhaul to connect wireless base stations. This is therefore a practical potential solution for the last-mile problem. The basic principles for designing a multichannel FSO system is that the effective bandwidths are increased with the number of channels. A 10×10 two-dimensional array should be capable of providing 100 Gbit/s in aggregate capacity with a 1 Gbit/s single transceiver system. The performance of this type of two-dimensional arrays with multiple transmitters and receivers will depend on the number of design factors such as the distance between the arrays, divergence angle of the light source, and package density of the transceivers on the surface of the array.

A multichannel system with $K+1$ simultaneously operating channels, in the performance FSO communication system analysis [10], with a Gaussian beam profile and on-off keying modulation for the transmitting information bits, a threshold value I_T for the receiver's output, a Gaussian shot noise with a power density N_0 were assumed. They modeled the array communication channel as a binary asymmetric channel and the capacity was estimated as a function of bit error probability. The channel capacity of their array considered the package density on each of the arrays, distance between arrays, and divergence angle of the transmitting light source. The array capacity C is given by [10].

$$C = \max_{p_1} H(\overline{p}_1 \overline{P}_e) - p_1 H(\overline{P}_e) \quad (3.1)$$

where H is the entropy, p_1 is the input symbol (ONE or ZERO) probability distribution, and P_e is the probability of error. The probability of error for multielement FSO communication is given by

$$p_e = Q\left(\frac{I_T}{2\sqrt{\frac{1}{2}\sum_{k=0}^{K-1} e^{-\left(\frac{4\theta_k}{\theta}\right)^2} + \frac{N_0 T_b}{4}}}\right) \tag{3.2}$$

where the signal from the transmitter was limited by the boundary of $(0 \leq t \leq T_b)$ for binary 0 or binary 1. In Eq. (3.2) θ_k is the angle of transmission from the *kth* interferer and θ is the divergence angle of the beam. Q is the usual Q-factor in a Gaussian distribution. When including OOC to minimize the interarray interference for the communication system, for an orthogonal code of weight N and length F, the probability of error is then given by

$$p_e = Q\left(\frac{I_T}{2\sqrt{\left(\frac{N^2}{2F}\right)\left(1-\frac{N^2}{2F}\right)\sum_{k=0}^{K-1} e^{-\left(\frac{4\theta_k}{\theta}\right)^2} + \frac{N_0 T_b}{4}}}\right) \tag{3.3}$$

The capacity of the FSO communication system for the array and including OOC can be calculated from Eq. (3.1) for capacity. Their results [10] show that an array without OOCs can be used for shorter distances ($\sim$150 m) with single laser, and the arrays can be used over longer distance ($\sim$500 m) using VCSELS laser array. It was also shown that by proper design of the pattern for the transceivers packing and implementing optical orthogonal codes on these arrays, capacity close to 1 can be achieved. A helical array design has shown to achieve better capacity. The multichannel FSO communication systems with two-dimensional arrays backhaul can therefore provide a practical solution to offer hundreds of Gbit/s in solving the last-mile problem.

3.2.2.3.1.3 OFDM-FSO System for the Last-Mile Problem. As mentioned earlier, FSO is an emerging solution for the last-mile broadband connectivity where deployment of fiber is expensive and impractical. The last-mile bottleneck arises when the high bandwidth demands in metropolitan area networks require flexible and cost-effective service. The technology of FSO communication can address any connectivity needed in optical networks, such as core, edge, or access networks. Severe atmospheric effects such as turbulence can degrade the performance of FSO communication links including in clear air turbulence conditions. Orthogonal frequency-division multiplexing (OFDM) enables robust transmission over optical channels in presence of atmospheric turbulence-induced scintillation [11,12]. In OFDM the high data rate is divided into multiple low data rates, which are then transmitted in parallel. The objective of this scheme is to lower the symbol rate and provide high tolerance to deep fades caused by atmospheric turbulence. The OFDM-based FSO communication system will have the advantages of both OFDM and FSO so that it is a good candidate for a last-mile solution for broadband connectivity in presence of atmospheric turbulence [12].

OFDM is a very well-established modulation scheme for modern broadband communication systems for FSO communications. In some instances, low-density parity-check coded OFDM can provide an efficient coded modulation technique suitable for FSO transmission [13]. Their results show that under strong turbulence case, comparing the BERs for coded OFDM and uncoded OFDM, the BER of convolution-coded OFDM have improved from a value of 10^{-2} to 10^{-3} for an SNR of 7.5 dB.

The performance of intensity modulation/direct detection-time division multiplexed-FSO (IM/DD-TDM-FSO) link recently has been compared with the IM/DD-OFDM-FSO link in terms of BER performance through turbulence following a gamma-gamma distribution. A gain of 14.7 dB in SNR and 7.2 dB improvement in receiver sensitivity with an OFDM-FSO link compared to a traditional TDM-FSO link is reported [14].

References

[1] I.I. Kim, E. Korevaar, Availability of free space optics (FSO) and hybrid FSO/RF systems, in: E. Korevaar (Ed.), Proc. SPIE, Optical Wireless Communications IV, November 27, 2002.

[2] I.I. Kim, H. Hkakha, P. Adhikari, E. Korevaar, A.K. Majumdar, Scintillation reduction using multiple transmitters, in: Free Space Laser Communication Technologies IX, Proc. SPIE, 2990, 1997, pp. 102–113.

[3] M. Tatarko, L. Ovsenik, J. Turan, Availability and reliability of FSO links estimation from measured fog parameters, MIPRO (2012) (MEET), Conference in Opatija, Croatia, May 21–25, 2012, 192–195.

[4] A.K. Majumdar, Free-space laser communication performance in the atmospheric channel, in: A.K. Majumdar, J.C. Ricklin (Eds.), Free-Space Laser Communications, Principles and Advantages, Springer Science LLC, 2008.

[5] A. Sevincer, M. Bilgi, M. Yuksel, N. Pala, Prototyping multi-transceiver free-space-optical communication structures, in: International Conference on Communications (ICC), 2010 IEEE, May 23–27, 2010.

[6] A.K. Majumdar, Advanced Free Space Optics (FSO): A Systems Approach, Springer Sience+Business Media, New York, 2015.

[7] A.K. Majumdar, T.M. Shay, Wide Field-of View Amplified Fiber-retro for Secure High Data Rate Communications and Remote Data Transfer, October 30, 2012. US Patent, No. 8, 301,032 B2, date of patent.

[8] https://www.techworld.com/data/what-is-li-fi-everything-you-need-know-3632764/.

[9] B. Ptanaik, P.K. Sahu, Optimized hybrid optical communication system for first mile and last mile problem solution of today's optical network, in: 2nd International Conference on Communication, Computing and Security [ICCCS-2012], SCiVerse Science Direct, 6, Procedia Technology, Elsevier, 2012, pp. 723–730.

[10] J. Akella, M. Yuksel, S. Kalyanaraman, multi-channel communication in free-space optical networks for the last-mile, in: Proceedings of the 2007 15th IEEE Workshop on Local and Metropolitan Area Networks, IEEE, 2007, 1-4244-1100-9/07.

[11] S.M. Rajendrakumar, M. Karruppaswamy, Analysis of link availability in FSO-OFDM system under various climatic conditions, Engineering Journal 19 (1) (January 30, 2015).

[12] H. Kaushal, V.K. Jain, S. Kar, Free Space Optical Communication, Springer, 2017.

[13] A. Sangeetha, B.M. Varghese, S. Khara, "Performance analysis of coded optical Ofdm over free space atmospheric link, International Journal on Recent Trends in Engineering and Technology 05 (02) (March 2011).

[14] P. Kumar, A. Srivastave, Enhanced performance of FSO link using OFDM and comparison with traditional TDM-FSO link, in: IEEE International Broadband and Photonics Conference, Bali, April 23–25, 2015.

CHAPTER 4

Fundamentals of Free-Space Optical Communications Systems, Optical Channels, Characterization, and Network/Access Technology

4.1 INTRODUCTION

This chapter discusses the basics of optical wireless communication (OWC) system and subsystem structure, which includes optoelectronic components (transmitter/receiver), OWC channel model, modulation for optical wireless channel, communication parameters for evaluating performance, link budget, and OWC performance: bit-error-rate (BER), and signal-to-noise ratio (SNR). The chapter then presents OWC network and access technology, providing an up-to-date introduction to optical networks, and introduces optical wireless technology, specifically addressing the network aspect into the global communications infrastructure. The chapter explains how OWC systems play an important role as broadband access technology where directly propagating an optical beam from fiber termination points through free-space using full-OWC transceivers. The technology explains how at the receiving end a corresponding transceiver can directly couple the free-space propagated optical beam into either an optical fiber connection port or another free-space optical (FSO) port similar to a relay system. The global connectivity can thus be developed. The chapter presents a basic design concept of full-OWC link with a potential of offering stable, error-free transmission at multiple Gb/s data rates globally. Topics discussed in this chapter include:

- Fundamentals of optical wireless network
- Optoelectronic components, including optical fibers, optical transmitters and receivers, and optical multiplexers and amplifiers
- Various types of optical networks
- Modulation and multiple access techniques
- Satellite communications (fixed and mobile services)

The chapter describes the basics of indoor OWC using multiple sources including visible light communication (VLC)-based multiple-input–multiple-output (MIMO) technique to provide reliability improvement and bandwidth

Optical Wireless Communications for Broadband Global Internet Connectivity.
https://doi.org/10.1016/B978-0-12-813365-1.00004-7

efficiency increase. Communication blocks in an indoor OWC system include intensity modulation/direct detection (IM/DD), optical transmitters and receivers (VLC and LED arrays and Fly Eye receivers), optical wireless channel propagation characteristics, and channel modeling: channel impulse response, system performance (BER) and outage for indoor OWC systems, orthogonal frequency division multiplexing (OFDM) for indoor OWC and aircraft (commercial airplanes) on-board wireless communication. This chapter also addresses fundamentals of coding concepts specifically to deal with atmospheric turbulence and scattering channels. Some of these channels include terrestrial optical links and others may include ground-aerial links (e.g., ground-to-airplane/balloons/low earth orbit [LEO] satellites) wherever the communication channel has some part of the turbulent/scattering medium. The chapter discusses basics and general design implementations of low-density parity-check for repetition MIMO, and also space-time coding with applications. Another excellent candidate for optical coding, OFDM is explained with applications in the atmospheric channels. This chapter explains channel capacity of hybrid FSO wireless channel with basics and calculation of channel capacity relevant to OWC systems that have to operate under different atmospheric conditions. It provides information about improving the optimum power adaptation that maximizes the total channel capacity. A block diagram of the system concept is discussed. Multiple access techniques in an OWC system is also considered, where the need of multiple access technologies for multiuser OWC systems is explained and addresses specifically the optical code-division multiple access (OCDMA) technique and the basic architecture of an OCDMA/OWC system. The chapter illustrates and analyzes the OCDMA/OWC in both indoor and outdoor FSO links. Overall FSO link performance, theory and analysis, is provided based on the fact that the physics of optical wave propagation are entirely different for different OWC scenarios and environments. Providing a comprehensive theory of optical propagation, this chapter shows some typical examples that are useful for OWC engineers who want to develop end-to-end system design for achieving Internet connectivity. Mitigation techniques for designing an atmospheric OWC system are also discussed. Eye-safety issues for limiting power of a source (transmitter) for indoor and outdoor environments are clearly pointed out.

4.2 FREE-SPACE OPTICAL COMMUNICATION SYSTEMS: VARIOUS TYPES FOR DIFFERENT OPTICAL NETWORK ARCHITECTURE

FSO communication systems are where free space acts as a communication channel between transceivers that are line-of-sight (LOS) for successful transmission of optical signals. The channel can be atmosphere, space, or vacuum, whose characteristics determine the transmission and reception of optical signals for designing reliable and efficient communication systems. Using FSO technology data is transmitted by propagation of light through atmospheric or space communication channels, allowing optical connectivity. FSO communication offers a high data rate to meet the tremendous increasing demand of broadband traffic mostly driven by Internet access and HDTV broadcasting services. Compared to fiber optics technology, FSO offers much more flexibility in designing optical network architectures at very high speeds, at tens and hundreds of Gbit/s rates. However, FSO communication is affected by atmospheric effects, which limits sensitivity and achievable data rates with acceptable BER. Some of these degradations are turbulence, absorption, and scattering, and various mitigation techniques exist for reliable and efficient data transmission [1] and to increase the communication performance. Both point-to-point, point-to-multipoint, multipoint-to-point, and multipoint-to-multipoint FSO communications are possible, depending on the different scenarios of establishing optical links. FSO communication is the most practical alternative to solve the bottleneck broadband connectivity problem. The data rates provided by FSO links continue to increase in both long- and short-range applications. FSO will be one of the most unique and powerful tools to address connectivity bottlenecks that have been created in high-speed networks during the past decade due to the tremendous success and continued acceptance of the Internet. The next generation of Internet connectivity will push the limits of existing infrastructure with high-bandwidth applications such as videoconferencing, streaming multimedia content, and network-enabled portable devices. Clearing these bottlenecks is crucial for the future growth and success of the contemporary Internet society. The bandwidth of optical communications access and edge networks will be needed to satisfy these demands. Communication systems are concerned with the transmission of *information* from a source to a user. The purpose of a communication system is therefore to transfer information. A very basic block diagram of any communication system (optical or radiofrequency (RF)) is shown in Fig. 4.1.

Fig. 4.1 shows a single point-to-point system, whereas in a multiplexed system there may be multiple input and output message sources and users (also called destinations). Fig. 4.2 shows other possible configurations and links for multipoint connections.

OWC is the next frontier for high-speed broadband connection and offers the following unique features and advantages: high bandwidth/capacity, ease of deployment, compact size, low power, and improved channel security. OWC can transmit and exchange voice and video communication data through the atmosphere/free-space at the rates of tens of Gbit/s and much more.

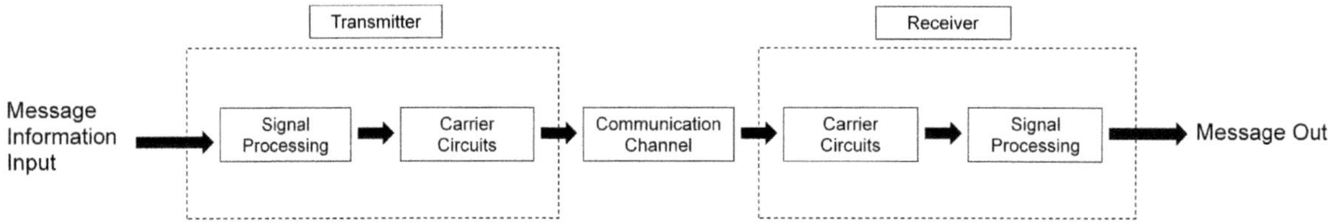

FIGURE 4.1
Basic communication system.

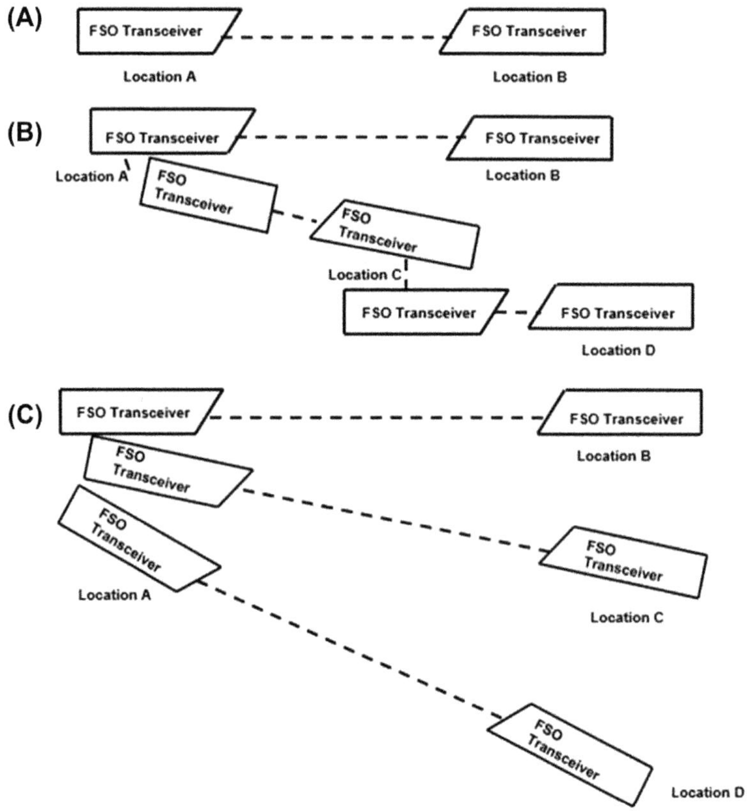

FIGURE 4.2
Point-to-point and multipoint connections and links: (A) point-to-point architecture, (B) mesh architecture, and (C) point-to-multipoint architecture.

FSO communications can provide high data rates in Gbits/s ranges through the atmosphere for ranges from a few hundreds of meters to a few kilometers. FSO links include the following: (1) chip-to-chip communication, (2) indoor infrared (IR) or VLC, (3) interbuilding communication, and (4) free-space laser communications including airborne, space borne, and deep space missions.

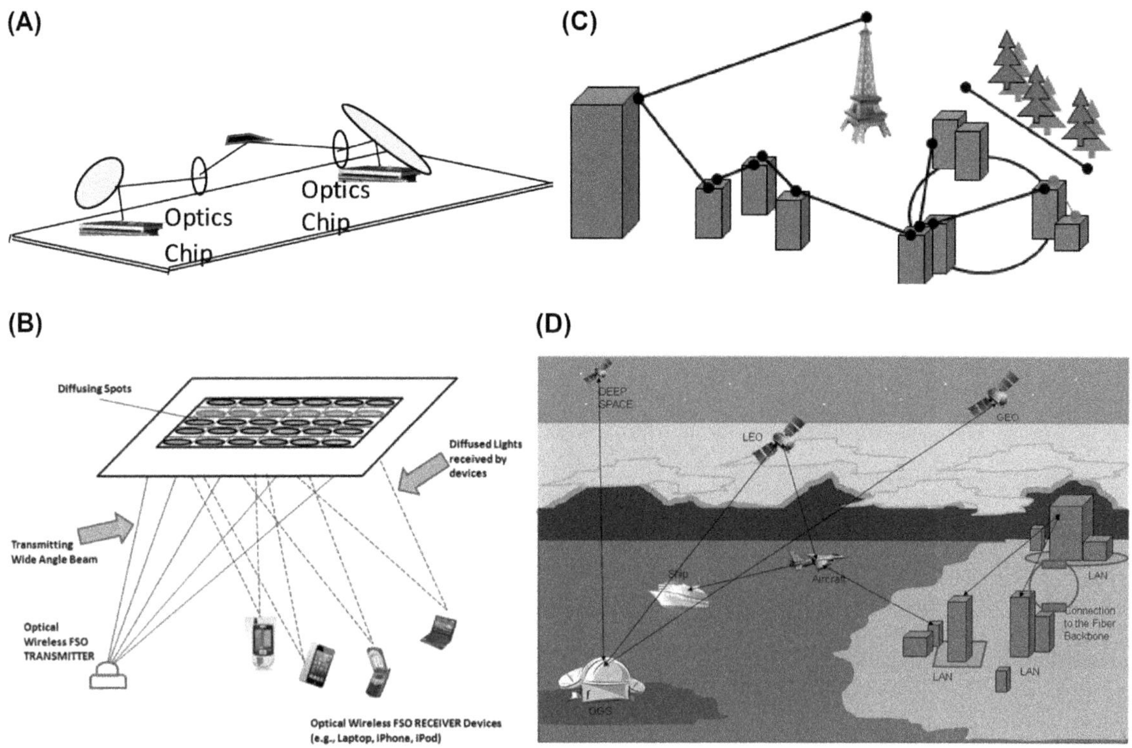

FIGURE 4.3
Wireless optical communications: (A) chip-to-chip, (B) indoor infrared or visible light communication, (C) interbuilding communication, and (D) airborne/space platform free-space laser communications.

Fig. 4.3 shows various FSO communication link applications and scenarios for different categories of transmission ranges. For very short distances such as chip-to-chip or board-to-board interconnections FSO links are also used to address the complex communication requirement in high-speed, compact optoelectronic devices and lower power consumption to make them compatible with other interconnected network devices. Other communication links are discussed next.

4.3 POINT-TO-POINT AND POINT-TO-MULTIPOINT COMMUNICATIONS BETWEEN BUILDINGS (SHORT RANGE)

Terrestrial FSO communications have been developed for the last many years and commercially available systems are now readily available. Fig. 4.4 shows an FSO terrestrial link between two buildings with rooftop transceivers establishing full-duplex communication links. These types of FSO links are suitable in

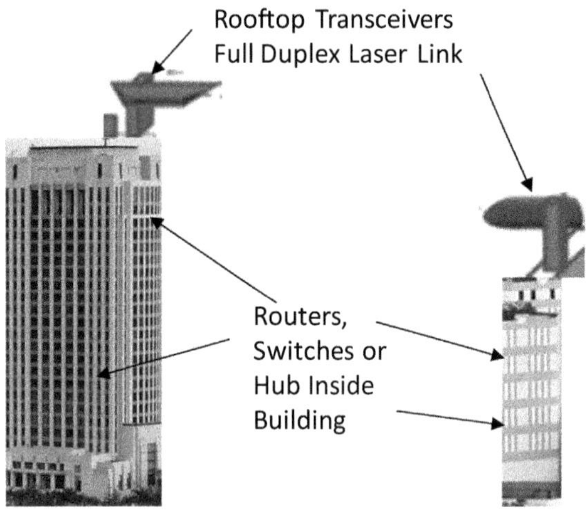

FIGURE 4.4
A typical free-space laser full duplex communication link between two office buildings.

urban areas where it is difficult to rig the roads for broadband fibers. Interconnecting local area network segments inside a building in the campus can provide high-speed connectivity to end users. FSO links can reach existing metropolitan area fiber rings to extend the FSO services to core infrastructure as well as to solve the last-mile problem. Terrestrial links include communication between building-to-building, mountain-to-mountain, or ground-to-mountain (slant path) links. FSO network for this type of link can be point-to-point or point-to-multipoint, or ring or mesh topology as shown in Fig. 4.5.

To design multibeam FSO communication links, a hybrid wavelength division multiplexing (WDM)/multibeam architecture can be used as shown in Fig. 4.6. Multibeam hybrid WDM can be used for FSO communications to improve data rate in the presence of atmospheric turbulence. The transmitter consists of a continuous wave laser, Mach-Zehnder modulator, data generator, and a demultiplexer. A number of continuous WDM channels spaced by a set of a number of channels is integrated to the demultiplexer and then transmitted to the destination through the atmospheric channel. In the receiver, all beams are collected and separated into a single beam profile using a demultiplexer. An avalanche photo detector (APD) photodiode is used to convert optical signal into electrical signals and is filtered to recover the data. This technique can offer a potential solution to the last-mile problem to achieve high capacity, increased transmission distance, and high data rate in multiples of Gbit/s range.

The transmitter consists of laser diodes at operating at 1550 nm wavelength, one optical WDM multiplexer, one optical splitter, and spatially diverse lenses.

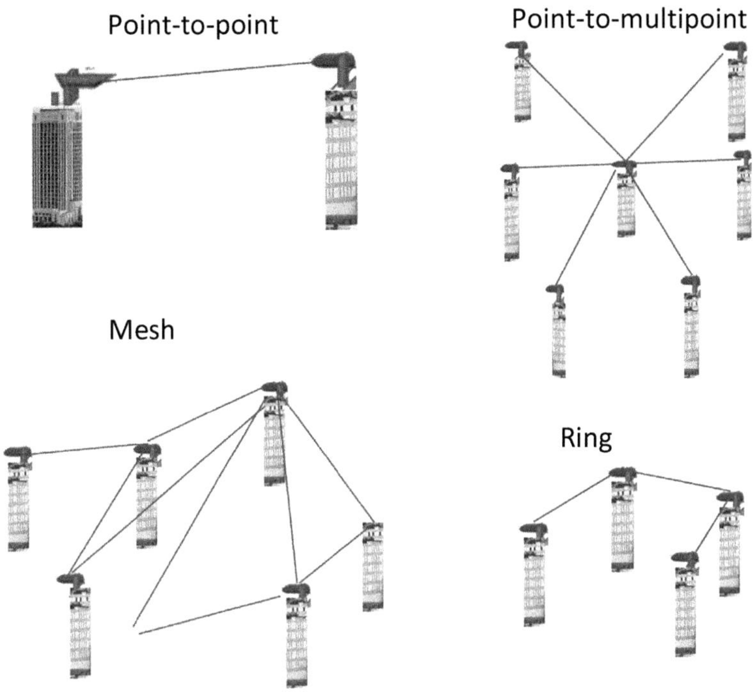

FIGURE 4.5
Terrestrial FSO links, point-to-point, point-to-multipoint, and mesh and ring topologies.

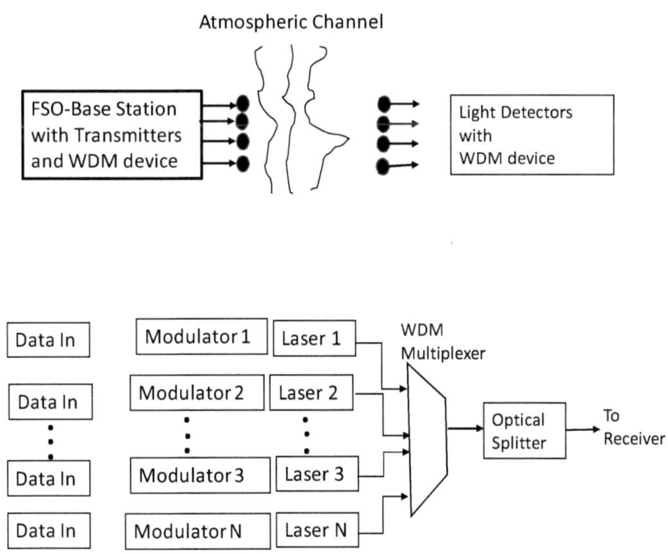

FIGURE 4.6
The architecture of hybrid wavelength division multiplexing/multibeam free-space optical communication link.

The transmitting beams are sent to the FSO channel and received at the receiver end. Multibeam architecture can improve the FSO communication performance for terrestrial links under adverse weather conditions.

4.4 INDOOR COMMUNICATION LINK

The design and performance factors of long-distance FSO communication systems apply equally to short-distance systems such as indoor optical communication but with one obvious exception: atmospheric loss has no effect for all indoor systems. The communication power link budget is therefore determined almost entirely by the transmitter launch power, free-space loss, and receiver sensitivity. Indoor communications have been described in the literature; see for example [1]. A base station for each cell has a number of terminals connected by short-range FSO communication links. These links are basically IR links using LEDs or diode lasers at visible wavelengths offering VLC. The details of the indoor communication system with IR and visible light/Li-Fi will be discussed in the next chapter. Propagation modeling for indoor OWC is analyzed in terms of impulse responses for an indoor channel with an LOS and non–line-of-sight (NLOS) path [1] to evaluate received power and the SNR of the indoor communication system.

4.5 TERRESTRIAL FREE-SPACE OPTICAL LINKS

FSO networks establish a point-to-point and LOS optical wireless connection between two transceivers through outdoor atmospheric channels. Because of the limitations imposed by the atmosphere on optical propagation and available communications systems performance, the LOS requirement allows the distances from hundreds of meters up to tens of kilometers. FSO-based optical telecommunication networks offer important means for establishing broadband Internet access. Optical wireless terrestrial networks are used to bridge existing high-data-rate networks where there are no optical fibers laid or it is impractical to ditch. FSO links provide solutions for the last-mile problem and offer high-bandwidth connection over a large distance for remote end-users.

OWCs have been used in a number of most active development efforts; for example, in terrestrial FSO communications, VLCs, and space FSO systems. Wireless FSO communications systems can easily provide connectivity over 2–3 km at data rates up to several Gb/s. Commercially available systems are already available to provide this service in a metropolitan city between two buildings without any costly fiber optics.

Fig. 4.7 shows a terrestrial FSO application in a city indicating various possible weather effects and building sway for tall buildings.

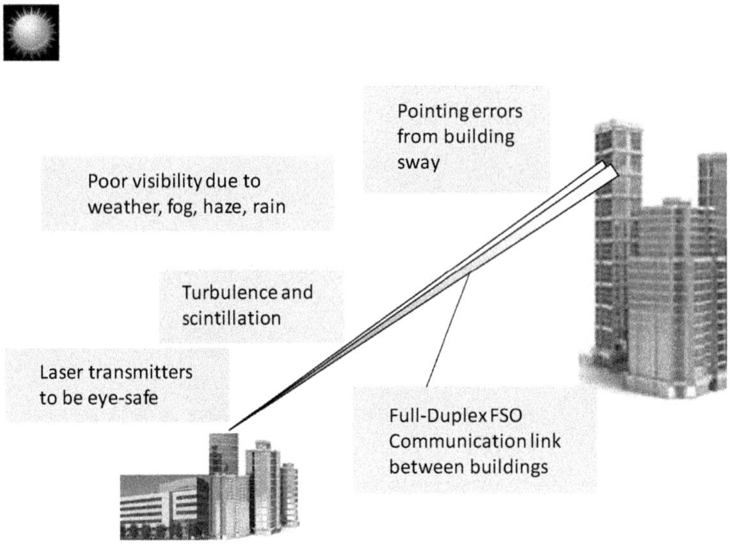

FIGURE 4.7

Terrestrial free-space optical application in a city influenced by weather effects and building sway.

4.6 LONGER RANGE (HORIZONTAL AND SLANT PATH) FREE-SPACE OPTICAL LINKS: VARIOUS FREE-SPACE OPTICAL COMMUNICATION SCENARIOS

The FSO is an outdoor wireless communication technology with a potential high data rate and bandwidth per link exceeding 10 Gb/s or even 50 or 120 Gb/s per link. Fig. 4.8 depicts the different FSO communication scenarios,

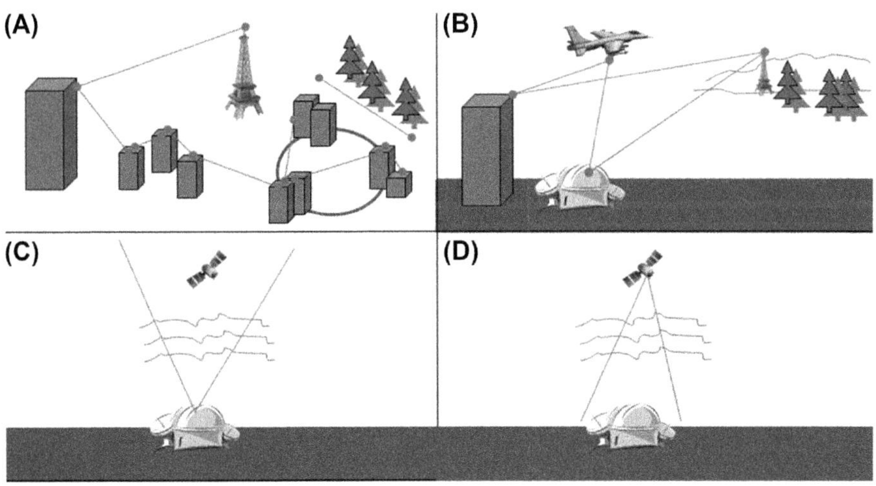

FIGURE 4.8

Various free-space optical communication scenarios, including near the ground and tropospheric atmosphere communication channels: (A) horizontal link, (B) slant path FSO link, (C) uplink, and (D) downlink.

each of which are described as follows [1]. FSO communication scenarios include both near the ground and tropospheric atmosphere communication channels. The atmospheric effect near the ground is the most severe on FSO communication performance, and the effects are less in the free space.

Data rates up to 100 Gb/s over several km under clear weather condition has been tested. Using the atmospheric mitigation techniques to combat atmospheric and weather effects, accurate statistical channel modeling and MIMO processing optimized relaying and link robustness can be increased for terrestrial FSO systems. FSO has the potential of providing backhaul technology for 5G wireless systems or a last-mile solution for accessing networks.

- *Horizontal link*: Such as from rooftop-to-rooftop of tall buildings, from window-to-rooftop, or from window-to-window, and some FSO nodes may be mounted on tall poles. Some of these FSO links may include a base transceiver station, backhaul, and cell sites. The range of this horizontal link is typically up to a few kilometers, connecting office buildings in a metro area (Fig. 4.8A). In computing the fluctuating intensity of the received signal, the strength of the turbulence is usually assumed uniform for the horizontal link. Physical obstructions such as birds, insects, tree, limbs, or other factors can temporarily or permanently block the laser LOS. Furthermore, building motion due to wind, differential heating and cooling, building sway, or ground motion over time can result in serious misalignment of a fixed-position FSO communication system.
- *Slant path FSO link*: The slant path situation can arise when the FSO path is not quite horizontal and the turbulence strength is not uniform along the propagation path (i.e., variable). Some of the examples to be treated in other chapters of the book include unmanned air vehicle (UAV) (or airborne) or top-of-a-mountain to a ground link (Fig. 4.8B).
- *Uplink*: Typical example for an uplink scenario is a ground-to-space propagation path, such as ground station to a satellite. It is necessary to consider the altitude profile of the turbulence strength parameters in order to compute the fluctuations of the signal intensity received at the receiver. (Fig. 4.8C).
- *Downlink*: From space-to-ground is an example for a downlink FSO communication, such as a satellite-to-ground link. The altitude variation of the turbulence strength must be considered to compute the received FSO signal intensity fluctuations (Fig. 4.8D).

4.6.1 Space Link

For space-based applications, optical communications between a satellite and ground system offer unique opportunity with potential data transfer and high bandwidth as well as to provide global Internet connectivity in remote locations. Some of the atmospheric communication links are discussed next.

Recent demonstrations show research and development for a broad variety of possible applications for FSO links from stratospheric platforms to ground, aircraft to ground, and satellites to ground. Space links include ground-to-satellite and satellite-to-ground links and intersatellites. Some of the examples are links between ground and LEO/ geostationary equatorial orbit (GEO) and between LEO and GEO as shown in the Fig. 4.3D. Some of the interesting and promising applications for FSO links is the interconnection of airships or high altitude platforms (HAPs) placed in the stratosphere for establishing an Internet connection to users on the ground and optical terminals to build a network among them offering broadband data access to remote locations where only terrestrial infrastructure exists or where they would be very expensive to build. HAPs, being above the heights of clouds, have no link blockings for inter-HAP links due to clouds. HAPs can thus be used as a data relay for optical downlinks from satellites; for example, an LEO can dump its data via a high data rate optical link to the HAP.

4.7 AIRBORNE-TO-GROUND COMMUNICATION SYSTEM

UAVs as airborne platforms offer the emerging technology for UAV-based FSO communication links.

UAVs are a possible future application for both civil and military use. The large amount of data generated by UAVs requires high data rate connectivity, thus making FSO communication very suitable architectures. Research in the area of FSO communications is usually based on point-to-point link or long-range link for space applications. Mobility (i.e., relative movement of either the transmitting or receiving terminal, or both) is the greatest challenge in FSO communication technology. By including UAV to provide a communication terminal, a number of flexible and practical applications can be developed. There is increasing interest in UAVs for many applications, especially in the area of surveillance, due to zero risk of human casualty. In order to meet the increasing demand, efficient ways to communicate with UAVs with FSO optics are needed since the UAVs are moving platforms. Many commercial FSO links operate at 1–2 Gbps over ranges of 1–3 km.

4.7.1 UAV Scenarios for Free-Space Optical Communication Links

There are three basic FSO communication link scenarios with UAV platforms: ground-to-UAV, UAV-to-ground, and UAV-to-UAV or between UAVs (UAV swarm).

4.7.1.1 Ground-to-Unmanned Air Vehicle Mobile Free-Space Optical Channel

Fig. 4.9A shows the ground-to-UAV mobile FSO link where the speed of the UAV can reach several hundred meters per second. Tracking in this scenario is typically accomplished by mechanical components such as a two-dimensional rotating gimbal oriented on global positioning system (GPS) data from the UAV. The small divergent angle of the transmitting beam makes the tracking task difficult. The received SNR variation can be caused by misalignment of the transmitted beam and the detector due to mechanical issues. Pointing uncertainty errors and GPS positioning errors limit the data rates of such a link. An efficient tracking method is therefore necessary to improve the data rate.

4.7.1.2 Unmanned Air Vehicle-to-Ground Free-Space Optical Communication Link

Fig. 4.9B illustrates the concept of a bidirectional daytime and nighttime optical communication link from a UAV to a stationary ground station. One of the tasks the UAV might have is to take science images over desired targets and then download the images via the optical communication channel.

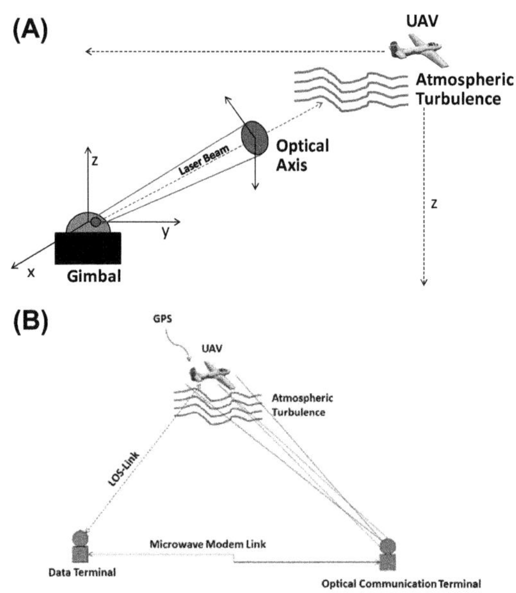

FIGURE 4.9
Unmanned air vehicle (UAV) mobile communication links: (A) ground-to-UAV, (B) UAV-to-stationary ground station.

An optical communication terminal receives a command from ground via RF links, which also receives continuous updates of the GPS information collected by the UAV GPS receiver. Simultaneously the UAV provides the optical communication terminal with its GPS information. The optical communication terminal uses updated information from both the UAV GPS receiver and the ground location to blind point the gimbal by sending a beacon signal to initiate the communication signal. Data communication starts when the signal terminal tracks on the beacon signal. A demonstration of the FSO communication link at 2.5 Gbps is presented [2] for a UAV altitude of 15.8–18.3 km using a 200 mW downlink laser at 1550 nm. They reported BER of 10^{-9}, which needed the pointing requirement on the flight terminal of 19.5 μrad and a bias error of 14.5 μrad with a probability of pointing induced fades (due to turbulence) of 0.1%. Vibration uncertainties also need to be eliminated for establishing a high data rate communication link.

4.7.1.3 Unmanned Air Vehicle Swarms: Links Between Unmanned Air Vehicles

The use of UAVs for transferring high date rate information is very rapidly attracting attention. In some situations, it may be necessary to collect data for a defined area with a variety of sensors. When these UAVs are operating in swarm formation, the observation area can be increased and an occasional loss of a UAV will not completely stop the data transfer process using the highly efficient optical communication links. FSO communications offering high bandwidth can therefore provide high data rate connectivity required by the large volume of data in UAV swarms. Fig. 4.10 illustrates an FSO network for

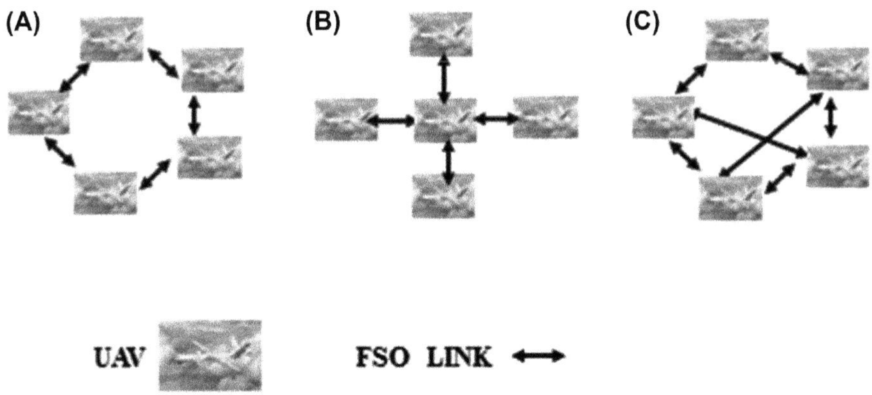

FIGURE 4.10

Free-space optical network for swarm unmanned air vehicles using three different architectures: (A) ring architecture, (B) star architecture, and (C) meshed architecture.

swarm UAVs using three different architectures: *ring architecture*, *star architecture*, and *meshed architecture*.

The key point here is to develop the UAV's capabilities to handle multiple sensor information in real time and in parallel so that a large amount of data can be transferred in real time with a goal to achieve a rate of 2.5 Gbps or higher.

In ring architecture mode, all FSO links are bidirectional and in case of a broken (or failure) link between any two UAVs, an indirect link may be used although the information is sent in the other direction where there is still a link [3]. In a star architecture, there is one UAV in the middle of the formation, which acts as an optical multipoint unit (OMU) [3] and is used as a repeater. All other UAVs equipped with transceivers are permanently connected with this OMU. A failure of the OMU can cause the breakdown of the whole configuration and a redundant OMU might be necessary. A meshed network can provide high reliability by combining the advantages of both the star and the ring architecture. Information flow from one UAV to another can be realized in different forms and can also be sent in the other direction of the ring network.

4.7.1.4 Free-Space Optical Between Unmanned Air Vehicles

Using multiple UAVs in cooperative swarm mode can survey a large area, collect a huge amount of data in short duration, and increase the robustness by FSO optical links to communicate between UAVs and communicating back to the command center on the ground station, all optically. Thus UAVs require high data rate connectivity. Data rates in the range of 100 Mbit/s to larger than 1 Gbit/s will be required to handle multiple sensor information and communicate with each other in real time and in parallel. UAVs equipped with FSO can have the technological capabilities to meet this high data rate requirement. With proper designs, the data rates can even go 2.5 Gbit/s or much higher using recent technological devices and optical networks.

4.7.2 Last-Mile Problem: Fiber Gateway Location-to-Optical Ground Station for Internet Connectivity

This section will be covered in the next chapter describing various architectures for FSO communication links between VLC and Li-Fi devices to connect between mobile devices such as a person carrying an iPhone sending/receiving messages on a street or in a car. Small-cell solutions to provide connectivity will also be explained, including optical-based small-cell concepts.

4.8 TERRESTRIAL AND SPACE PROPAGATION: POINT–POINT/AIRBORNE AND SPACE PLATFORMS TO GROUND PROPAGATION PATH: FREE-SPACE OPTICAL COMMUNICATION SYSTEMS PERFORMANCE

To understand how Internet connectivity is established by FSO communication systems with transceivers located at various platforms it is essential to understand propagation of optical beam through various media that are basically the communication channels. Atmospheric effects affect the propagating light and ultimately limit the communication system performance such as achievable data rates and communication capacity. FSO communication has increasingly attracted attention in the past decade for a number of applications for providing high bandwidth wireless communication links. Some of these applications include satellite-to-satellite cross-links, up-and-down links between space platforms and aircraft, ships, and other ground platforms, and among mobile or stationary terminals to solve the last-mile problem through the atmosphere. However, there are a variety of deleterious features of the atmospheric channel that may lead to serious signal fading, and even the complete loss of signal altogether. Absorption and scattering due to particulate matter in the atmosphere may significantly decrease the transmitted optical signal, while random atmospheric distortions due to optical turbulence can severely degrade the wavefront quality of a signal-carrying laser beam, causing intensity fading and random signal losses at the receiver. Atmospheric channel effects on free-space laser communication, and optical propagation relevant to FSO communications are explained and discussed in detail [1,4].

4.8.1 Free-Space Optical Transmitting Beam Types

Due to the tremendous growth into the telecommunications infrastructure using FSO communication links, there is increased need for computing the scintillation index (to be discussed later in this section) in turbulence with different kinds of laser beams such as plane, spherical, or Gaussian beam waves as depicted in Fig. 4.11. The object to search for the best (optimum) beam at the source (transmitter) is to minimize the degrading effects of turbulence in the atmospheric FSO links.

- Plane Wave
  A plane wave in this FSO context is defined as the type of electromagnetic wave when generated from the source point (transmitter end) propagates through the atmosphere in which the phase fronts are parallel planes.

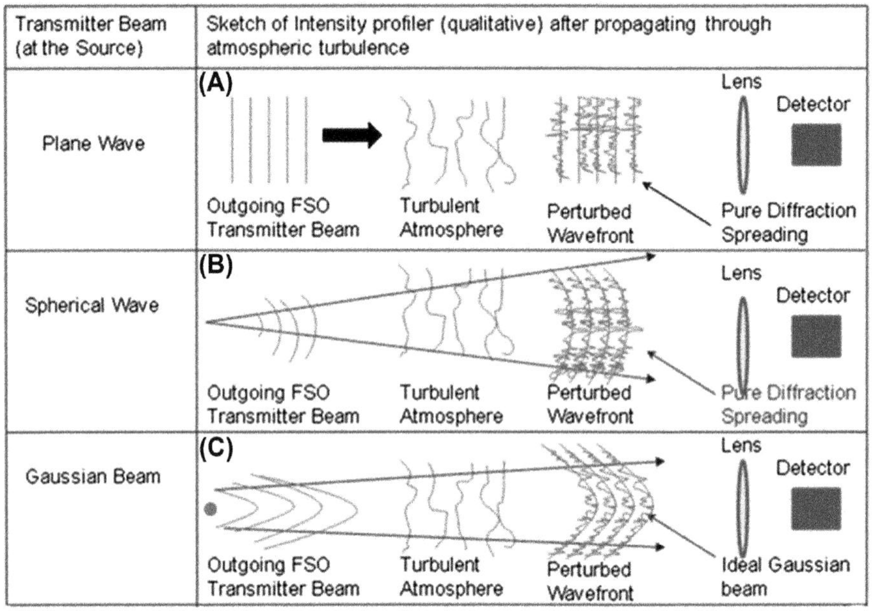

FIGURE 4.11

Sketches to illustrate intensity profiles of transmitting beams propagated through atmospheric turbulence: (A) plane wave, (B) spherical wave, and (C) gaussian beam wave.

Mathematical description:

$$U_0(r,z) = A_0 e^{i\phi_0 + ikz} \tag{4.1}$$

where U_0 is the complex amplitude at distance z from the transmitter wave propagating along the positive z-axis in free space, A_0 is a constant amplitude of the wave field, ϕ_0 is the phase, and k is the optical wave number related to the optical wavelength λ by $k = 2\pi/\lambda$.

Note that this formulation describes the plane wave propagation in free space. FSO communication has to be operated in a random atmospheric turbulent channel. When a plane wave propagates through this random medium, it undergoes both amplitude and phase fluctuations when the signal arrives at the detector surface. Fig. 4.11B sketches qualitatively how this happens.

- Spherical Wave

  A spherical wave is defined as one in which the phase fronts are spherical surfaces.

  Mathematical description:

  The complex wave amplitude is given by (at a distance, z)

$$U_0(r,z) = \frac{A_0}{4\pi z} \exp\left(ikz + \frac{ikr^2}{2z}\right) \tag{4.2}$$

The phase $\phi = k\left(z+\frac{r^2}{2z}\right)$ has a transverse radial dependency. Spherical waves are assumed when the source excitation for FSO communication can be considered as a point source. Fig. 4.11B sketches the spherical waves intensity profile propagated through atmospheric turbulence.

- Gaussian-beam wave

  A Gaussian beam is one whose transverse electric field and intensity (irradiance) distributions are approximated by Gaussian functions. Many lasers operating on the fundamental transverse mode, or TEM_{00} mode, of the laser's optical resonator emit an approximate Gaussian profile. The outgoing Gaussian beam passing through a lens also is transformed into another Gaussian.

  Mathematical description:

  The complex amplitude at distance z from the source is the Gaussian wave [5]:

$$U_0(r,z) = \frac{1}{1+i\alpha_0 z}\exp\left[ikz - \frac{1}{2}\left(\frac{\alpha_0 k}{1+i\alpha_0 z}\right)r^2\right] \tag{4.3}$$

where r is the radial distance from the beam center line, and $\alpha_0$is a complex parameter related to spot size and phase front radius of curvature as

$$\alpha_0 = \frac{2}{kW_0^2} + i\frac{1}{F_0} \tag{4.4}$$

W_0 is the radius at which the field amplitude falls to $1/e$ of that on the beam axis, and the phase front is taken to be parabolic with radius of curvature F_0. $F_0 = \alpha$ (collimated), $F_0 > o$ (convergent) and $F_0 < o$ (divergent). Fig. 4.9C sketches the Gaussian beam intensity profile distribution after propagating through turbulent atmosphere; see also [6].

4.8.2 Review of Optical Propagation Relevant to Free-Space Optical Communications Systems Analysis

Atmospheric channel, which in this case is random (stochastically dynamic), and the optical field originating from the optical transmitter undergoes various temporal and spatial changes in the communication signal before reaching the receiver. A communication scenario will then determine the FSO transceiver (both optical transmitter and receiver) design with the knowledge of the signal distortions due to atmospheric turbulence. The solution of optical wave equation (discussed later) provides the second- and fourth-order moment of the random optical field from which mean irradiance, scintillation index, and covariance function can be deduced. FSO communication parameters such as beam wander, angle-of-arrival fluctuations, correlation width, and so on can then be determined. Combined with scintillation statistics/probability density function and appropriate mitigation technique (such as aperture averaging, coding, adaptive optics, etc.) the FSO communication performance metrics such as BER, SNR, and probability of fading can be evaluated.

4.8.3 Review Physics of Optical Propagation Through Atmospheric Turbulence

The atmospheric index of refraction can be treated as a random field, and the index of refraction at any spatial point in the atmosphere at any given time can be considered a random variable. Specifically, the index of refraction is a random function of both space and time. The time dependency of the index of refraction is usually suppressed in optical studies, assuming the wave maintains a single frequency as it propagates. The index of refraction fluctuations caused by atmospheric turbulence can be written as

$$n(\vec{r}) = n_0 + n_1(\vec{r}) \tag{4.5}$$

where $\vec{r}$ is a point in space, $n_0 = \langle n(\vec{r})\rangle \cong 1$ is the mean value of the index of refraction of air at atmospheric pressure and $n_1(\vec{r})$ represents the random deviation of $n(\vec{r})$ from its mean value, with a zero average, $\langle n_1(\vec{r})\rangle = 0$. The refractive index of air at optical frequencies is

$$n(\vec{r}) = 1 + 79 \times 10^{-6}\frac{P(\vec{r})}{T(\vec{r})} \tag{4.6}$$

where P is the pressure in millibars and T is the temperature in degrees Kelvin. The index of refraction fluctuations within the visible and near-IR region of the spectrum are primarily due to random temperature fluctuations.

The covariance function of $n(\vec{r})$ can be written as

$$B_n(\vec{r}_1, \vec{r}_2) = B_n(\vec{r}_1, \vec{r}_1 + r) = \langle n_1(\vec{r}_1)n_1(\vec{r}_1 + \vec{r}) + n_0^2 \tag{4.7}$$

where $\vec{r}_1$ and $\vec{r}_2$ are two points in space, and $\vec{r} = \vec{r}_2 - \vec{r}_1$. If the media is assumed to be homogeneous and isotropic turbulent, the covariance function reduces to a function of one scalar distance $r = |\vec{r}_2 - r_1|$.

A locally homogeneous field is usually not characterized by the covariance function, but by the structure function given by

$$D_n(r) = \langle [n(r_1 + r) - n(r_1)]^2 \rangle = 2[B_n(o) - B_n(r)] \tag{4.8}$$

From Eqs. (4.7) and (4.8), the structure function of refractive index fluctuations can be determined and Kolmogorov-Obhukov two-thirds power law can be derived as

$$D_n(r) = C_n^2 \cdot r^{2/3} \quad l_o \ll r \ll L_o \tag{4.9}$$

where l_o (inner scale size) and L_o (outer scale size) were discussed earlier and C_n^2 is the index of refractions structure parameter. For FSO communications for horizontal links, C_n^2 is essentially constant; however, for uplink, downlink, and slant-path FSO scenarios, C_n^2 is a function of height. Typical values of C_n^2 range from $10^{-17}\,\mathrm{m}^{-2/3}$ (weak turbulence) and $10^{-12}\,\mathrm{m}^{-2/3}$ (strong/very strong turbulence). A similar concept of determining various frequency components in a time-varying electrical signal by Fourier transform method, the ability of different eddy sizes contributing to the refractive index of random medium can be applied. The three-dimensional spatial power spectrum of the random field $\Phi_n(\vec{k})$ form a Fourier transform pair with the covariance function (see Eqs. (4.14a) and (4.14b) where $\vec{k}$ is the wave number vector.

Using $D_n(r) = C_n^2\ r^{2/3}$, the wave number spectrum is (within the inertial range) [7]

$$\Phi_n(k) = 0.033C_n^2k^{-11/3} \quad 1/L_o \leq k \leq 1/l_o \tag{4.10}$$

To evaluate FSO communication performance, it is necessary to compute characteristics of the received signal on the detector such as intensity fluctuations, mean intensity, temporal power spectrum, and so on. The evaluation of these parameters will require a knowledge of the refractive index spectrum valid for the intended applications. Some of the commonly used spectrums are discussed next.

4.8.3.1 Tatarskii Spectrum

To include the inner or outer scale effects, the modified Kolmogorov power law spectrum is given by the Tatarskii spectrum, which uses a Gaussian function to extend coverage to the dissipation range ($k > 1/l_o$) where small eddies are influential:

$$\Phi_n(k) = 0.033C_n^2k^{-11/3}\exp\left(-k^2/k^2m\right) \tag{4.11}$$

where $k_m = 5.92/l_o$.

Von Karman Spectrum

This modified spectrum is valid over both the inner and outer scale parameters,

$$\Phi_n(k) = 0.033C_n^2\frac{\exp\left(-k^2/k_m^2\right)}{\left(k^2+k_o^2\right)^{11/6}} \tag{4.12}$$

where $k_o = 1/L_o$ (or $k_o = 2\pi/L_o$), $o \leq k < \propto$.

Hill Spectrum (Modified Atmospheric Spectrum) [2,8]

This spectrum includes a "bump" at high wave numbers near $1/l_o$:

$$\Phi_n(k) = 0.033C_n^2\left[1 + 1.802(k/k_l) - 0.254(k/k_l)^{\frac{7}{6}}\,\frac{\exp\left(-k^2/k_L^2\right)}{\left(k^2+k_o^2\right)^{11/6}}\right] \tag{4.13}$$

$o \leq k < \propto$

where $k_l = 3.3/l_o$.

4.9 EFFECTS OF OPTICAL PROPAGATION THROUGH ATMOSPHERIC TURBULENCE RELEVANT TO FREE-SPACE OPTICAL COMMUNICATIONS

For wireless optical communications or FSO communication links, atmosphere is not an ideal communication channel. Atmospheric turbulence can cause fluctuations in the received signal level as a result of a number of effects, which increase the bit errors in a digital communication link. In order to quantify the FSO performance limitations, a better understanding of these effects is needed. Known effects of atmospheric turbulence include:

1. Beam scintillation: Fluctuations in the spatial power density at the receiver plane. The scintillation process is low compared to the large data rates typical of FSO transmission.
2. Beam steering: Angular deviation of the beam from its original LOS causing the beam to miss the receiver.
3. Image dancing: The received beam focus moves in the image plane due to variations in the beam's angle of arrival (AOA) (the AOA of the phase fronts as a receiver fluctuates).
4. Beam spread: Increased beam divergence due to scattering causing a reduction in received power density.
5. Other deterioration includes
   a. deviations of the beam shape (from circularly symmetric Gaussian beam, for example) from those that are time dependent.
   b. wander of the centric of the beam.
   c. breakup of the beam into distinct patches of illumination whose shapes and locations fluctuate with time.
6. Spatial coherence degradation: Turbulence also induces losses in coherence across the beam phase fronts; can be deleterious for photomixing as incoherent receiver.

The random variations on the amplitude and phase of the propagating wave can be addressed theoretically, by solving the wave equation for the electric field and its respective statistical moments. The electric field for a propagating electromagnetic wave is derived from the wave equation

$$\nabla^2 E \neq k^2 n^2(r)E + 2\nabla[E\cdot\nabla\log n(r)] = 0 \tag{4.14a}$$

where $k = 2\pi/\lambda$ is the wave number of the electromagnetic wave, λ is the wavelength, $n(r)$ is the index of refraction whose time variations have been suppressed, and $\nabla^2 = \frac{\partial^2}{\partial x^2} + \frac{\partial^2}{\partial y^2} + \frac{\partial^2}{\partial z^2}$ is the Laplacian operation. The time variations in the refractive index are sufficiently slow that a quasi-steady approach can be used where n(r) is a function of position only. The third term on the left-hand side of Eq. (4.14a) is a depolarization term and can be neglected so that Eq. (4.14a) can be written as

$$\nabla^2 E + k^2 n^2 E = 0 \tag{4.14b}$$

Eq. (4.14b) for the vector E can be decomposed into three scalar equations, one of which can be denoted by a scalar component $U(r)$. This scalar component $U(r)$ is transverse to the direction of propagation along the positive z-axis. The scalar stochastic Helmholtz equation for $U(r)$ can then be written as

$$\nabla^2 U + k^2 n^2(r) U = 0 \tag{4.15a}$$

From Eq. (4.15a),

$$n(r) = n_o + n_1(r) \text{ with } n_o = \langle n(r) \rangle = 1, \ \langle n_1(r) \rangle = 0$$

In order to solve Eq. (4.15b) for weak turbulence case, Rytov approximation was used, which allows the field of the electromagnetic wave to be written as

$$U(r, L) = U_o(r, L)\exp[\phi(r, L)] \tag{4.15b}$$

where $U_o(r)$ represents the unperturbed field (i.e., an optical wave travelling through free-space at the receiver) and ϕ is a complex phase perturbation due to turbulence, which can be written as

$$\phi(r, L) = \phi_1(r, L) + \phi_2(r, L) + \ldots \tag{4.16}$$

where $\phi_1(r, L)$ and $\phi_2(r, L)$ are first- and second-order perturbations, respectively.

For FSO communication applications, what matters is the amount and quality of the received signal on the detector. The received signal in the form of optical field $U(r, L)$ on the detector located at a distance L from the transmitter end, which is then converted to electrical signal by the photodetector for further communications-related processing, r is a vector in the receiver plane transverse to the propagation axis. The coherent portion of the field is represented by the first moment $<U(r, L)>$, where $< >$ denotes an ensemble average. The mutual coherence function (MCF) of the wave defined by the second moment is given by

$$\text{MCF} \equiv \Gamma_2(r_1, r_2, L) = \in \langle U(r_1, L) U^*(r_2, L) \rangle \tag{4.17}$$

where r_1 and r_2 are two points in the receiver plane and $U^*(r, L)$ is the complex conjugate of $U(r_1, L)$. When the optical field in the receiver plane is detected at the same point, the mean irradiance is determined from the second moment. The fourth-order moment or the cross-coherence function of the field can be represented by

$$\Gamma_4(r_1, r_2, r_3, r_4, L) = \langle U(r_1, L)U^*(r_2, L)U(r_3, L)U^*(r_4, L)\rangle \tag{4.18}$$

From the fourth moment, the second moment of irradiance can be determined. The scintillation index σ_I^2 can then be evaluated from the second moment of irradiance and the mean irradiance. By setting $r_1 = r_2 = r_3 = r_4 = r$ for the same point in the receiver plane,

$$\langle I^2(r, L) = \Gamma_4(r, r, r, r, L) \quad \langle I(r, L)\rangle = \Gamma_2(r, r, L) \tag{4.19}$$

from which a theoretical expression for the scintillation index can be obtained as follows:

$$\sigma_I^2(r, L) = \frac{\langle I^2(r, L)\rangle}{\langle I(r, L)\rangle^2} - 1 \tag{4.20}$$

Weak and strong regimes of turbulence discussed below define various refractive index spectrums relevant to FSO communications. Weak or strong fluctuation theories are generally classified based on the values of Rytov variance, denoted by σ_R^2 (or σ_I^2):

$$\sigma_R^2 = 1.23 \cdot C_n^2 \cdot k^{7/6} L^{11/6} \tag{4.21}$$

where C_n^2 is the refractive-index structure parameter. The different ranges of the values of σ_R^2 specify the various turbulence conditions as follows:

$$\begin{aligned} &\sigma_R^2 < 1 (\text{weak fluctuations}) \\ &\sigma_R^2 \approx 1 (\text{moderate fluctuation}) \\ &\sigma_R^2 > 1 \ (\text{strong fluctuations}) \\ &\sigma_R^2 >> 1, \quad \sigma_R^2 \rightarrow \infty \ (\text{saturation regime}) \end{aligned} \tag{4.22}$$

For a Gaussian-beam the preceding criteria based on σ_R^2 is not adequate; the conditions for a Gaussian-beam are

$$\sigma_R^2 < 1 \text{ and } \sigma_R^2 \Lambda^{5/6} < 1 (\text{weak fluctuations})$$

where $\Lambda = \frac{2L}{kW^2}$, W = free-space beam radius at the receiver σ.

If either of these conditions fails, the fluctuations are classified as moderate to strong.

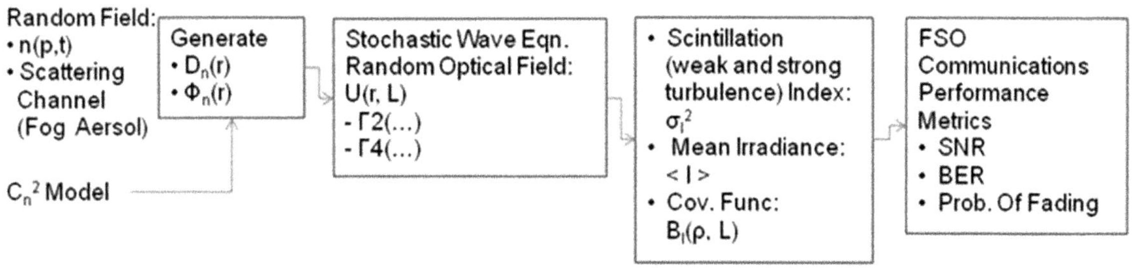

FIGURE 4.12
Flow chart for free-space opticalcommunication system performance evaluation.

This overall picture of the complete systems approach is shown in Fig. 4.12. The symbols used in the system blocks are identified and the functional relationships of the necessary parameters and how they are related to the solutions of the optical wave equation are discussed in detail in the next sections. The symbols used in the system blocks are the following and are related to the solutions of the optical wave equation [1]:

$n(r, t) =$ Random index of refraction (spatial and temporal)
$D_n(r) =$ Structure function of random refractive index
$\Phi_n(r) =$ Three-dimensional spatial power spectrum

The Gaussian-beam parameters can be characterized by [7,9]

$$\theta_o = 1 - \frac{L}{F_o}$$

and by the output parameter in the receiver's plane at $z = L$,

$$\Theta = 1 + \frac{L}{F} = \frac{\theta_o}{\theta_o^2 + \Lambda_o^2} \tag{4.23}$$

$$\Lambda = \frac{2L}{kW^2} = \frac{\Lambda_0}{\theta_o^2 + \Lambda_o^2} \tag{4.24}$$

where W and F are the beam radius and phase front radius at the receiver plane, respectively.

Mean irradiance and beam spreading are due to turbulence.

The results are derived and discussed in references [4,9] and are not repeated here. The following are only the essential results.

The mean from Eq. (4.19) gives

$$\langle I(r,L)\rangle = \Gamma_2(r,r,L) = \frac{w_o^2}{w^2}\exp\left(-\frac{2r^2}{w^2}\right)\exp\left[2\sigma_r^2(r,L) - T\right] \tag{4.25}$$

For a Kolmogorov spectrum (Eq. 2.60)

$$\Phi_n(k) = 0.033C_n^2k^{-11/3}$$

the quantity $\sigma_r^2(r,L)$ from Eq. (4.25) can be evaluated and is given by [10].

$$\sigma_r^2(r,L) = 0.066\sigma_R^2\Lambda^{5/6}\left[1 - {}_1F_1\left(-\frac{5}{6}; 1; \frac{2r^2}{w^2}\right)\right] \tag{4.26}$$

$$\cong 1.11\sigma_R^2\Lambda^{5/6}\frac{r^2}{w^2}, \quad r < w$$

where $\sigma_R^2 = 1.23C_n^2k^{7/6}\cdot L^{11/6}$ is the Rytov variance, and ${}_1F_1(a;\ c;\ x)$ is the confluent hypergeometric function.

Eq. (4.25) can then be approximated by the Gaussian function as follows:

$$\langle I(r,L)\rangle \cong \frac{w_o^2}{w_{LT}^2}\exp\left(-\frac{2r^2}{w_{LT}^2}\right)\left[w/m^2\right] \tag{4.27}$$

where W_{LT} is a measure of the effective or, long-term beam spot size given by

$$W_{LT} = W\sqrt{1+T} = W\sqrt{1 + 1.33\sigma_R^2\Lambda^{5/6}} \tag{4.28}$$

Mean irradiance and beam spread for strong fluctuations [9].

The Rytov variance criterion for strong fluctuation condition is

$$\sigma_R^2 \gg 1 \text{ or } \sigma_R^2\Lambda^{5/6} \gg 1$$

$\Lambda = \frac{2L}{kw^2}$, where w is the spot size of the Gaussian beam parameter at the receiver. The mean irradiance in this case is given by

$$\langle I(r,\ L)\rangle = \frac{w_o^2}{w_{LT}^2}\exp\left(-\frac{2r^2}{w_{LT}^2}\right) \tag{4.29}$$

where w_{LT} is the effective or long-term spot radius.

$$W_{LT} \cong W\sqrt{1 + 1.63\sigma_R^{12/5}\Lambda} = W\sqrt{1 + 4q\Lambda/3} \tag{4.30}$$

$$L \ll z_i = \left(C_n^2k^2l_o^{5/3}\right)^{-1} \quad q = 1.22\sigma_R^{12/5}$$

This is valid for the propagation length $L \ll z_i$, which is valid under most practical situations.

4.9.1 Scintillation Index

Atmospheric turbulence can ultimately limit the FSO communications performance. Scintillation index is therefore very critical in determining the communications system performance when data with high bandwidth is transferred through atmospheric turbulence. The notation and the development of the mathematical formulations used in this section are taken from Andrews and Phillips (2005) [9]. Results are provided for both weak and strong turbulence regimes.

4.9.1.1 *Case of Strong Turbulence*

The scintillation results described earlier were limited to weak fluctuations as defined by $\sigma_R^2 < 1$. Strong fluctuation situations can be attained when the strength of the turbulence parameter C_n^2 is high and for long-range communication links (e.g., ≈ 5 km or longer). The scintillation index for strong turbulence regime is typically $\sigma_R^2 \approx 1$ to $\sigma_R^2 > 1$. The saturation regime is defined when $\sigma_R^2 \rightarrow \propto$. The extended Rytov theory has been recently applied to explain the irradiance fluctuations applicable for moderate-to-strong turbulence regimes. The extended theory assumes an effective random part of the refractive index term as follows:

$$n_{1,e}(r) = n_x(r) + n_y(r) \tag{4.31}$$

where $n_x(r)$ and $n_y(r)$ are large-scale and small-scale in-homogeneities, respectively. The received irradiance of an optical wave is the resultant of the small-scale (diffractive) fluctuations multiplicatively modulated by random large-scale (refractive) fluctuations.

The perturbed optical wave can be written in this case as

$$U(r) = \mathrm{U}_o(r, L)\exp\left[\phi_x(r, L) + \phi_y(r, L)\right] = \mathrm{U}_o(r, L)$$
$$\exp[\phi_x(r, L)]\cdot\exp\left[\phi_y(r, L)\right] \tag{4.32}$$

The extended Rytov theory also assumes the form of the atmospheric spectrum as an effective spectrum given by

$$\Phi_{n,e}(k) = \Phi_n(k)\cdot G(k, l_o, L_o) = 0.033C_n^2\cdot k^{-11/3}\cdot G(k, l_o, L_o) \tag{4.33}$$

where $G(k, l_o, L_o)$ is an amplitude spatial filter.

With the assumptions of effective random refractive index term and effective spectrum, Andrew and Phillips (2005) [9] have derived the scintillation index for moderate-to-strong-to saturation regimes. A detailed explanation on the derivation of the scintillation index equations can be found in Andrews and Phillips (2005) [10]. The following is a summary of their results for plane wave, spherical wave, and Gaussian-beam wave models [9].

4.10 EFFECTS OF FREE-SPACE OPTICAL COMMUNICATION SIGNAL TEMPORAL FREQUENCY SPECTRUM

For FSO communications, the irradiance fluctuations resulting from the propagation of optical beam through the atmospheric turbulence are a main noise. It is therefore necessary to formulate the temporal power spectrum of irradiance fluctuations, both on the optical axis and in points located a certain distance from it. Quantitative analysis of the temporal behavior of the irradiance provides significant merit figures in the optical link. Without the detailed temporal information, the overall performance of an FSO communication link is essentially incomplete. The mean values of the SNR or mean BER (to be discussed in another chapter) are not the relevant parameters, but the instantaneous values are important. In order to evaluate the burst error rate and the overall availability of the FSO communication system, knowledge of the statistics of the instantaneous values is necessary [11]. Also, the temporal characteristics of the FSO system are necessary to develop and design optimal schemes for detection and coding. Also see the book by H.R. Anderson, *Fixed Broadband Wireless System Design* (Wiley & Sons, West Sussex, England, 2003).

Tatarskii [12] and Ishimaru [13] defined the temporal spectrum of irradiance fluctuations, or power spectral density $S(\omega)$ by the Fourier transform of the temporal covariance function as follows.

$$S(\omega) = 2\int_{-\infty}^{\infty} B_I(\tau, L)e^{-j\omega\tau}d\tau = 4\int_{-\infty}^{\infty} B_I(\tau, L)\cos(\omega\tau)d\tau \tag{4.34}$$

where $B_I(\tau, L)$ is the temporal covariance function. The general fourth-order cross-coherence function was defined earlier by Eq. (4.18) and the MCF.

4.11 EFFECTS OF PROBABILITY DENSITY FUNCTION MODELS FOR INTENSITY FLUCTUATIONS RELEVANT TO FREE-SPACE OPTICAL COMMUNICATION SYSTEMS

FSO communication systems have recently attracted considerable attention for a variety of applications because of the high bandwidths and the ability to transfer a high data-rate of information. The performance of a lasercom system operating in the atmosphere is reduced by irradiance scintillations and phase fluctuations due to optical turbulence effects at the optical wavelength. Fading or scintillation impair and degrade FSO system performance particularly for link ranges of 1 km and above even in clear sky conditions. The reliability of an FSO communication system depends on the probability of detection, miss and false alarm, and the probability of fade, all of which require an accurate knowledge of the probability density function (PDF) for the received optical power. An accurate mathematical model for the exact PDF of the received signal under all intensity fluctuation regimes is difficult. From the probability theory, random variable is a function whose integral over an interval determines the probability that the random variable attains a certain value in that interval.

In an FSO communication system, the PDF of the received power at the photodetector is assumed to come from the same family as the PDF of irradiance at a point in the pupil plane of the receiver [10]. Over the last decades, various statistical models for irradiance PDF have been proposed for different degrees of turbulence regimes. Depending on weak or strong irradiance fluctuations and the propagation distances, these proposed PDFs have different forms. Log-normal distribution is the most widely used model for the probability density function of the irradiance due to its simplicity: this PDF, however, is only applicable to weak turbulence regime. As the turbulence strength increases, multiple scattering effects become important and significant deviations from log-normal statistics are exhibited in experimental data. Log-normal PDF underestimates the behavior in the tails as compared with measurement results. For communication systems, accurate detection and fade probabilities primarily depend on the tails of the PDF, hence underestimating this region significantly affects the accuracy of the communication performance. In a recent series of paper a gamma-gamma PDF has been proposed [14,15] based on work by Nakagami [16]. This model is a two-parameter distribution based on a doubly stochastic theory of scintillation, and assumes that small-scale irradiance fluctuations are modulated by large-scale irradiance fluctuations of the propagating wave, both governed by independent gamma distributions. The gamma-gamma distribution can be directly related to the atmospheric parameters defining turbulence conditions. In this chapter, we will consider mainly the two PDFs, namely, *log-normal* and *gamma-gamma*. Log-normal distribution is applicable to a weak turbulence case and the gamma-gamma distribution is assumed to be valid for both weak and strong turbulence regimes. These two PDFs, which are used in this chapter for the computational purpose of evaluating communication performance, are given by the following equations:

Log-normal distribution:

$$p(I) = \frac{1}{I\sigma_I(D)\sqrt{2\pi}} \exp\left\{ -\frac{\left[\ln\left(\frac{I}{<I>}\right) + \frac{1}{2}\sigma_I^2(D)\right]^2}{2\sigma_I^2(D)} \right\}, \quad I > 0 \tag{4.35}$$

where $\sigma_I(D)$ is the aperture-averaged scintillation index as defined earlier, and $<I>$ is the mean intensity.

Gamma-gamma probability density function:

$$p(I) = \frac{2(\alpha\beta)^{(\alpha+\beta)/2}}{\Gamma(\alpha)\Gamma(\beta)} I^{(\alpha+\beta)/2-1} K_{\alpha-\beta}\left(2\sqrt{\alpha\beta I}\right), \tag{4.36}$$

where $K_a(.)$ is the modified Bessel function of the second kind of order a. Here α and β are the effective number of small-scale and large-scale eddies of the turbulent environment. These parameters are directly related to the atmospheric turbulence conditions as follows [14,15]:

Plane wave with aperture-averaged scintillation index:

$$\alpha = \left[\exp\left[\frac{0.49\sigma_1^2}{\left(1+0.65d^2+1.11\sigma_1^{12/5}\right)^{7/6}}\right]-1\right]^{-1} \tag{4.37}$$

$$\beta = \left[\exp\left[\frac{0.51\sigma_1^2\left(1+0.69\sigma_1^{12/5}\right)^{-5/6}}{1+0.90d^2+0.62d^2\sigma_1^{12/5}}\right]-1\right]^{-1} \tag{4.38}$$

Spherical wave with aperture-averaged scintillation index:

$$\alpha = \left[\exp\left[\frac{0.49\sigma_2^2}{\left(1+0.18d^2+0.56\sigma_2^{12/5}\right)^{7/6}}\right]-1\right]^{-1} \tag{4.39}$$

$$\beta = \left[\exp\left[\frac{0.51\sigma_2^2\left(1+0.69\sigma_2^{12/5}\right)^{-5/6}}{1+0.90d^2+0.62d^2\sigma_2^{12/5}}\right]-1\right]^{-1} \tag{4.40}$$

where the parameters σ_1, σ_2, and d were defined earlier.

4.11.1 Received Signal-to-Noise-Ratio and Bit Error Rate

In almost every area of measurements, the ultimate limit to the detectability of a weak signal is set by noise—unwanted signals that obscure the desired signal. The same is true for free-space laser communication systems where the shot noise, background noise, and thermal noise contribute to the total noise at the receiver. The goal of a lasercom system for digital communication is to transmit the maximum number of bits per second over the maximum possible range with the fewest errors. Electrical data signals are converted to optical signals via a modulator. A 1 is transmitted as a pulse of light while a 0 has no light output. The number of 1's and 0's transmitted per second determines the speed of the link (bit rate). At the receiving end of the link, the optical signal is detected by an optical-to-electrical converter (e.g., a photo detector). A decision circuit then identifies the 1's and 0's in the signal, and thus recovers the information sent.

For digital communication systems, information is sent over an optical link as digital symbols. This is accompanied by encoding the source information into binary symbols (bits) and transmitting the bits as some type of coded optical field, for example, by encoding on a bit-by-bit basis (binary encoding). Each bit is then sent individually by transmitting one of two optical fields to represent each bit. In this chapter, we will consider only a direct-detection (DD) system in which the standard binary procedure is to pulse an optical source (e.g., a laser or an LED) on or off depending on data bit. This encoding is referred to as on-off keying (OOK). At the receiver OOK decoding is based on a decision as to whether or not the pulse slot time has high enough field energy. The selected threshold determines the best performance in decoding the correct signal with the lowest probability of making a bit decision error and thus the BER can be obtained. The other modulation format of pulse position modulation (PPM) will also be considered.

4.11.2 Relationship Between Signal-to-Noise Ratio and Bit Error Rate

The performance of the lasercom system can be predicted by computing the BER of the system, which depends on the modulation format, and the SNR, where the noise contributions come from all possible sources, which include signal shot noise, dark current noise, thermal/Johnson noise in the electronics following the photodetector, and the background noise. Assuming a Gaussian distribution of noise, the SNR at the output of the photodetector in the absence of turbulence is given by Ref. [14]

$$\mathrm{SNR}_0 = \frac{P_S}{\sqrt{\left(\frac{2h\nu B}{\eta}\right)(P_S + P_B) + \left(\frac{h\nu}{\eta e}\right)^2 \left(\frac{4kT_N B}{R}\right)}} \tag{4.41}$$

where P_S is the signal power of the optical transmitter and P_B is the background noise, both in watts; η is the detector quantum efficiency; e is the electronic charge in coulombs; h is the Planck's constant; ν is the optical frequency in hertz; k is the Boltzmann constant; B is the bandwidth of the (detector) filter; T_N is the effective noise temperature; and R is the effective input resistance to the amplifier of the detector. For shot noise limited operation, the background noise and the thermal noise can be ignored.

In presence of turbulence, the SNR is a fluctuating term (i.e., an instantaneous value) and the average (mean) value is to be taken. The mean SNR can be expressed as [14]

$$\langle \mathrm{SNR} \rangle = \frac{\mathrm{SNR_0}}{\sqrt{\frac{P_{S0}}{\langle P_S \rangle} + \sigma_I^2(D)\mathrm{SNR}_0^2}} \tag{4.42}$$

where $\mathrm{SNR_0}$ is the SNR in the absence of turbulence defined earlier, P_{S0} is the signal power in the absence of atmospheric effects, the mean input signal power $\langle P_S \rangle$(i.e., the mean of the instantaneous input signal power P_S), and $\sigma_I^2(D)$is the aperture-averaged scintillation index previously discussed.

Let us now understand the communication performance criteria defined by the parameter BER. The signal entering the decision circuit fluctuates due to the various noise mechanisms. The BER is the probability of incorrect bit identification by the decision circuit. The design of optical communication systems relies on the estimation of performance figure of merit such as BER. One of the most commonly used parameters is the receiver sensitivity defined as the minimum averaged received optical power required to achieve a given BER. In order to simulate the characteristics of optical links or subsystems, it is therefore essential to be able to compute the BER in an effective and reliable way. It is important to investigate the degree to which the optical communication may be expected to be degraded by atmospheric turbulence, by calculating the expected increase in bit error probability due to turbulence, including the effect of a finite aperture receiver.

We also assume the Gaussian distribution for both noise and signal plus noise; the probability of error in absence of turbulence is given by

$$\mathrm{BER_0} = \frac{1}{2}\left(\frac{\mathrm{SNR_0}}{2\sqrt{2}}\right) \tag{4.43}$$

In presence of turbulence, the probability of error is given by [14,17]

$$\mathrm{BER} = \frac{1}{2}\int_0^\infty p_I(s) erfc\left(\frac{\langle SNR \rangle s}{2\sqrt{2}\langle i_S \rangle}\right) ds \tag{4.44}$$

where $p_I(s)$ is the probability distribution of irradiance, i_S is the instantaneous signal current whose mean value is given by $\langle i_S \rangle = \eta e <P_S>/h\nu$, $<P_S>$ is the mean signal power, and $\langle \mathrm{SNR} \rangle$ is the mean SNR in the presence of turbulence defined earlier. Depending on the level of turbulence and the communication link scenario, various forms of the probability distributions can be modeled. In this chapter, we will compute BER using lognormal and gamma-gamma probability distributions. Note that the probability distribution contains the parameter s, which is related to the aperture-averaged scintillation index. Some of the numerical results showing BER computations will be presented later in the chapter.

4.11.3 Probability of Fade

The object of designing any communication system is to obtain uninterrupted communications between the transmitter and the receiver during the exchange of information of sending and receiving data. We have discussed earlier that because of the presence of randomly varying atmospheric channel, the received signal fluctuates so that it is possible for the received signal to fall below an acceptable detection level and error rate. The fading probability for an optical communication link can be determined from the knowledge of statistics of intensity fluctuations (PDF) and the scintillation index. The reliability of a communication system thus depends on the fading probability. Let us denote $I(t)$ as the instantaneous value of irradiance normalized to its mean. We need to obtain the fraction of time $I(t) \geq I_T$, where I_T is the threshold level of intensity specified by the communication system. Because of the assumption of an ergodic process for the statistics of optical propagation in our case, the time averages are equal to ensemble averages, so that we can also write the fraction of time of a fade where $I \leq I_T$ as follows:

$$\begin{aligned} \text{Fraction } (I \leq I_T) &= CP_I(I \leq I_T) \\ &= \int_0^{I_T} p_I(I)dI \end{aligned} \tag{4.45}$$

where CP_I is the cumulative probability for irradiance and is assumed to be either a lognormal (for weak turbulence) or a gamma-gamma distribution (for weak to strong turbulence) model, for the numerical examples described in this chapter. The fade threshold parameter F_T of a signal (in decibels) can be defined by [14,18]

$$F_T = 10 \log_{10}(\langle I(0,L)\rangle / IT) \tag{4.46}$$

where $<I(0, L)>$ is the on-axis mean intensity at a communication distance L. We can also write

$$\ln\left[\frac{I_T}{\langle I(0,L)\rangle}\right] = -0.23F_T \tag{4.47}$$

In the numerical examples described in the following section, we show the probability of fade as a function of the threshold parameter F_T (given in dB below the mean on-axis intensity at the receiver) for various communication scenarios, such as downlink, uplink, and horizontal links and for two model PDFs (lognormal and gamma-gamma).

4.11.4 Airborne-Ground Propagation Path: Slant Propagation

The propagation path characteristics for airborne platforms such as airplane and UAV are basically a slant path contributing atmospheric turbulence close to the ground to the FSO communication system performance. This was discussed earlier in this chapter.

4.12 SATELLITE-BASED FREE-SPACE LASER COMMUNICATION SYSTEM

Optical communication utilizing satellite platforms has the potential to extend the reach of establishing Internet connectivity from terrestrial limits of ~200 km to global scales. This way, optical connectivity to remote locations worldwide is possible. Optical propagation from ground-to-satellite (uplink) and satellite-to-ground (downlink) effects limit the data transfer rate as well as the availability and fading statistics relevant directly to global FSO communication system performance. The performance of LEO satellites for downlink and uplink scenarios is important to design optical network architectures by optimum ground-based locations and the systems for seamless connectivity *anytime, anywhere*. Optical global hubs at the optical ground stations and the satellite will both require to design both uplink and downlink transceivers for two-way communications; transmitters, receivers, and optical telescopes need to be designed properly for exchanging very high data rate information with the required communication parameters such as BER, SNR, and probability of fade. The optical propagation characteristics and the models help to resolve important design considerations such as operating wavelength, type and specifications of sources and detectors, telescope designs, specific orbits, and ground station locations for addressing overall system performance. Limitations of laser satellite communication are primarily from weather effects (close to the ground) and pointing requirements. Absorption due to clouds, fog, and turbulence presents significant loss of laser signals, which could cause black-out periods without any connectivity. The locations of ground stations are therefore very important due to these weather condition criteria. Due to the low divergence of laser beam width, the FSO communication link to be established requires the satellite or other spacecraft to maintain accurate pointing at the ground station. Fig. 4.13 shows system architecture consisting of a downlink laser beam (e.g., 1550 nm) as primary downlink method, and an uplink (e.g., 850 nm) beacon for acquisition and tracking. The angular pointing error, distance from the satellite to the ground station, and the elevation angle from the ground are need to be evaluated. High data-rate FSO communication from space is realized by the narrow beam width the laser communication provides. Better pointing accuracy in the presence of atmospheric absorption, scattering, and turbulence is therefore a direct indication of higher data rate communication capability for an FSO communication system. A brief theory of optical propagation as it relates to laser satellite communication is outlined next, with emphasis on the optical propagation parameters that directly affect BER, SNR, and probability of fades FSO communication systems performance.

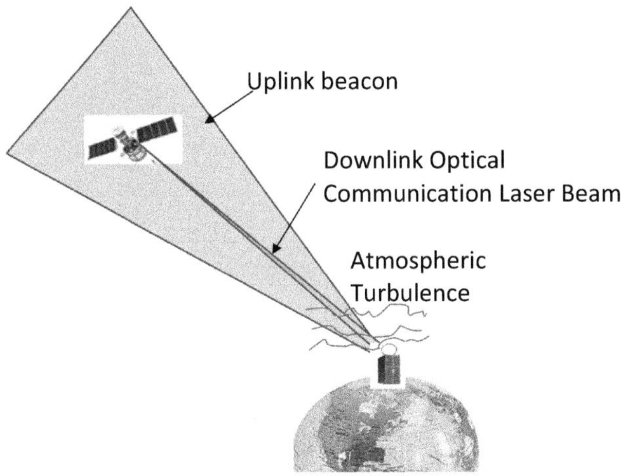

FIGURE 4.13
System architecture consisting of a downlink laser beam (e.g., 1550 nm) as primary downlink method, and an uplink (e.g., 850 nm) beacon for acquisition and tracking.

4.13 ESSENTIAL RESULTS OF ATMOSPHERIC OPTICAL PROPAGATION RELEVANT TO LASER SATELLITE COMMUNICATION CHANNEL

Basic concepts and the results of effects of atmosphere on satellite-ground propagation path (uplink and downlink) are summarized next. The following analysis applies to a slant path connecting a GEO satellite, LEO satellite, and a ground or airborne platform; for example, a HAP or a UAV. All the platforms can be used to provide high data rate Internet backbone. The intensity fluctuation statistics for uplink and downlink optical propagation are quite different and are not symmetric. The effects of turbulence near the transmitter for the uplink laser beam and near the receiver for the downlink beam for treating the path are not the same. The refractive index turbulence strength profiles are also dependent on the zenith angle of the platform location, which also needs to be taken into account. The expressions for the scintillation index, beam spreading, and spatial coherence radius are therefore different for uplink and downlink propagation. The FSO communication systems performance such as BER, SNR, and probability of fade for the two paths are therefore affected accordingly.

4.13.1 Uplink and Downlink Optical Wave Models for Free-Space Optical Communication System Performance

Most of the expressions necessary to define the FSO communication system have already been derived and discussed in detailed analysis in the excellent book by Larry C. Andrews and Ronald L. Phillips [11], and are not repeated here. Only the essential results are outlined and summarized here. In order to analyze the slant path optical propagation characteristics, the important point to remember is that the altitude-dependent atmospheric turbulence model for different zenith angle should be taken into account in the calculation of SNR and BER for communication system performance.

We start with the Gaussian-beam wave with unit amplitude. The Gaussian-beam parameters can be characterized by Ref. [11] and are repeated here for reference to our slant path propagation analysis:

$$\theta_o = 1 - \frac{L}{F_o}$$

$$\Lambda_0 = \frac{2L_0}{kW_0^2}$$

where W_0 and F_o are the beam radius and phase front radius of curvature at the transmitter output aperture.

The output parameter in the receiver plane at $z = L$,

$$\Theta = 1 + \frac{L}{F} = \frac{\theta_o}{\theta_o^2 + \Lambda_o^2}$$

$$\overline{\Theta} = 1 - \Theta$$

$\Lambda = \frac{2L}{kW^2} = \frac{\Lambda_0}{\theta_o^2 + \Lambda_o^2}$ where W and F are the beam radius and phase front radius at the receiver plane, respectively.

4.13.1.1 Mean Irradiance and Beam Spreading Due to Turbulence

The mean irradiance is given by

$$\langle I(r,\ L)\rangle = \Gamma_2(r, r, L) = \frac{w_o^2}{w^2}\exp\left(-\frac{2r^2}{w^2}\right)\exp\left[2\sigma_r^2(r, L) - T\right] \tag{4.48}$$

For a Kolmogorov spectrum

$$\Phi_n(k, h) = 0.033C_n^2(h)k^{-11/3}$$

where $C_n^2(h)$ is the refractive index turbulence strength, which is a function of the altitude h. A number of profile models are discussed for the last many years for the refractive-index structure parameter $C_n^2(h)$; see, for example, Refs. [1,4]. MCF discussed earlier is closely related to the wave structure function, which is dominated by the phase structure function. For a Gaussian-beam wave for a downlink path is given by the equation [11]:

$$D(\rho, L) = 2.91k^2\rho^{5/3}\mu_0 \sec(\zeta) \tag{4.49}$$

where $\mu_0 = \int_{h_0}^{H} C_n^2(h)dh$,

where h_0 is the height above ground level of the uplink transmitter and/or downlink receiver, and $H = h_0 + L\,cos(\zeta)$ is the satellite altitude and ζ is the zenith angle. The other quantity of interest to the satellite-to-ground propagation path is the *spatial coherence radius*, which is the distance ρ where $D(\rho, L) = 2$, which after substitution in the preceding equation is given by

$$\rho_0 = \left(\frac{\cos\zeta}{1.45\mu_0 k^2}\right)^{3/5} \tag{4.50}$$

Another parameter r_0, called the Fried's parameter, is defined by the atmospheric coherence width $r_0 = 2.1\rho_0$.

$$r_0 = \left[0.42\sec(\zeta)k^2 \int_{h_0}^{H} C_n^2(h)dh\right]^{-3/5} \tag{4.51}$$

For an uplink propagation path, the spatial coherence radius is given by [11]

$$\rho_0 = \left(\cos\zeta|1.46\,k^2\,(\mu_{1u} + 0.622\,\mu_{2u}{}^{11/6})\right)^{3/5} \tag{4.52}$$

where Λ, the Gaussian-beam parameter, was defined earlier and

$$\mu_{1u} = \int_{h_0}^{H} C_n^2(h)\left(+\,^{-}\left(\frac{h-h_0}{H-h_0}\right)\right)^{5/3} dh \tag{4.53}$$

and

$$\mu_{2u} = \int_{h_0}^{H} C_n^2(h)\left(1-\left(\frac{h-h_0}{H-h_0}\right)\right)^{5/3} dh \tag{4.54}$$

Angle of arrival fluctuations is an important parameter for both uplink and downlink paths. For a receiver aperture of diameter $2W_G$, the variance of the angle-of-arrival fluctuations is given by [11]

$$\text{Downlink}: \beta_a^2 \cong 2.91\,\mu_0\,\sec(\zeta)(2W_G)^{-1/3} \tag{4.55}$$

$$\text{Uplink}: \beta_a^2 \cong 2.91\left(\mu_{1u} + 0.62\mu_{2u}^{11/6}\right)\sec(\zeta)(2W_G)^{-1/3} \tag{4.56}$$

where the parameters are defined earlier.

Turbulence-induced beam wander: When the size of turbulence eddies are larger than the transmitting beam size, random deflection of the beam from the on-axis occurs when propagating through the atmospheric turbulence. The phenomena is termed as beam wander and will lead to link failure. This is more prominent for uplink and will result in beam displacement by several hundred meters. The root-mean-square (rms) beam wander displacement is given by [14]

$$\sigma_{BW}^2 \approx 0.54(H - h_0)^2 sec^2(\zeta) \left(\frac{\lambda}{2W_0}\right)^2 \left(\frac{2W_0}{r_0}\right)^{5/3} \tag{4.57}$$

where h_0 is now the altitude of the transmitter.

Mean irradiance at the pupil plane of the receiver can be approximated by a Gaussian spatial profile and is expressed as [11]

$$\langle I(r,\ L)\rangle \cong \frac{w_o^2}{w_{LT}^2} \exp\left(-\frac{2r^2}{w_{LT}^2}\right) \left[w/m^2\right] \tag{4.58}$$

where w_{LT} is the effective or long-term spot size of the Gaussian beam in the presence of atmospheric optical turbulence and is given by

$$W_{LT} = W\left[1 + 4.35\ \mu_{2d}{}^{\frac{5}{6}} k^{7/6}(H - h_0)^{\frac{5}{6}} sec^{\frac{11}{6}}(\zeta)\right]^{1/2} \tag{4.59}$$

The downlink parameter μ_{2d} can be written as

$$\mu_{2d} = \int_{h_0}^{H} C_n^2(h) \left(\frac{h - h_0}{H - h_0}\right)^{5/3} dh \tag{4.60}$$

4.13.2 Scintillation Index

The variance of intensity fluctuation of satellite optical communication determines the SNR and therefore BER to determine the performance of satellite FSO communication system performance. When the satellite altitude H is much greater than 20 km, the irradiance fluctuations scintillation index $\sigma_I^2(r, L)$can be simplified for satellite-to-ground path, taking account of the zenith angle as follows [11]:

$$\sigma_I^2(r, L) = 2.25\ k^{7/6} sec^{\frac{11}{6}}(\zeta) \int_{h_0}^{H} C_n^2(h)(h - h_0)^{5/6} dh \tag{4.61}$$

The aperture average factor can be included in the scintillation index expression [4,11] when the receiver aperture is larger than the transverse irradiance correlation width. In addition, various adaptive optics techniques [1] can be applied to minimize the scintillation index to increase the received SNR and lowering the BER in the performance of the satellite laser communication system performance.

4.13.2.1 Uplink Scintillation Index and Beam Wander

The long-term spot size for a ground-to-satellite optical path is useful in designing the receiver aperture and detection methods and can be written as

$$W_{LT} = \sqrt{1 + \left(\frac{D_0}{r_0}\right)^{5/3}} \quad \text{for } 0 \leq \frac{D_0}{r_0} < 1 \tag{4.62}$$

$$= W\left[1 + \left(\frac{D_0}{r_0}\right)^{\frac{5}{3}}\right]^{\frac{3}{5}} \quad \text{for } 0 \leq \frac{D_0}{r_0} < \infty \tag{4.63}$$

where r_0 is the atmospheric coherence width and $D_0 = 8w_o^2$.

For an uplink path, much more beam spreading occurs close to the transmitter at the ground location and therefore requires proper design so that the maximum intensity of the upward beam reaches the receiver located at the satellite.

For an uplink scenario, beam wander, which is the angular deviations of the beam from the boresight, is an important factor to be taken into account. Beam wander can cause time-varying power fades, which at times, can be large enough to lose the Internet connectivity. The transmitting beam type and size such as of a collimator beam from ground to satellite can be selected and designed by calculating the rms angular beam wander. The uplink scintillation index, the pointing error variance, and covariance function of irradiance for an uplink satellite laser communication is described elsewhere [11].

4.13.3 Fluctuation Statistics and Probability of Fade

For a downlink channel the satellite FSO communication system performance depends on the PDF of irradiance fluctuation of the laser beam, which can be modeled as lognormal PDF for a Gaussian-beam wave for weak regime. The lognormal PDF can be written as [11]

$$p_I(I) = \frac{1}{I\sigma_I(r,L)\sqrt{2\pi}} \exp - \left(\frac{-\left[\ln\left(\frac{I}{< I(r,L) >}\right) + \frac{1}{2}\sigma_I^2(r,L)\right]^2}{2\sigma_I^2(r,L)} \right)$$

$$= \frac{1}{I\sigma_I(r,L)\sqrt{2\pi}} \exp - \left(\frac{-\left[\ln(\frac{I}{< I(0,L) >} + \frac{2r^2}{w_{LT}^2} + \frac{1}{2}\sigma_I^2(r,L)\right]^2}{2\sigma_I^2(r,L)} \right) \tag{4.64}$$

A gamma-gamma PDF distribution, defined earlier in this section, is valid for weak to strong turbulence regime and therefore can be used by substituting the irradiance I to normalized: $I = \frac{I}{\langle I(r,L) \rangle}$ in the previous equation of $p_I(I)$.

4.13.3.1 Satellite-Based Free-Space Optical Communication Performance Parameters Relevant to Internet Connectivity

The parameters that determine the FSO communication system performance are *received power* (evaluated from the link equation), *SNR* in the presence of atmospheric turbulence, *probability of fade (or miss)*, determined by a threshold irradiance I_T below the mean on-axis irradiance in decibels (dB). These were explained and discussed earlier.

Cloud blockage: For a satellite-to-ground or ground-to-satellite optical link occasionally opaque clouds may block the signals, causing LOS laser communications to be totally lost. Using optical sites and various ground stations can provide solutions for mitigation applying spatial and temporal diversity techniques. The database from the International Satellite Cloud Climatology Project can provide a database for weather conditions to select optimum optical ground sites and develop an optical space network.

4.13.3.2 Recent Design Technology Concepts for Gigabit Laser Internet Connectivity in Space/Airborne Path

Gigabit laser in space at 0.6 Gbit/s over 45,000 km has already been achieved with GEO satellite. In order to overcome the old solution of storing the data on the LEO satellite and then sending the data down to the earth as the LEO passes over the base station has been solved by using a combination of LEO and GEO connected by optical laser pipes. This means the laser propagation effects for both GEO-to-LEO and LEO-to-ground locations are important and need to be addressed simultaneously. The overall Internet connectivity at high data rates are thus affected, and the European Data Relay Systems plans to use a combination of LEOs and GEOs where LEO sends the collected data immediately up to a stationary higher-altitude GEO satellite, which in turn can then immediately *relay* the data to earth for processing. This is close to almost real-time gigabit-level Internet connectivity possible data exchange with ground stations and Internet users. Furthermore, instead of LEOs any airborne manned or unmanned platforms such as airborne, UAV, and balloon platforms can be used. The theory of optical propagation effects for both uplink and downlink paths as outlined earlier in this chapter can be directly applicable to analyze and evaluate any FSO communication systems independent of where they are located in air or space.

4.13.4 Intersatellite Links

The propagation path for intersatellite links such as GEO-GEO, LEO-GEO, and LEO-LEO optical links do not suffer the atmospheric absorption, scattering, turbulence effects, and cloud outages. Since the two satellites move with different relative velocity, the acquisition and tracking can be one of the major challenges to maintain the two satellites for LOS links. Other challenges for establishing optical link connectivity are discussed [19]: point-ahead-angle, doppler shift, satellite vibration and tracking, and background noise sources. The transmission and detection schemes need to be power efficient because of the large propagation distance, and coherent detection is more efficient for intersatellite links for optimum SNR. Gbps data rates reported recently are data rates of 5.6 Gbit/s between LEO and LEO [6,10] and 1.8 Gbit/s between GEO and LEO [20].

4.13.4.1 Orbital Angular Momentum for Atmospheric Free-Space Optical Communication System: Potential Scheme for Gigabit Range Data Rate

Optical angular momentum (OAM) is associated with the helicity photon wavefront and related to spatial distribution. Application of OAM beams in high data rate FSO communication has been explored recently [21,22]. In order to fully utilize the terabit capacity of the optical carrier for potential FSO communication systems, research has been devoted to OAM-multiplexed free-space laser communication links. Increased capacity and spectral efficiency for high data rate communication are being studied using information carrying OAM beams, specifically for long distance, deep space, or near Earth optical communication for future potential technology. OAM for FSO applications is discussed in Ref. [19]. As described in Ref. [19], the OAM transmitted beam can be expressed as follows:

$$U(r,\theta,t) = S(t)\cdot A(r)\cdot \exp(il\theta) \tag{4.65}$$

where $A(r)$ is the amplitude of the waist of the Gaussian beam, $\exp(il\theta)$ is the azimuthal phase term, r is the radial distance from the center axis of the beam, and $S(t)$ is the signal containing the data to be transmitted. This is for a single OAM beam, and when multiplexing N such information carrying OAM beams, the total multiplex OAM beam can be written as

$$\text{Transmitted OAM beam: } U_{mux}(r,\theta,t) = \sum_{m=1}^{N} S_m(t)A_m(r)\exp(il_m\theta) \tag{4.66}$$

Received demultiplexed can then be obtained by multiplying an inverse azimuth phase term as follows.

Received demultiplexed OAM beam:

$$U_{Rx}(r,\theta,t) = \exp[(-\,il_n\theta)]\sum_{m=1}^{N} S_m(t)A_m(r)\exp(il_m\theta) \tag{4.67}$$

The data rates will depend on the number of light beams and the modes of the resultant OAM beam and the types of modulation used. Terabit free-space data transmission employing OAM multiplexing for short distances in the presence of turbulence have been reported [23,24], which show the future potential of achieving a very high data rate *at least* in the multigigabit ranges between satellites and satellite-to-ground links for near-Earth FSO communications even in the presence of some level of turbulence to make high data rate Internet connectivity possible.

4.13.5 Atmospheric Propagation Characteristics of Different Backhaul Free-Space Optical Communication Links and Network

The need for gigabit capacity low-cost backhaul solutions is increasing because of demands of very high data rates to satisfy today's Internet-based optical links and networks. Terrestrial FSO communications are limited because of maximum range limitations at very high data rates due to atmospheric effects such as turbulence and scattering. The other viable possibilities are to use airborne and space platforms, which can provide optical links for multigigabits data rate, and the propagation paths are less affected by atmospheric effects. In addition to fixed backhaul, mobile backhauls are urgently needed to connect with a large number of cell sites, which in turn require more bandwidth and a large number of backhaul connections per cell site for potential 5G Internet connectivity. Integration of space-based high data rate backhauls and mobile cell sites can provide an ultimate broadband Internet global connectivity solution anytime, anywhere with all-optical technology networks. Different high bandwidth backhaul terminals are part of different propagation scenarios and therefore it is important to understand the basic physics of optical interactions for various atmospheric and space propagation effects relevant to FSO communication systems performance. This section discusses the optical propagation effects for HAP, UAV/drones, balloons, small satellites/constellation of satellites, and the effects on communication systems performance such as SNR, BER, and probability of fades for these different scenarios.

4.13.6 High-Altitude Platform System Architecture

HAPs are located at altitudes of about 17–25 km in the atmosphere where the propagation path is less severely affected than the close-to-ground path. There are two types of optical communication links possible: (1) integrating both HAP to satellite and HAP to ground, and (2) directly HAP-to-ground applicable to establish connectivity in remote areas. Fig. 4.14 shows the HAP system architecture for establishing an optical link.

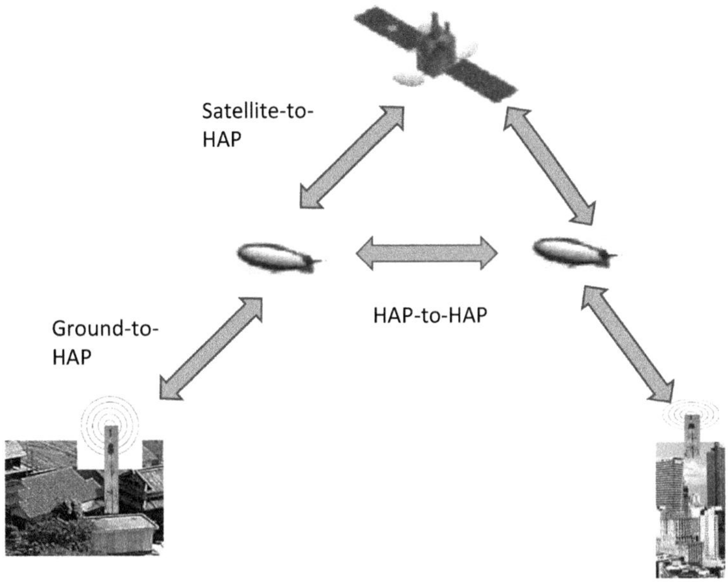

FIGURE 4.14
High-altitude platform as free-space optical communication link connection platform.

4.13.7 Satellite-to-High-Altitude Platform Propagation and Communication Performance

Optical propagation and communication from or to high-altitude platforms and aircrafts that are situated well above clouds at typical heights of 17–22 km will suffer severe atmospheric effects than close-to-ground propagation. The optical link between satellite and HAP can potentially provide as broadband backhaul communication channels if information data from communication terminals onboard the HAP is to be transmitted to a satellite. Optical links can serve this purpose effectively. HAPs can also be used as a data relay station when receiving information from a satellite via optical ink. Not limited to HAPs only, conventional airplanes flying at altitudes up to 10 km can serve as platforms to accommodate laser communication terminals. Channel propagation characteristics and effects on the communication performance parameters were discussed earlier in this section. Mathematical channel models based on the analytical descriptions earlier can be applied to investigate the propagation and communication parameters for HAP and satellite path. The analysis can include the reduced amount of atmosphere at high altitudes, which impacts the effect of pointing errors and beam wander in the case of strong atmospheric turbulence as well as to include high HAP moving speeds. The essential parameters relevant to calculate the link budget for communication systems performance can be addressed for different optical modulation formats, DD receiver concepts, and noise sources. The link budget can determine the available power margin given a certain required BER and system outage probability [1,11,17]. Recently [16] the effect of atmospheric turbulence on a free-space laser communication system for data rates up to 10.7 Gbit/s and using forward error correction (FEC) for link scenarios including HAP and satellite platforms is analyzed using single-mode coupled optically preamplified receivers, allowing for efficient suppression of background noise and highly sensitive detection. The transmitting wavelength of 1550 nm was used in the analysis. Analysis for the HAP path includes [16] the following.

Absorption and scattering: For the HAP-to-satellite propagation path, the variation of atmospheric attenuation with zenith angle ζ, which is the angle between zenith and LOS, between the transmit and receive telescope on the HAP can be accounted for by the approximation $a_{\mathrm{atm}}(\zeta) = a_{\mathrm{atm}}(0)\sec(\zeta)$, where $a_{\mathrm{atm}}(0)$ is the atmospheric loss of about 0.2 dB at the HAP height of 20 km and ζ is the zenith angle. For HAP-to-satellite transmission the atmospheric loss is typically 0.2−0.8 dB.

Beam spread and beam wander: The concepts of beam spread and beam wandering were discussed earlier. From the analysis for HAP-to-GEO uplink propagation, the maximum beam spread loss is shown to be only 0.03 dB for different atmospheric turbulence conditions.

Scintillation index: For HAP the analysis [16] reports for HAP-to-satellite scenarios, the typical value of intensity fluctuations variance, σ_I^2, to be about 0.025.

Phasefront distortions: The phasefront distortion impacts the coupling efficiency into a single-mode fiber, which for a downlink satellite-to-HAP scenario the coupling loss is smaller than 1.2 dB. For uplink HAP-satellite, the phasefront disturbance is negligible.

4.13.8 Performance Parameters for High-Altitude Platform-Satellite Propagation and Communication Link

In order to establish broadband/high data rate backhaul links, optical satellite-HAP links can provide the needed backhauls and HAP to globally receive and distribute data to HAPs. LEO satellites can range from 400 km whereas GEO satellites are about 35,786 km above the ground. From the link budget calculation [17] recently reported [16] a data rate of 1 Gbit/s with a transmitting power of 40 dBm required a link margin of 9.7 dB at BER $= 10^{-9}$. If FEC is applied, a positive link margin can be achieved for a higher data rate of 10 Gbit/s.

System outage probability:

As mentioned earlier, the outage probability is an important parameter for characterizing system performance and is defined as the probability that the communication link cannot be closed. The outage probability was defined as the probability by which BER is greater than a specified threshold value. A short-term power loss has to be compensated for to achieve the required BER. For a data rate of 10.7 Gbit/s and for a BER $= 10^{-9}$ short-term power loss was shown to be about -3 dB for a scintillation index σ_I^2 of 0.1 and about -10 dB for σ_I^2 of 0.3 using wavelength of 1550 nm, lognormal probability density function for fluctuation statistics at HAP platforms and optically preamplified receiver [16].

Theoretical analysis of the optical link between HAPs and other satellite platforms in performing the link analysis uses transmitter and receiver parameters and atmospheric effects. HAP-GEO optical links performance analysis under weak turbulence conditions are reported [25]. Other recent studies on HAP-based optical link connections include [26] HAP-GEO-HAP link [18] and HAP-LEO scenario [27] using differential phase shift keying modulation, communication between an aircraft and GEO satellite, theoretical analysis of HAP-HAP links [28] at 20 km altitude assuming a distance of 500 km using communication wavelength of 1550 nm, intensity modulation, and a quadrature APD receiver, which achieved a BER of 10^{-6} at a data rate of 384 Mbit/s and a transmit power of 800 mw, optical backbone networks between HAPs [29] for all-optical routing in the network using dense wavelength division multiplexing. The HAP links can offer very high data rates and have very limited beam spreading, causing negligible interference to any other terrestrial or orbit system. However, the main disadvantage is that the system needs complex pointing, acquisition, and tracking units. These stratospheric vehicles can offer a terrestrial backbone network.

For the ground-to-HAP communication area, development of an optical terminal designed for UAV was reported [30] for a planned project to fly at 18 km for 2.5 Gbit/s data rate and a distance of 50 km at wavelength of 1550 nm. A unidirectional optical downlink from a stratospheric balloon to a ground station within the EU project Capanina [8] was performed where the terminal on the balloon was equipped with a transmitter at 1550 nm to achieve 1.25 Gbit/s data rate. A BER of 3×10^{-7} and a path over 64 km was reported. Later, another project involved unidirectional downlink tests of the terminal over a distance of 40 km using a 1550 nm laser beam at 1550 nm from the flight terminal and 1590 nm from the ground station [31].

The recent experimental and theoretical studies in the HAP platform optical communication links therefore have the potential to establish practical and

Table 4.1 Satellite and Airborne Platform Altitudes for Free-Space Optical Communication Links

FSO Communication Platforms	Altitude Above the Ground (km)
Geostationary equatorial orbit	35,786
Low earth orbit	160–2000
High-altitude platform	17–25
Balloons (stratospheric)	18–37
Unmanned air vehicle	6–10
Aircraft	~10

high-capacity FSO communication terminals for HAP-to-HAP and HAP-to-satellite links with up to 10 Gbit/s using the all-optical technology currently available.

Table 4.1 summarizes the typical altitudes of space and air platforms relevant to integrating FSO communication systems for Internet connectivity.

4.13.9 Unmanned Air Vehicle-To-Ground and Between Unmanned Air Vehicles Free-Space Optical Propagation Characteristics and Communication System Performance Analysis

UAVs/drones can play a major role in a digital connected society, for example, Internet of Things (IoT). All two-way communications from drone to existing infrastructure can be accomplished via secure interfaces on mobile networks using existing networks. Establishing connectivity with drones will also enable pilots to avoid collisions with other drones and aircrafts.

Optical channel characterization and analysis for communication system performance:

4.13.9.1 Links Between UAVs and Ground Stations

FSO technology for deploying Internet connectivity can be used for slant paths between UAVs and fixed ground station. The communication systems performance for UAV mobile links along slant paths can be evaluated in terms of the two basic parameters, average received SNR and turbulence-induced BER at a high data rate such as a few Gbit/s. Theory of optical propagation through atmospheric turbulence relevant to FSO communications and systems performance in the atmospheric turbulence channel are discussed in detail [1,17]. Atmospheric turbulence models, the link analysis, and the scintillation index due to intensity fluctuations are also explained in the references. This section discusses the laser propagation links at different UAV altitudes and communication parameters for different distances. The degradation for the BER for a slant path is quite different than a horizontal path. BER performance versus SNR and horizontal distances has been presented [1,17] for an IM/DD OOK modulation format for plane wave, spherical wave, and Gaussian wave models. The quality performance of an FSO link has been studied [32] for the mobile slanted path communication optical link between a UAV and a fixed ground station. In their paper, the slant path is divided into small intervals and simultaneously changing parameters along the slant path is taken into account for each interval. In their communication system design erbium doped fiber amplifiers are used to increase the optical output power and simultaneously limiting the amplified spontaneous emission (ASE) noise to improve BER.

4.13.9.2 Atmospheric Turbulence Modeling for Unmanned Air Vehicle Slant Path and Average Signal-to-Noise Ratio, Bit Error Rate Expressions

The atmospheric turbulence modeling for the slant path is considered by dividing into small intervals; the Rytov variance (σ_R^2) for a Gaussian beam wave for each interval given by [32]

$$\sigma_{R,ij}^2 = 1.23c_{n,i}^2 k^{7/6} L_j^{11/6} \tag{4.68}$$

where k is the wave number. This model takes into account of the turbulence strength profile $c_{n,i}^2$ for each ith intervals along the slant path using the Hufnagel-Valley model (HV-21 model) [4] and for jth path length of path L_j. Assume the UAV cruising at a fixed altitude h, the propagation distance $L_j = h/\sin\theta_j$ where θ is the time-varying angle between the transmitter and the receiver. The longitudinal component for the large-scale and small-scale log-irradiance variance is given by (for a Gaussian beam wave) [9,32]

$$\sigma_{I,ij}^2 = \exp\left[\frac{0.49\sigma_{\mathrm{B},ij}^2}{\left(1+0.56\sigma_{\mathrm{B},ij}^{12/5}\right)^{7/6}} + \frac{0.51\sigma_{\mathrm{B},ij}^2}{\left(1+0.69\sigma_{\mathrm{B},ij}^{12/5}\right)^{5/6}}\right] - 1 \tag{4.69}$$

where

$$\sigma_{B,ij}^2 \simeq 3.86\sigma_{R,ij}^2 \begin{cases} 0.4\left[(1+2\Theta_i)^2 + 4\Lambda_i^2\right]^{5/12} \\ \cos\left[\dfrac{5}{6}\tan^{-1}\left(\dfrac{1+2\Theta_i}{2\Lambda_i}\right)\right] - \dfrac{11}{16}\Lambda_i^{5/6} \end{cases} \tag{4.70}$$

Θ_i Λ_i are the output plane beam parameters; these Gaussian beam parameters are already defined in reference [1,9]. In order to calculate the SNR in presence of turbulence, first the SNR in the absence of turbulence $\mathrm{SNR}_{0,i}$ is determined assuming that all the noise sources are uncorrelated, and is given by [32]

$$\mathrm{SNR}_{0,i} = i_{s,i}^2 \Big/ \sigma_{N,i}^2 \tag{4.71}$$

The SNR and BER in the presence of atmospheric turbulence is calculated by taking the average value of the time varying SNR, which can be expressed as [9,32].

$$\langle \mathrm{SNR}_{ij} \rangle = \frac{\mathrm{SNR}_{0,i}}{1 + 1.63\sigma_{R,ij}^{12/5} \cdot \Lambda_i + A \cdot \sigma_{I,ij}^2 \cdot \mathrm{SNR}_{0,i}} \tag{4.72}$$

$$\langle \mathrm{BER} \rangle = \frac{1}{2} erfc\left(\frac{1}{2\sqrt{2}\sqrt{\sum_{ij} 1/\langle \mathrm{SNR}_{ij} \rangle}} \right) \tag{4.73}$$

The simulation results reported in Ref. [32] are the following:

The transmitting wavelength considered was 1.55 μm, laser power of 100 mW, the beam waist radius 2 cm, and the receiver aperture diameter 10 cm. The visibility was 2 km for haze and 10 km for clear sky. For the UAV altitude of 3 km and clear sky visibility of 10 km, a difference of about 16 dB between the slant path and horizontal average SNR was estimated for a distance greater than 9 km. The values of BER for the slant path of UAV and ground station, and for a clear sky, were better than 1.5e-13 in achieving a reliable link for distances up to 43.2 km. For haze conditions the BER was estimated to be 6.3e-15 for a distance only up to 8.7 km. There is degradation of the BER over a slant path compared to a horizontal path at the same condition, which has to be taken into account for establishing a communication link with a moving platform like UAV. Their simulation results show that a reliable communication link for UAV can be possible for variable ranges from 1 to 35 km under clear sky conditions.

4.13.9.3 FSO Communication Links Between Flying UAVs: BER and Adaptive Beam Divergence Analysis

This section gives an example of a novel solution for FSO communication links between UAVs showing the viability of establishing multi-Gbit/s optical communication links in the presence of atmospheric turbulence. Optical communication offers the advantages of sensor information exchanges at high data rates as well as secure communications needed for a number of tactical applications. There is an increasing interest in UAVs for critical civil and military applications especially in the area of surveillance of a defined area with a variety of sensors. Using multiple UAVs in cooperative swarm mode can survey a larger area, collect a huge amount of data in short duration, and increase the robustness by sharing information between UAVs and sending information back to the command center on the ground. Thus UAVs require high data rate connectivity. Data rates in the range of 100 Mbit/s to larger than 1 Gbit/s will be required to handle multiple sensor information in real time and in parallel. UAVs equipped with FSO can have the technological capabilities to meet this high data rate requirement [1,17]. With proper designs, the data rates can even go 2.5 Gbit/s or much higher using recent technological devices and optical networks. FSO communication between UAVs is also secure. This section discusses the free-space laser communication link analysis between two UAVs and communication performance analysis in terms of achievable BER in the presence of atmospheric turbulence in between the UAVs. Finally, the effects of varying beam divergence to keep the connectivity between the UAV platforms are also described to suggest developing adaptive beam divergence for maintaining the communication links.

Some of the main challenges involved in establishing and maintaining the LOS between two UAV terminals are the effects of atmosphere and the effective divergent loss consisting of geometrical and pointing error losses for inter-UAV communication links. The communication channel for UAV FSO communication is the atmosphere where the beam attenuation is due to the atmosphere on one side and the alignment of the FSO units on the other side. In order to achieve an LOS communication link, the alignment becomes a technical challenge in tracking and acquisition in the presence of the atmosphere. Here we address and analyze the communication system performance for a link between two UAVs separated by a fixed distance.

The UAV communications system performance will depend on the atmospheric propagation path between UAVs. Optical signals propagating in the earth's atmosphere experience degradation due to absorption, scattering, and turbulence [1,4,17]. The horizontal link described in this paper is defined when both of the communication terminals communicate via a horizontal path in space. For a specific modulation scheme such as OOK modulation, and knowing the parameters such as wavelength, UAV height, transmitter divergence angle, and a data rate, the UAV communication performance BER can be computed [1]. For the horizontal case the value of C_n^2remains constant along the propagation path. The value of C_n^2at the altitude of the UAVs needs to be accounted for when estimating intensity fluctuations or beam wander. For establishing a successful FSO communication link between two UAVs, the most important criteria to start with is to make sure that the mechanical gimbal can accurately track the moving UAV in presence of atmospheric turbulence and absorption. From the UAV side, it is important that the minimum transmitting beam divergence is such that the probability of the fading signal reaching the receiver due to beam wandering caused by atmospheric turbulence is below a required threshold. Adjusting the transmitting beam divergence by adaptive means may help in solving the problem of pointing ability to establish and maintain the LOS between two UAVs.

4.13.9.4 Atmospheric Propagation Characteristics Between Two Unmanned Air Vehicles: Bit Error Rate and Signal-To-Noise Ratio Calculations for Unmanned Air VehicleLink

Assuming a collimated optical beam transmitted from one UAV and for this FSO system (plane wave) the weak and symmetrical atmospheric turbulence, the so-called log-irradiance variance is given by [33]

$$< \chi^2 >= 0.31 C_n^2 k^{7/6} L^{11/6} \tag{4.74}$$

Assuming the only dominant noise source is the atmospheric turbulence between the UAVs, χ represents the fluctuations of the log of the amplitude of the field, $A_n(r)$ is the amplitude of turbulence-induced noise, and $A_0(r)$ is the amplitude of the laser beam without turbulence.

$$\chi = ln\left[\frac{A_n(r)}{A_0(r)}\right] = \ln(1+\varepsilon) \tag{4.75}$$

where $\varepsilon = \frac{A_n(r)}{A_0(r)}$.

The performance of the communication system between UAVs is evaluated generally in terms of the two parameters, SNR and BER. The object is to increase the average received power and to reduce the effects of atmospheric turbulence by applying a number of various techniques for the optimum design. The SNR in the presence of turbulence is given by

$$\text{SNR} = \frac{\langle A_0(r)^2\rangle}{\langle A_n(r)^2\rangle} = \left[\langle \varepsilon^2\rangle\right]^{-1} \tag{4.76}$$

At the UAV altitude, we assume a weak turbulence model, ε is therefore very small, and

$$\ln(1+\varepsilon) \approx \cdot \varepsilon$$

The SNR can be therefore be written as for weak turbulence,

$$\text{SNR} = \langle \chi^2\rangle^{-1} = \left(0.31 C_n^2 k^{\frac{7}{6}} L^{\frac{11}{6}}\right)^{-1} \tag{4.77}$$

For the FSO link between UAVs with OOK modulation scheme the BER can be written as [34]

$$\text{BER} = \frac{\exp^{\left(-\frac{\text{SNR}}{2}\right)}}{(2\pi \text{SNR})^{0.5}} \tag{4.78}$$

From the preceding equations, the UAV FSO communication systems can be evaluated for a typical turbulence strength, C_n^2, optical wavelength, λ (and thus wave number $k = 2\pi/\lambda$),and path length, L.

If the beam spreading effect is included in the SNR calculation, the effective SNR can be defined as [9,33,35]

$$\mathrm{SNR_{eff}} = \frac{\mathrm{SNR_0}}{\left[1 + 1.33\,\sigma^2_{lnIR}\left(2L/kw_L^2\right)^{5/6}\right] + F\cdot\,\sigma^2_{lnIR}\cdot SNR_0} \tag{4.79}$$

where σ^2_{lnIR} is the log amplitude variance, $\chi = 0.5\,\sigma_{lnIR}$, F is the aperture averaging factor, which is the ratio of normalized intensity variance of the signal at a receiver with diameter D to that of a point receiver $D = 0$ [33]:

$$F = \frac{\sigma^2_{lnIR}\,(D)}{\sigma^2_{lnIR}(D = 0)} \tag{4.80}$$

The value of the aperture averaging factor depends on the wavelength, turbulence strength, path length, and the wave number defined before.

The data rates can go 2.5 Gbit/s or much higher using recent technological devices and optical networks. The achievable data rate depends on the laser transmitter power, receiver, range, and atmospheric transmission and turbulence characteristics [17]. Achievable data rate is calculated based on the two UAVs link scenario with the following parameters: a laser transmitter power = 50 mW, wavelength = 1550 nm, propagation path between two UAVs = 700 m, receiver aperture diameter = 10 cm, transmitting angle = 10 mrad, receiver sensitivity = 1568 photons/bit, atmospheric attenuation factor = 0.2 dB/km (clear atmosphere), 4 dB/km (haze condition), turbulence-induced loss = 6 dB (weak turbulence), 10 dB (moderately strong turbulence). Data rates for different atmospheric conditions are shown in Table 4.2.

The data rates in the order of a few Gbit/s can be achieved using a much lower power transmitter down to only 10 mW.

Table 4.2 Data Rates for Different Atmospheric Conditions Between Unmanned Air Vehicles

Communication Range = 700 m			
Clear Atmosphere: $\alpha = 0.2$ dB/km		Haze: $\alpha = 4$ dB/km	
With No Turbulence-induced Loss	With turbulence-induced loss 1. 6 dB loss 2. 10 dB loss	With no turbulence-induced loss	With turbulence-induced loss 1. 6 dB loss 2. 10 dB loss
27.6 Gbit/s	1. 6.93 Gbit/s 2. 2.76 Gbit/s	14.99 Gbit/s	1. 3.76 Gbit/s 2. 1.499 Gbit/s

4.13.10 Adaptive Beam Divergence Technique for Inter-Unmanned Air Vehicle Free-Space Optical Communication Link

For establishing inter-UAV FSO communications, in general the beam divergence present in the FSO link should be sufficient to offset any error introduced into the alignment and tracking algorithm by the gimbal with a very low probability of signal fade for the UAV optical link. A new technique based on adaptive beam divergence is presented [36,37] for inter-UAV FSO under varying distance conditions. A single beam divergence employed in the link of optical communications may limit the transmission distance between UAVs. An adaptive beam divergence can improve the free-space communication link performance and provides advantages over a fixed beam divergence in the inter-UAV FSO. The collimated beam output can therefore be altered to provide adaptive beam divergence adjustment in order to improve the FSO communication performance under varying distance conditions. If high amplitude jitter is present, the adaptive beam divergence method can be used to suppress the jitter.

Fig. 4.15 shows the BER performance calculated for the inter-UAV optical communication in the presence of atmospheric turbulence at the height of the UAV (5 km). The BER is calculated using the effective SNR_{eff}. From the calculations, we have also found that for BER, there is marginal improvement for SNR compared with SNR_{eff}. In other words, there is no appreciable effect of beam spreading at this short range. The strength of turbulence, Cn^2, was

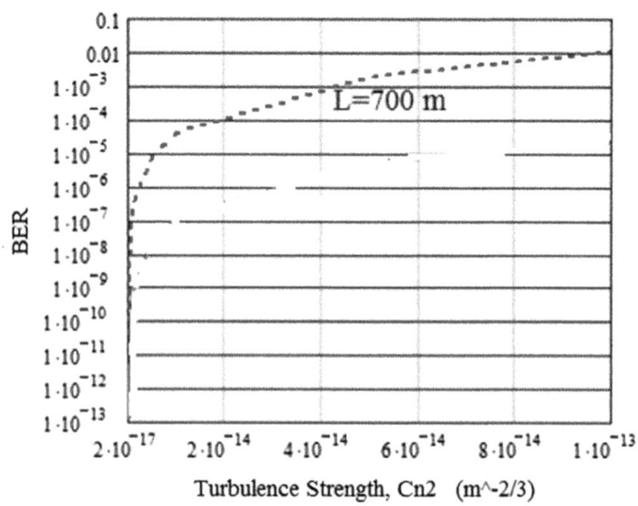

FIGURE 4.15

Calculated bit error rate expected for free-space opticalcommunication link between two unmanned air vehicles separated by 800 m for different atmospheric turbulence strength, Cn^2.

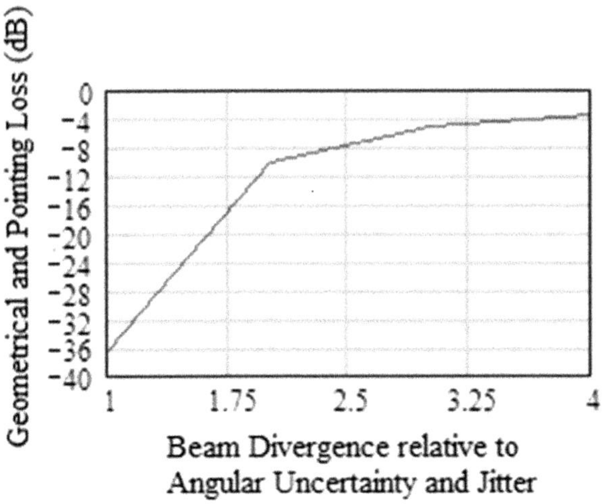

FIGURE 4.16
Geometrical and pointing loss at the edge of the uncertainty area due to beam divergence relative to the angular size of uncertainty area and jitter.

assumed from the range of 2.10^{-17} to 10^{-13} $m^{-2/3}$, which is still in the weak turbulence regime so that Eqs. (4.7) and (4.9) could be applied under the approximation described in the text to calculate SNR and then BER. The results of BER are shown for the communication range of 700 m. For the communication range of 700 m, the BER will increase little from 10^{-12} for very weak turbulence to about 10^{-2} for somewhat moderate turbulence strength. The typical BER of 10^{-6} can be easily accomplished for weak turbulence strength of about 5.10^{-17}.

Fig. 4.16 shows the geometric and pointing loss in dB as a function of beam divergence relative to angular uncertainty and jitter. From Fig. 4.16, for a value of $x = 1$ (relative beam divergence of 1), the combined loss is about −36 dB and increasing x to 2.5 or 3.5, the losses can be less by about −8 and −4 dB, respectively. This loss can be compensated either by adjusting the beam divergence adaptically or sometimes increasing the laser input power. The acceptable communication capability can be improved if automatic adaptive beam divergence is implemented for inter-UAV FSO communications.

4.13.11 Some Issues and Comments in Establishing Unmanned Air Vehicle Communication Links

- Networks with UAV swarm
  Recent operation of the UAVs in a swarm scenario and potential applications need a high data rate communication infrastructure between the individual UAVs belonging to that swarm. We have

mentioned the FSO propagation characteristics for UAV in the presence of atmospheric turbulence. The UAV in a swarm formation can cover a wider range of geographical territories and the individual UAVs need to communicate with each other at very high data rate. Optical communication networking scenarios can thus be established with different network configurations to optimize the communication link performance. Typically ring, star, and the meshed swarm formation are used [38]. All the UAVs have bidirectional links for a ring architecture connecting one to the next and one to the previous UAV. In a star architecture, one UAV is chosen as the OMU, which provides the link to the ground station. Each of the UAVs are equipped with individual transceivers. The meshed architecture has the most reliable network, which combines the advantages of both star and ring architecture and offers the most secure links. The transmitter wavelength can be 1.55 μm and the link distances can be a few km, 4–8 km. The atmospheric turbulence model at the heights of the UAVs can be applied; an example was already discussed in the previous section for achieving SNR and BER for multigigabit links.

- FSO links combined with OFDM modulation between UAVs and ground station
  An extremely reliable FSO communication link in the presence of atmospheric turbulence can be established between UAVs and between UAV and ground station. OFDM modulation technique provides a practical solution for establishing the high bandwidth optical communication links offering multipath mitigation in the turbulent atmosphere [39]. Coherent optical OFDM offers the most performance in receiver sensitivity and spectral efficiency by overlapping subcarrier spectrum and using coherent detection and signal set orthogonality to avoid interference. The technique can be appropriately applied to turbulent atmosphere between UAVs and between UAV and ground station. The system includes a pseudo-random binary sequence generator for bit sequences with specific bit rate, and the bit generator is connected with quadrature amplitude modulator (QAM) sequence generator that generates QAM data signals. These data signals are then OFDM modulated [39]. They reported mathematical analysis of OFDM transmitter, optical up-conversion for finally transmitting optical power of laser beams through the path of the mobile FSO link. The atmospheric propagation parameters include the attenuation and turbulence model, which was modeled as gamma-gamma distribution, and the Hufnagel-Valley 21 model. Their simulation shows that the maximum range that is available by the system for a reliable link of BER equal to 1e-6 for clear sky, for hazy weather is 12.5 km and for foggy weather is 3 km. The results also show BER versus optical SNR indicating higher birates to be

more sensitive to BER improvement as the optical SNR increases. For birates equal to 20 Gbps, there is no BER improvement, indicating the system performance limit of the link [39]. The higher order of QAM introduced in the system will significantly increase the received power for specific transmitted power. Because the turbulence levels are high at lower altitudes, improvement of each subsystem is required in order to overcome any turbulence and scintillation effects for UAVs flying at low altitudes.

- Optimal UAV height for minimum outage probability of an air-to-ground link communication network
  A very large network of UAVs has the potential of providing broadband connectivity to remote areas. Furthermore, aerial base stations can provide a low latency and large coverage networks with communication capacity enhancements. UAVs also have the potential to be a part of IoT to assist low power transmitters for sending their data to a destination, achieving energy-efficient communication. A cellular network can also be integrated with the UAV as low-altitude platforms to compensate cell overload or site outage. UAVs can therefore be very effectively used to improve the capacity of the cellular network. A recent paper [40] discusses the analysis of deriving the height of the UAV for minimum outage probability at every user location on the ground and the optimum altitude resulting in maximum coverage radius. The paper considered a UAV as an aerial base station to serve ground nodes in a downlink scenario. An altitude-dependent path loss exponent and fading function together with air-to-ground channel models are considered. In their air-to-ground wireless networking using a UAV in the presence of randomly distributed ground relays over the coverage region that includes both ground destination as well as cooperative relays was considered. Thus UAV placed in an adjustable altitude communicates with a ground node either directly or through a terrestrial relay. Their results show that the optimal altitude for maximum reliability is determined by the direct link of communication in an air-to-ground cooperative system. The relaying communication can treat a wide range of UAV altitudes. The optimal UAV altitude for a specific scenario of 1000 m communication range using low transmit power is shown to be 1300 m for direct communication. For a relaying communication, a range of 700–2000 m was for that optimal UAV altitude.

4.13.12 Airborne Lasercom Terminal Network Link

An airborne platform can be used to provide a node in a network comprised of other airborne, space-borne, and ground-based platforms/terminals. The airborne platforms will play a role in establishing Internet connectivity in global optical networks that require data transfer at a very high rate to locations on the ground. If there is a cloud or other LOS blockage the data can be routed through air-to-air-to-ground or air-space-to-ground FSO links. Development of a second-generation optical communication demonstrator intended for airborne applications like air-to-ground and air-to-air optical links at JPL was reported [41]. The development described the capability for unidirectional high data rate (2.5 Gbit/s) transmission at 1550 nm with the ability to receive an 810 nm beacon to aid acquisition, pointing, and tracking. Airborne platforms as an important element of the evolving network centric surveillance architecture is also described [42], showing the various applications of lasercom terminals.

4.13.13 Atmospheric Propagation Characteristics Relevant to Airborne Laser Communication Systems Performance

The atmospheric path characteristics and effects of atmospheric absorption, scattering, and turbulence for airborne terminal to ground location can analyzed using the similar theoretical concepts for slat path as previously discussed. The impact of the atmospheric channel on overall link budget performance is discussed in the previous sections. Near the ground location fog, rain, dust, snow, smoke, molecular absorption, and aerosols particulate matter all attenuate the signal-carrying laser beam from the airborne platform. The attenuation of the signal to a certain extent can be compensated for by increasing the signal gain or some atmospheric mitigation technique. Random fluctuations in the atmosphere's refractive index severely degrade the wavefront quality of a signal-carrying laser beam, causing intensity fading and random signal losses at the receiver. This results in increased system BERs along a slant propagation path between an airborne terminal to the ground terminal. Results from numerical simulations are presented [13] for analyzing communication performance for downlink and uplink, which are relevant to the airborne lasercom. The average BER of M-ary PPM optical communication systems was recently analyzed [43] under the combined influence of atmospheric turbulence and aero-optic effects. In their analysis, the atmospheric turbulence is modeled by the exponential Weibull distribution to describe the PDF of the irradiance under weak to strong conditions in the presence of aperture averaging. This PDF model can be accurate for the combined effects of atmospheric turbulence and aero-optics that arise because of the distortions generated close to the aircraft for affecting the optical propagation. Their research shows the BER performance for different receiver aperture sizes to improve the system performance significantly in the presence of atmospheric turbulence and aero-optics induced distortions. This compensation technique is important in developing an airborne optical communication system for high data rate transmission.

4.13.13.1 Constellation of Satellites/Small Satellites

In order to provide global broadband telecommunications service, it is possible to use a massive satellite constellation where each satellite is equipped with a laser transceiver transferring information between them as well as from the satellite belonging to the constellation to ground station. In the next few years, satellite entrepreneurs backed by billions of dollars from investors plan to cover the earth with thousands of miniature satellites to beam low-cost, ubiquitous broadband service. Some of the advantages of using these small satellites are that they are much less costly to launch, largely mass-produced, and can satisfy demand for high-speed Internet connectivity everywhere including any remote area worldwide and at anytime. Networks of satellites for high-speed broadband access around the world have already been proposed by companies like SpaceX, OneWeb, and Boeing, as well as by Facebook Inc. for bringing Internet to poorly connected areas. Constellation of satellites approach to provide high-speed Internet services is growing as consumers need the high-speed service not well-served by broadband fiber or cable including remote areas where digging cables are impractical and expensive. Boeing proposed a 2900-satellite constellation to provide high-speed broadband access. OneWeb expects to launch the first 10 satellites of a constellation that eventually will be 900 spacecraft.

There are also some efforts to use a constellation of satellites of several hundred LEO satellites to provide a global, broadband *Internet-in-the-sky*; for example, Teledesic proposed to provide affordable access to fiber-like telecommunications capability anywhere in the world similar to a local service provided through a global network.

A reliable Internet connection to remote areas is essential so that universal 5G services could be integrated with IoT developments across all industries and in all regions. Satellite communications can provide a practical resource to extend and complement existing terrestrial networks to meet the demanding 5G requirements both in terms of flexibility and global connectivity. Technology for future Internet connectivity everywhere is now available and will be more affordable with time. Based on all-optical technology, it will be possible to design an Internet gateway constellation of satellites to provide backhaul connectivity to terrestrial 5G relay nodes that can even create an on-ground 5G network, which requires much more bandwidth than an IoT sensor that is read every hour, and more bandwidth is needed during school time than at home. Atmospheric turbulence effects close to the ground and Doppler shifts related to satellite channels will be challenging to terrestrial-based systems using satellites in the constellation system. Another important factor to be considered is satellite handover so that all the transmitting and receiving nodes in the system should pass through the handover process. A 5G-based LEO megaconstellation architecture to foster the integration of satellite and terrestrial networks in 5G I was recently proposed [44]. The impact of typical satellite channel impairments as large delays and Doppler shifts on network layer procedures are also discussed in their paper.

FSO communications can provide size, weight, and power (SWaP) advantages since the optical devices used for transmitter, receiver, and optics are very small, compact, and fast, requiring very small power to operate. SWaP can be critical in applying FSO communications to small satellite platforms and terminals with Gbps data rates. Network systems architectures and requirements provide the capability of acquisition, establishment, and maintenance of FSO communication links. It is important to address the efficient design concepts to develop techniques such as modulation formats and capacity for high data rate transmission.

4.13.13.2 Global Coverage for Free-Space Optical communication Between Constellation of Low Earth Orbit Small Satellites and Earth Station

Laser communication between satellites and also between any transmission satellite and a ground station in the range of multiples (preferably hundreds) of Gbit/s is needed in order to establish global Internet connectivity. A high data rate laser link will require high accuracy pointing capability and a wide band optical earth receiver in conjunction with a laser beacon and two-way data link. A microsatellite platform can be used for the transmission satellites. A laser communication system can be designed to provide a high data rate laser link to the earth from LEO satellites, with a high accuracy of pointing capability. The effects of atmospheric turbulence and pointing limitations limit each satellite's access to the ground station. This will require a relatively large number of satellites at low altitudes. Within the orbital plane to secure constant communication a Walker constellation in a "string of pearls" formation, where each satellite can easily access the two satellites closest to it (the one ahead and the one behind it), can be used. This will minimize the total number of satellites and required orbital planes required. Atmospheric effects such as turbulence will affect the received BER and probability of fades of the signal to a tolerable amount sufficient for a reliable link. One of the key factors in a system constellation is the maximum angle relative to the zenith when the transceiver may point toward the receiver and transmit the data. The quality of the communication link is satisfied as long as the maximum angle is not reached. The angle is zero when the satellite passes exactly above the ground station. Since the path is longer as the angle increases, the turbulence effects on the communication system performance also increases. *A zenith angle no larger than 30° permits that BER is not deteriorated beyond a required BER needed for successful high-quality communication.* When LOS is blocked by clouds or severely affected by fog and rain, redundant transmission to several terminals in different locations will be needed. A constellation of 27 satellites at LEO to provide constant communication has been reported from the constellation size analysis [45]. Mobile ground stations can be useful in reducing the number of satellites and therefore cost in order to increase the number and quality of new applications.

It is therefore obvious that the satellite market is shifting toward smaller (micro- and nanosatellite), lower mass and increased performance platforms applicable to FSO communications possible everywhere in the global world. The impact of the telecommunication market has been to provide unprecedented worldwide coverage using constellations with less than 1000 satellites.

A laser communication system concept for small satellites has been recently reported [46], which presents the C3PO system with advanced concept for laser uplink/downlink communications with space objects describing some key technologies. The project reported designs a communications system that uses a ground-based laser to illuminate a satellite, and a modulated retroreflector [1] to return a beam of light modulated by data to the ground. The system therefore requires no laser on the satellite and is well-suited to small satellites to provide high data rate in Gbit/s range for very light on-board mass. For the C3PO system, each ground-based station is equipped with components used for both uplinks and downlinks, a tracking system for satellite acquisition and pointing, and a ground-based laser for downlink with a modulated format for uplink communication. A modulated retroreflector [1,46] is integrated at the space-based terminal and uplink communications will use an embedded optical receiver on the satellite receiving the modulated signal from the ground.

4.13.13.3 Recent Updates Regarding Constellation of Small Satellites for High Data Rate Laser Communications to Remote Places

Two demonstration satellites of its proposed network of satellites planned by SpaceX, which have the shape of a box 3 ft × 2.2 ft × 2.2 ft will be a part of the first phase of testing for SpaceX's so-called constellation, which will initially consist of 4425 satellites. The constellation is intended to provide broadband Internet service at fiber-like speeds for individual households and small businesses. SpaceX plans to start launching operational satellites in phases and will reach full capacity by 2024. After the launch of 800 satellites commercial service could start. As mentioned earlier OneWeb and Boeing XCo. plan to launch constellation of hundreds or even thousands of small satellites to provide Internet access, especially to remote global locations where fiber optic cables and cell towers do not reach. Small satellite constellations can also offer more competition where terrestrial Internet access is already available.

4.13.13.4 New Concept Technology for Free-Space Optical Communications Using a Constellation of Small Satellites

Based on this book's description and discussions of global Internet connectivity by all-optical technology, it is expected that every small satellite in the constellation can be equipped with very compact, fast laser transceivers based on development of advanced optical devices so that multi-Gigabit/s Internet access *anytime, anywhere* will be practically possible in the near future.

4.13.13.5 Balloons for Establishing Free-Space Optical Communications in Remote Places

FSO communication has the potential of building a high bandwidth backhaul network with wireless optical communication systems providing throughput of 10 Gb/s or more using a balloon mesh network. FSO communication for mobile nodes in three dimensions thus has the potential of offering Internet access to remote areas. The balloons are launched at high altitude and can drift in space due to winds causing random movements that will lead to misalignment of LOS FSO communications between adjacent balloons and between a balloon and ground station. It is therefore important to keep the network intact in spite of the shift and movement of the balloons due to wind. An algorithm to track the drifting balloon and evaluate the tracking system by state estimation of the drifting balloon's position with respect to another nearest balloon is reported [47]. In case of misalignment they proposed a backup optical link to keep the network intact based on Kalman filtering–based state estimation technique for three-dimensional movement and speed.

To provide high-speed Internet to inaccessible areas and other regions of the world that are not connected, Facebook and Google are considering various aerial devices. Some of the projects include Facebook's Aquila, the WI-Fi lasering, solar-powered drone, and Google's Project Loon Internet balloons, which plan to beam Internet from the sky. Google's Project Loon's plan to beam wireless Internet from the skies through gigantic floating balloon hotspots started launching about 30 of their inflatable envelop-type aerial vehicles from New Zealand [48]. The balloons travel about 20 km above the Earth's surface and are solar-powered. Google plans to launch its (literal) around-the-world coverage launching more than 300 of these balloons in a continuous string around the globe. This ensures the connectivity is maintained by one balloon when the neighboring balloon floats away. The balloons are 12 m high and 15 m wide when inflated and are made from lightweight polyethylene plastic. Each balloon provides connectivity to an area of 80 km in diameter underneath it. The connected devices with the balloons have the capability of 4G-type speed and the balloons can now last a couple of months. A sketch of establishing connectivity with balloons is depicted in Fig. 4.17.

With Facebook's Internet drone project Aquila leading the way, Google has now launched another project, Skybender [48], to deliver 5G Internet to remote

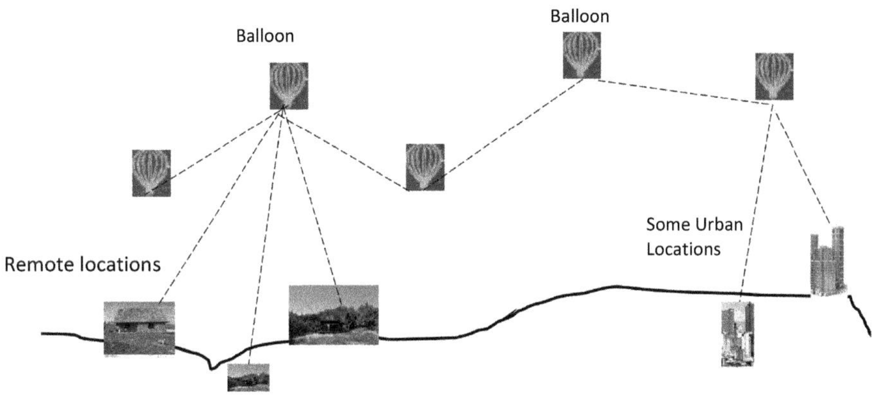

FIGURE 4.17
A sketch of the connectivity with the balloons.

parts of the world and increase Internet speeds in cities. The drones beam down the backbone of 5G networks and the solar-powered drones hover at heights of between 18.5 and 27.5 km, above commercial aircraft altitudes, and will help to avoid problematic weather conditions.

4.14 CONCLUSIONS

This chapter discusses the fundamentals of FSO systems required for terrestrial, airborne, and space-based platform and optical propagation channel models for understanding and evaluating system performance. SNR and BER for optical signals in presence of atmospheric turbulence are analyzed for airborne, UAV, HAP, balloons, and satellite platforms equipped with all-optical transceivers to provide potential high data rate Global Internet connectivity anytime, anywhere particularly in remote locations. Different optical propagation paths such as slant uplink and downlink, and between UAVs or satellites are considered to analyze the different communication system performance parameters affected by different scenarios. With recent development of compact and fast optical and electrooptical devices and advanced optical networks, there is definitely a future potential to design and develop high data rate global Internet connectivity based on all-optical technology.

References

[1] A.K. Majumdar, Advanced Free Space Optics (FSO): A Systems Approach, Springer, New York, 2015.

[2] C. Lee, C. Shen, C. Cozzan, R.M. Farrell, J.S. Speck, S. Nakamura, B.S. Ool, S.P. DenBaars, Gigabit-per-second white light-based visible light communication using near-ultraviolet laser diode and red-, green-, and blue-emitting phosphors, Optics Express 25 (15) (July 24, 2017).

[3] H. Haas, Wireless Data from Every Light Bulb, TED Global, Edinburgh, July 2011.

[4] J.C. Ricklin, S.M. Hammel, F.D. Eaton, S.L. Lachinova, Atmospheric Channel Effects on Free-Space Laser Communication, Free-space Laser Communications: Principles and Advances, Springer, New York, 2008.

[5] H. Le-Minh, D. O'Brien, G. Faulkner, A Gigabit/s indoor optical wireless system for home access networks, in: CSNDSP, 2010.

[6] B. Smutny, H. Kaempfner, G. Muehlnikel, U. Sterr, B. Wandernoth, F. Heine, et al., 5.6 Gbps optical intersatellite communication link, in: Proc. SPIE, Free Space Laser Comm. Tech. XXI, vol. 7199, 2009.

[7] V.I. Tatarskii, The Effects of the Turbulent Atmosphere on Wave Propagation, Available from: U.S. Department of Commerce, Springfield, VA, 22151, 1971. Translated by IPST Satf.

[8] J. Horwath, M. Knapek, B. Epple, M. Brechtelsbauer, B. Wilkerson, Broadband backhaul communication for stratospheric platforms: the stratospheric optical payload experiment (STROPEX), in: Proc. SPIE, vol. 6304, 2006, 63041N-1–63041N-12.

[9] L.C. Andrews, R.L. Phillips, C.Y. Hopen, Laser Beam Scintillation with Application, SPIE Press, 2001.

[10] B. Smutny, R. Lange, H. Kämpfner, D. Dallmann, G. Mühlnikel, M. Reinhardt, K. Saucke, U. Sterr, B. Wandernoth, R. Czichy, In-orbit verification of optical inter-satellite communication links based on homodyne BPSK, in: Proc. SPIE, Free-Space Laser Communication Technologies XX, vol. 6877, 2008.

[11] L.C. Andrews, R.L. Phillips, Laser Beam Laser Beam Propagation through Random Media, second ed., SPIE Press, 2005.

[12] H. Kaushal, V.K. Jain, S. Kar, Overview of Wireless Communication Systems, Chapter, January 2017, https://doi.org/10.1007/978-81-322-3691-7_1. ResearchGate: https://www.researchgate.net/publication/312133486.

[13] A.K. Majumdar, J.C. Ricklin, Effects of the atmospheric channel on free-space laser communications, in: D.G. Voelz, J.C. Ricklin (Eds.), Free-Space Laser Communications V., Proc. of SPIE, vol. 5892, SPIE, Bellingham, WA, 2005, 58920K-1–58920K-16.

[14] H. Kaushal, V. Kumar, A. Dutta, H. Aennam, H. Aennam, V. Jain, S. Kar, J. Joseph, Experimental study on beam wander under varying atmospheric turbulence conditions, IEEE Photonics Technology Letters 23 (22) (2011) 1691–1693.

[15] J.A. Anguita, M.A. Neifeld, B.V. Vasic, Turbulence-induced channel crosstalk in an orbital angular momentum multiplexed free-space optical link, Applied Optics 47 (13) (2008) 2414–2429.

[16] F. Fidler, Optical backhaul links between HAPs and satellites in the multi-gigabit regime, in: GLOBECOM Workshops, 30 Nov.-4 Dec. 2008, IEEE, 2008, 978-1-4244-3062-8/08.

[17] A.K. Majumdar, J.C. Ricklin, Free-Space Laser Communications: Principles and Advances, Springer, New York, 2008.

[18] S. Betti, E. Duca, I. Toselli, HAP-GEO optical links: performance analysis under weak turbulence conditions, in: Proc. Symp. Commun. Syst., Netw. Digit. Signal Process, 2008, pp. 77–81.

[19] H. Kaushal, G. Kaddoum, Optical communication in space: challenges and mitigation techniques, IEEE Communications Surveys and Tutorials (May 28, 2017), https://doi.org/10.1109/COMST.2016.2603518.

[20] H. Zech, F. Heine, D. Tröndle, S. Seel, M. Motzigemba, R. Meyer, S. Philipp-May, LCT for EDRS: LEO to GEOoptical communications at 1.8 Gbps between Alphasat and Sentinel 1a, in: Proc. SPIE, Unmanned/Unattended Sensors and Sensor Networks XI and Advanced Free-Space Optical Communication Techniques and Applications, vol. 9647, 2015.

[21] I.B. Djordjevic, Deep-space and near-Earth optical communications by coded orbital angular momentum (OAM) modulation, Journal of the Optical Society of America 19 (15) (2011) 14277–14298.

[22] T. Lei, M. Zhang, Y. Li, P. Jia, G.N. Liu, et al., Massive individual orbital angular momentum channels for multiplexing enabled by dammann gratings, Light: Science and Applications (2015).

[23] J. Wang, J.Y. Yang, I.M. Fazal, N. Ahmed, Y. Yan, H. Huang, Y. Ren, Y. Yue, S. Dolinar, M. Tur, A.E. Willner, Terabit free-space data transmission employing orbital angular momentum multiplexing, Nature Photonics 6 (7) (2012) 488–496.

[24] H. Huang, G. Xie, Y. Yan, N. Ahmed, Y. Ren, Y. Yue, D. Rogawski, M. Tur, B. Erkmen, K. Birnbaum, S. Dolinar, M. Lavery, M.J. Padgett, A.E. Willner, 100 Tbit/s free-space data link using orbital angular momentum mode division multiplexing combined with wavelength division multiplexing, in: OFC/NFOEC Tech. Digest, 2013.

[25] S. Betti, E. Duca, I. Toselli, HAP-GEO optical links: performance analysis under weak turbulence conditions, in: Communication Systems, Networks and Digital Signal Processing. CNSDSP 2008. 6th International Symposium on Date of Conference: 25–25 July 2008 in Graz, Austria, 2008.

[26] F. Fidler, M. Knapek, J. Horwath, W.R. Leeb, Optical communications for high-altitude platforms, IEEE Journal of Selected Topics in Quantum Electronics 16 (5) (September/Oeptember 2010).

[27] M. Antonini, S. Betti, V. Carrozzo, E. Duca, M. Ruggieri, Feasibility analysis of a HAP-LEO optical link for data relay purposes, in: Presented at the Aerosp. Conf., Big Sky, MT, 2006.

[28] E. Katimertzoglou, D. Vouyioukas, P. Veltsistas, P. Constantinou, Optical interplatform link scenarios for 20 km altitude, in: Proc. 16th IST Mobile Wireless Commun. Summit, 2007, pp. 1–5.

[29] C. Fortuna, A. Vilhar, J. Horwath, M. Mohorcic, Wavelength requirements in optical transport networks based on high altitude platforms, in: Proc. Symp. Commun. Syst., Netw. Digit. Signal Process, 2008, pp. 87–91.

[30] A. Biswas, Airborne optical communications demonstrator design and pre-flight test results, in: Proc. SPIE Free-Space Laser Commun. Technol. XVII, vol. 5712, 2005, pp. 205–216.

[31] J. Horwath, C. Fuchs, Aircraft to ground unidirectional laser-comm. terminal for high resolution sensors, in: Proc. SPIE, vol. 7199, 2009, 7199-1–7199-8.

[32] A. Hatziefremidis, K.E. Zarganis, H.C. Leligou, N. Pleros, K.E. Zoiros, Performance evaluation of optical channel transmission between UAVs and Ground Stations, in: MATEC Web of Conferences 41, 01005, 2016, EDP Sciences, 2016, https://doi.org/10.1051//matecconf/2016 41 01005.

[33] L.C. Andrews, R.L. Phillips, C.Y. Hopen, Laser Beam Scintillation with Applications, SPIE Press, Philadelphia, PA, 2001.

[34] X. Guoliang, Z. Xuping, W. Junwei, F. Xiaoyong, Influence of atmospheric turbulence on FSO link performance, in: Proc. SPIE, Optical Transmission, Switching, and Subsystems, vol. 5281, 2004, pp. 816–823.

[35] I. Toseli, L.C. Andrews, R.L. Phillips, V. Ferrero, Free space optical system performance for a Gaussian beam propagating through non-Kolmogorov weak turbulence, in: IEEE Trans. Antennas Propag, vol. 57, 2009, pp. 1783–1788.

[36] K.H. Heng, N. Liu, Y. He, W.D. Zhong, T.H. Cheng, Adaptive beam divergence for inter-UAV free space optical communications, in: IEEE Conference IPGC, 2008.

[37] S.G. Lambert, W.L. Casey, Laser Communications in Space, Artech House, 1995.

[38] S.S. Muhammad, T. Plank, E. Leitgeb, A. Friedl, K. Zettl, T. Javornik, N. Schmitt, Challenges in establishing free space optical communications between flying vehicles, in: CSNDSP08 – Proceedings, IEEE, 2008, 978-1-4244-1876-3/08.

[39] K.E. Zarganis, A. Hatziefremidis, Performance analysis of coherent optical OFDM applied to UAV mobile FSO systems, American Journal of Optics and Photonics 3 (1) (2015) 5–12, https://doi.org/10.11648/j.ajop.20150301.12.

[40] M.M. Azari, F. Rosas, K.-C. Chen, S. Pollin, "Ultra reliable UAV communication using altitude and cooperation diversity, IEEE Transactions on Communications 66 (1) (January 2018).

[41] A. Biswas, N. Page, J. Neal, D. Zhu, M. Wright, G.G. Ovtiz, W.H. Favr, H. Hernnzati, Airborne optical communications demonstrator design and preflight test results, in: Proc. SPIE 5712, Free-Space Laser Communication Technologies XVII, April 18, 2005, https://doi.org/10.1117/12.600834.

[42] J.A. Maynard, D. Begley, Airborne laser communications: past, present and future, in: D.G. Voelz, J.C. Ricklin (Eds.), Free-space Laser Communications V, Proc. of SPIE, vol. 5892, SPIE, Bellingham, WA, 2005.

[43] J. Zhao, S. Zhao, W. Zhao, J. Cai, Y. Liu, X. Li, BER performance analysis of M- ary PPM over exponentiated Weibull distribution for airborne laser communications, Journal of Optical Technology 84 (10) (2017) 658–663.

[44] O. Kodheli, A. Guidotti, A. Vanelli-Coralli, Integration of Satellites in 5G through LEO Constellations, June 30, 2017 arXiv:1706.06013v2 [cs.NI].

[45] E. Pinsky, D.L. Barnett, A. Oren, Global Coverage For Fast Response Communication Between Constellation Of Leo Small Satellites And Earth Stations, August 2008. ResearchGate article: https://www.researchgate.net/publication/258302604.

[46] Y. Gouy, E. Steck, C. Sanchez, G. Faulkner, D. O'Brien, F. Sproll, P. Wagner, D. Hampf, W. Riede, M. Salter, Q. Wang, D. Platt, D. Jakonis, X. Piao, M. Karlsson, O. Oberg, I. Petermann, A. Michalkiewicz, J. Krezel, A. Debowska, B. d'Humières, Y. Thueux, THE C3PO project: a laser communication system concept for small satellites, in: Proc. Vol. 10096, Free-Space Laser Communication and Atmospheric Propagation XXIX, SPIE LASE, 2017.

[47] M.B. Awan, S. Mohan, Balloon mesh free space optical communication with tracking and link Switching, in: ICTON, IEEE, 2015, 978-1-4673-7880-2/15.

[48] Google's Projects Loon and Skybender Are Taking Internet Connectivity to Greater Heights, Autonomous Vehicles Drones IoT Megatrends, March 10, 2016.

Further Reading

[1] D.C. O'Brien, G. Faulkner, H. Le-Minh, O. Bouchet, M.E. Tabach, M. Wolf, J.W. Walewski, S. Randel, S. Nerreter, M. Franke, K.-D. Langer, J. Grubor, T. Kamalakis, Gigabit optical wireless for a home access network, in: Proc. PIMRC, 2009, pp. 1–5.

[2] H. Le Minh, Z. Ghassemlooy, D. O'Brien, G. Faulkner, Indoor gigabit optical wireless communications: challenges and possibilities, in: The 12th International Conference on Transparent Optical Networks (ICTON), 27 June- 1 July 2010, Bavarian Academy of Sciences and Humanities, Munich, Germany, 2010.

[3] A. Gupta, N. Sharma, P. Garg, M. Alouini, Cascaded FSO-VLC communication system, IEEE Wireless Communications Letters 1 (2017), https://doi.org/10.1109/LWC.2017.2745561.

[4] E. Katimertzoglou, D. Vouyioukas, P. Veltsistas, Optical interplatform links scenarios for 20 km altitude, in: Mobile and Wireless Communications Summit, 1–5 July 2007.

[5] S. Kidder, T.H. Vonder Haar, A Satellite Constellation to Observe the Spectral Radiance Shell of Earth, 2004.

CHAPTER 5

All-Optical Broadband Global Communications for Internet Connectivity: Free-Space Optic Links and Optical Network Architectures

5.1 INTRODUCTION

Previous chapters discussed the basics and fundamentals of Internet connectivity and various types and methods of network architectures using optical links. As mentioned earlier, free-space optical (FSO) links can be used to set up a complete FSO network, which is supplement to conventional radio frequency (RF) links combined with fiber-optic links. This chapter will introduce a big picture for establishing global Internet connectivity using *all-optical* connectivity, which includes both FSO and some portion with broadband fiber optics [1,2]. Both FSO and fiber-based optical connectivity use optical wavelength technology as a means of transmitting and transferring messages and detecting signals. The major difference in optical communing using FSO and fiber optics is the two different communication channels in the two cases. FSO channel suffers through atmospheric effects such as turbulence and scattering whereas fiber-optic channels are not affected by atmospheric effects.

This chapter also identifies a number of recent technological developments that have shown promise for enhancing and developing innovative optical wireless communication (OWC)/FSO systems. These include (1) modulating retroreflector (MMR)-based FSO duplex communications (includes fiber-based amplified retromodulators and microelectromechanical modulators (MEMS) deformable mirror retroreflector modulator), and (2) Wi-Fi/Li-Fi FSO optics via LED lights. The chapter explains the physics of atmospheric reciprocity leading to *reciprocity-enhanced optical communication* through atmospheric turbulence. The fiber-array–based (multiple channel) amplified retromodulator is discussed by a recent patent by the author and his colleague to provide a pixelated fiber array system for both incoming and outgoing optical beams, eliminating the need of complex aligning and pointing in FSO communication at multigigabit data rate. The link budget analysis for a satellite-based system laser interrogator and low-power consuming Gb/s amplified fiber retromodulator

Optical Wireless Communications for Broadband Global Internet Connectivity.
https://doi.org/10.1016/B978-0-12-813365-1.00005-9

shows the feasibility of space-to-ground communication. Newly developed technology, Wi-Fi/Li-Fi, to increase the bandwidth of conventional Wi-Fi systems several times using FSO transmission via LED lights is described, where up to 100 Mb/s rate data can be sent to *each user* for an indoor system. The increased bandwidth should eliminate problems like video streaming that stalls and buffers in home applications; for example, gaming systems or watching movies with tablets/computers. There is the potential to achieve up to 1000× higher capacity per area than traditional Wi-Fi systems. The chapter discusses the latest advancement of achieving Li-Fi Internet breakthrough for 224 Gb/s connection broadcast with an LED bulb for ~3 m wireless link. The chapter discusses how a transceiver unit be integrated into every smart phone, tablet, and laptop to form part of the fourth- and fifth-generation (5G) mobile network infrastructure to use this Li-Fi technology. OWC/FSO features create this huge technological opportunity.

The breakthrough technology developed initially for defense purposes and later commercialized for civilian use has created various applications today. FSO is one of the recent examples—a way to use lasers through free space instead of fiber-optic cables to transmit data. For example, from connected homes to cheap smart phones in everybody's pockets, there is skyrocketing growth in the number of network-connected devices coupled with advances in optical technology. These are driving fundamental changes in the way the very technology will design global network infrastructure. The basic idea of building a whole new global network backbone has taken a totally new concept, which may consist of satellites that beam data down to earth using lasers, 5G wireless networks, and intercontinental submarine cables, capable of handling unprecedented amount of bandwidth only possible using optical wavelength. Internet connectivity in remote locations where there is no connectivity at all now as well as high bandwidth can thus be established with this *global backbone*. This newly designed backbone will enable a tremendous amount of data to come in from the *edge of the network* than outward, from centralized computing hubs. Eventually this will provide a total global coverage capable of bringing connectivity literally anywhere in the world; for example a constellation of 8–12 laser-enabled satellites, called HALO, that will circle the planet and thus be able to combine terrestrial networks to create hybrid high-capacity network.

The most practical way to design and develop this global coverage concept will be to create the primary hubs at the *equinox* data centers where the bulk of the world's networks interconnect and will be able to distribute data from space to terrestrial networks at various locations. A number of ground nodes comprising each hub will be equipped with laser heads to link with a number of different laser-enabled satellites so that there will be always an optical link connectivity with any of the satellites at that location and time. The network structure can thus be very flexible to send laser beams to any of the laser-enabled satellites to establish instant data transfer between any remote points on the earth. Furthermore, submarine optical cable landing stations can also be included to other ground-node locations to maximize the possibilities for transferring data across the oceans and to distribute it almost anywhere on the ground. Data rates of 100 Gbit/s between ground nodes and satellites, and 200 Gbit/s from satellite to satellite—about 100 times or faster than radio links used in satellite communication today—are predicted [3].

5.2 VARIOUS TYPES OF FREE-SPACE OPTICAL SYSTEMS FOR DIFFERENT NETWORK ARCHITECTURES: SATELLITE COMMUNICATIONS, FREE-SPACE OPTICS (TERRESTRIAL), AND WIRELESS LOCAL AREA NETWORKS COMBINED

The diverse range of FSO communication applications range from indoor communications between various devices such as smart phones, PCs, laptops, and the like, to outdoor interbuilding links (short- and long-distance terrestrial links), ground-to-air/air-to-ground terminals (which can include airborne platforms like unmanned air vehicles [UAVs], balloons, high-altitude platforms [HAP], etc.), satellite uplink/downlink (including small satellites), and other terminals involving intersatellite and deep space probes. By connecting FSO links and networks to the backbone realized with optical fibers, the FSO networks and links can constitute a complete global *all-optical* network around the world, which can provide global Internet connectivity in remote places. Various FSO networks involving optical wireless satellites, terrestrial, and local area networks (LANs) can be integrated and operated as a whole is shown in Fig. 5.1 with a simplified schematic of the fiber optic connectivity to connect from the house/office base station to the fiber backbone and thus eventually to the Internet server using FSO communications.

5.2.1 Integrated All-Optical Satellite/Airborne, Terrestrial Local Area Network and Home Network Conceptual System Scheme

This section discusses various FSO system architectures that interconnect satellites with airborne and ground-based platforms via laser links. In order to support worldwide broadband access to the Internet, global backbones capable of delivering terabits of data transfer need to be established. Depending on the applications, various FSO link system architectures need to design laser/optical links in a variety of environments, including the previous types of FSO communication systems.

It is important to understand the functions of various FSO systems to design full-OWC links that can operate directly propagating an optical beam from a

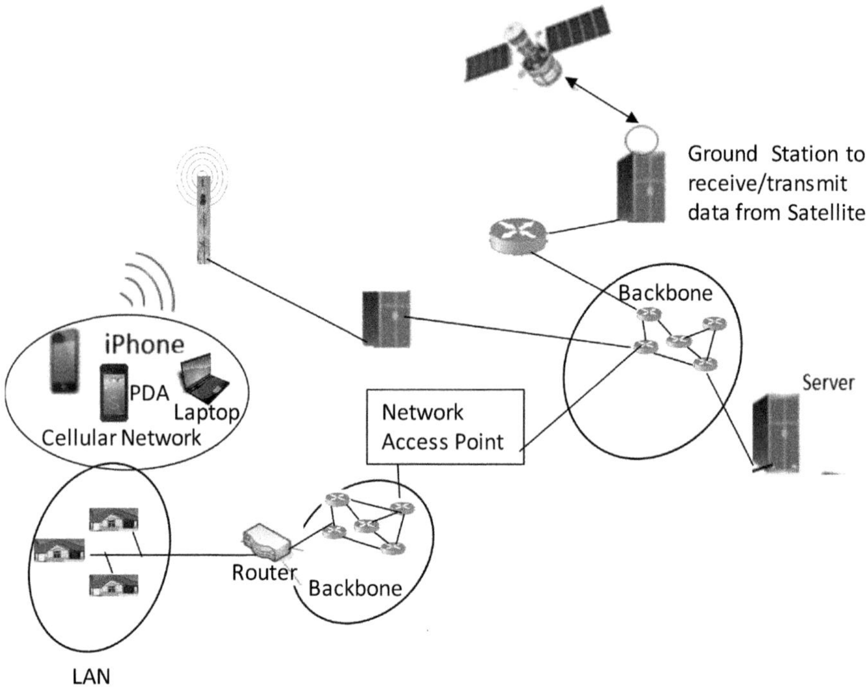

FIGURE 5.1

Various free-space optical networks involving optical wireless satellites, terrestrial, and local area networks integrated with a fiber-optic connectivity to the Internet server.

fiber termination point through a free-space channel using an OWC transceiver. A corresponding transceiver at the receiving end can directly couple the free-space propagated optical beam into an optical fiber connection port. The FSO system links will include terrestrial FSO (e.g., between buildings in metropolitan office areas, a campus, or hospitals; between mobile platforms), horizontal and slant path terrestrial optical links, airborne systems (e.g., between aircraft, UAVs, drones, HAPs, balloons such as Project Loon, satellites). Indoor FSO communications (within a building among various users in a room or elsewhere in the building) will also be a part of the global Internet communications link at one extreme end. Basically the functions of the various FSO optical communication systems are (1) the FSO-based LAN will provide the connectivity to fixed and mobile users in a network cell; (2) terrestrial FSO will establish a broadband data link between two fixed locations addressing the issues like the *last-mile problem*, and also connecting to an optical ground system (OGS), which eventually establishes connectivity between the laser-based satellites and the OGSs; and (3) satellites (including a constellation of a few satellites) can provide the optical connectivity between remote locations globally, almost *anytime, anywhere*.

5.2.2 Major Challenges for Efficient System Integration of All the Systems

All the systems mentioned earlier must work together, which requires seamless coupling of a free-space propagated optical beam to a fiber so that ultimately the overall global system interoperability is accomplished. As shown in Fig. 5.1, based on the availability of today's advanced high-speed, compact, light-weight, and low-power consumption optical technology it is possible to develop broadband global Internet connectivity by connecting FSO links and networks to fiber-optic and FSO to the global optical *backbone*. Fundamental challenges involve the availability of FSO links and the reliability of the individual FSO networks (and the individual components) integrated for designing the overall system. The enormous complexity of building an externally reliable global optical system arises from two reasons: (1) the characteristics of the physical links, which are part of an integrated network with great geographic extent and large numbers of uses, must be well matched to the network architecture; and (2) the propagation medium is different for each layer protocols. For example, for an intersatellite link, the propagation medium has practically no atmosphere (i.e., like a vacuum link) and the only effect to channel will come from propagation loss (due to unavoidable diffraction-limited beam spread). On the other hand, for terrestrial communication links involving atmosphere, when multiple users share the same optical medium, a channel's randomness interacts with the upper layer protocol. As a result, throughput performance of the network is reduced drastically. In short, all optical global networks therefore must be designed to account for multiple-network layers. Reliability of FSO terminals made of a number of individual FSO systems with FSO transmitting and FSO receiving terminals will depend upon the reliability of all the optical components and the interfaces to the network associated with the components. Different availabilities and reliabilities will determine the different network architecture to work under different weather conditions.

Some of the other challenges include acquisition, tracking, and pointing (ATP) in the presence of different atmospheric propagation channels and to design the network terminals capable of mitigating atmospheric effects to transmit and receive broadband data.

5.3 CONCEPT ARCHITECTURES FOR ALL-OPTICAL FREE-SPACE OPTICAL SYSTEMS TO ACHIEVE GLOBAL INTERNET CONNECTIVITY

This section starts with indoor communications of the previously mentioned FSO networks system followed by longer range links of outdoor, airborne, and satellite links to support global Internet connectivity solutions.

5.3.1 Indoor Optical Communications: Visible Light Communication and Li-Fi

Indoor communications have been described in the literature extensively [4,5]. Indoor FSO networks are desirable for wireless broadband communications inside houses and offices. The optical wireless home and office networks are used to construct a LAN comprised of various cells that divide various sections within the building. A base station for each cell has a number of terminals connected by short-range FSO communication links. These links are basically infrared (IR) links using LEDs or diode lasers at visible wavelengths offering visible light communication (VLC). These wireless optical cells in a given section of the building or a room are connected and integrated with a broadband infrastructure. Indoor FSO optical links can be line-of-sight (LOS), or non—line-of-sight (NLOS), which is established by reflection, scattering, or diffused mechanisms inside the room. LOS link requires a direct path between a transmitter and a receiver whereas for a NLOS link lights from a diffused source undergo multipath propagation due to reflection and scattering by the walls, ceilings, floor, and furniture or objects in the room. Because of a better power budget the LOS links achieve higher capacity supporting higher data rate compared to NLOS links; however, the NLOS links are more robust to support mobile terminals. The diffused system for NLOS suffers multipath dispersion, causing pulse spreading and severe intersymbol interference (ISI), thus offering a lower data rate. Since fundamentals and different architectures of indoor communications are discussed in the literature [4,5], some specific areas of indoor communications, namely VLC and introductions to the Li-Fi for broadband communications, will be emphasized here. Fig. 5.2 shows visible light spectrum of electromagnetic waves for selecting LED/LD emitters for visible light communication applications.

5.3.2 Infrared and Visible Light Communication High Data Rate Indoor Communications

Indoor OWCs include two main technologies: IR and VLC. IR technology is well developed with various applications and devices conforming to Infrared Data Association standards. A bidirectional Gb/s IR OWC in home-access networks (HANs) is reported in [6] where the gigabit IR links provide high speed wireless transmission to data hotspots within the home. The method in their

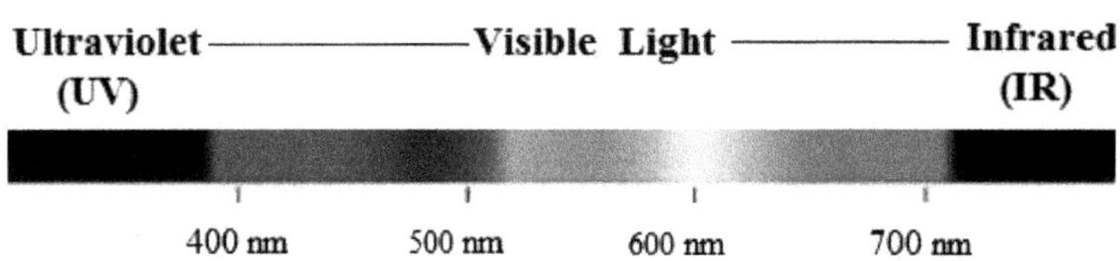

FIGURE 5.2
Visible light spectrum of electromagnetic waves for selecting LED/LD emitters for visible light communication applications.

research involves increasing the FOV of an indoor high-speed IR link by adopting a cellular communication scheme in which multiple narrow FOV links are combined together to form a wider FOV. The cellular IR system consists of a base station (BS) and a mobile terminal within the BS coverage area with identical transceiver components for both mobile terminal and BS. Thus, there is a multiple element transmitter at the BS and the mobile terminal has a number of angle diversity receivers. An LOS communication link is thus accomplished for any transmitter and a receiver. The transmitter source is an 825 nm laser diode (LD) operating at a measured power level of 50 mW (within eye safety regulation). The receiver is 0.5 mm (diameter) Avalanche Photo Diode (APD) with integrated preamplifier with a sensitivity of about −35 dBm at 1.25 Gbit/s for a bit error rate (BER) of 10^{-9}. The full field-of-view (FOV) is 30 degrees with a minimum range of 3 m. A bidirectional gigabit Ethernet OWC system is reported also in [7], where with a similar increasing FOV method a measured BER of 10^{-11} is reported for 1.25 Gbit/s non—return-to-zero (NRZ) on-off-keying (OOK) link over a 3 m range with a coverage area of 1.3 m × 0.45 m. Another experimental realization of a wide FOV IR transparent-wireless-fiber link with a capacity over 100 Gbit/s is reported [8] by combining a narrow beam (±0.15 degrees) link with a wide-angle holographic beam splitting, at both transmitter and receiver, using liquid crystal spatial light modulators (SLMs). In their work an optimum operating wavelength of 1550 nm was used and a seven-channel Nyquist wavelength division multiplexing (WDM) transceiver was used and the system operation at 3 m provided a coverage area up to 3.4 m in diameter in the receiver plane, establishing a digital coherent WDM transmission at 112 Gbit/s and 224 Gbit/s for a full angle of 60 and 36 degrees, respectively. Since the present book focuses on how to establish the Internet connectivity from a transceiver terminal (from the house or office, street, mobile platforms, or practically anywhere) to ultimately an Internet server, broadband connectivity is essential. Fig. 5.3 shows a schematic of a very high data rate ~100 Gbit/s indoor optical communication using an optical wireless module containing a holographic beam steering device and angle magnification, as well as connectivity with the fiber backbone, which can be integrated with optical ground stations to transmit and receive optical communications with a satellite at a high rate.

A VLC system for high-speed indoor VLC networks requires an efficient design of both the illumination and communication aspects using the same visible source. For illumination, white LEDs are used either by combining separate RGB (red-green-blue) emitters or by using a blue emitter in combination with yellowish phosphor. In order to achieve an acceptable data rate and link quality in a VLC system, received power and signal-to-noise ratio (SNR) level at any location inside the house or office must be analyzed, including indoor VLC systems for supporting mobility. Specifically, the influence of FOV angle on the connectivity performance in the practical indoor scenarios is an important consideration in design such as for a VLC system.

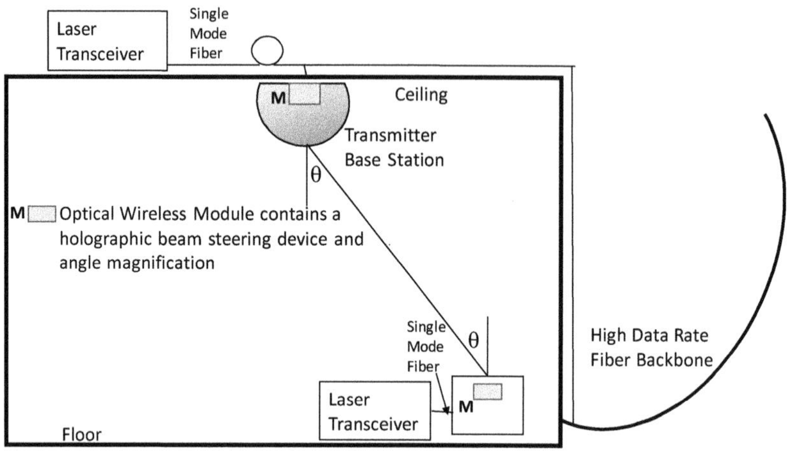

FIGURE 5.3
A schematic of the fiber-optic connectivity from very high data rate ~100 Gbit/s indoor optical wireless communication to the Internet server via fiber backbone.

For mobility scenarios connectivity is one of the most vital points to consider since it ultimately defines the communication coverage of the system viewing by the FOV angle of the receiver in the receiver plane. Ideally full connectivity for continuous data connection in all places in the room is desired. Changing the receiver plane affects the connectivity as well changing the lighting area of an LED and view area of the receiver. Properly designing the lighting layout can therefore adjust the connectivity within a range of receiver plane and FOV angle.

Recently a design of an indoor VLC system was reported by studying the effect of setting lighting positions and setting FOV angle on connectivity and link switching performance [4]. Two LEDs were used (which could be generalized to any number of multiple-input–multiple-output (MIMO) systems) and installed on the ceiling of a room that was 5 m × 2.5 m × 2.5 m; the distance between the LEDs was 2.5 m, half-power angle of the LEDs was 70 degrees, transmitted optical power of 72 W with a center luminous intensity of 2628 and a reflectance factor of the walls of 0.8. PD detector area of 1.0 cm^2 with VLC receiver is a mobile platform, OOK modulation was used, the received power was −1 to 4 dBm in most of the places in the room, at the highest FOV of 48 degrees, the average received power of about 2.8 dBm, and average SNR of about 31 dB. A comprehensive lighting configuration is thus reported in their work analyzing various relationships of various parameters in a VLC system to achieve the highest link quality and optimum connectivity supporting the link switching process.

A visible communication system based on white light generated using a near-ultraviolet LD pumping RGB-emitting phosphors was recently demonstrated [9]. The transmitter used was a III-nitride LD on a semipolar (2021) substrate. The transmitter wavelength was 410 nm and an avalanche photodetector was used. They reported at the data rate of 1.06 Gbit/s, BER (approximately) of $10^{-3.8}$ for a received optical power of −5 dBm. Using quadrature amplitude modulated (QAM) orthogonal frequency division multiplexing (OFDM) modulation scheme the researchers predicted achieving a much increased data rate using a higher speed photodetector (>GHz).

Another VLC communication system demonstration [10,11] involved use of LDs with transmitting powers up to 50−100 mW at wavelengths between 421 and 429 nm, achieving data rates of 3.4 Gbit/s showing the potential of GaN laser diodes for high-speed free-space VLC over short distances such as indoor scenarios. Other researchers published [10,12] demonstrations of potential VLC data rates at over 100 Gbit/s using off-the-shelf LDs in a number of scenarios with illumination constraints. The three-color LDs used to generate white light for their demonstration were a red LD operating at 658 nm, a green LD at 520 nm, and a blue LD at 450 nm, and the collimated outputs were combined using beam splitters, mirrors, and optical filters. The modulation scheme for communication was an *M*-ary QAM symbol. The researchers demonstrated the communication parameter results for the following scenarios of illumination: array of RGB used for complete indoor illumination achieving overall communication data rate of 14 Gbit/s and an average BER of 6.07×10^{-4} at a distance of 30 cm from the transmitter; using entire visible light spectrum employing multiple LD transmitters with 12 triplets of LDs (36 streams total) delivering an illumination of approximately 871.28 1× to achieve a data rate of 105.41 Gbit/s with an average BER of 2.04×10^{-3}.

A technique using beam steering and computer-generated holograms has been described to develop a 20 Gb/s mobile indoor VLC system [13]. Their system employs beam steering of part of a VLC system using adaptive finite vocabulary of holograms in conjunction with an imaging receiver and delay adaptation technique (DAT) to enhance SNR and to mitigate the impact of intersatellite link (ISL) at a high data rate of 20 Gbit/s. The adaptive system provided a BER of 10^{-9} at all receivers' locations when operated at 20 Gbit/s in a harsh room environment. This research is an important accomplishment of indoor VLC systems operated at 20 Gbit/s using a simple modulation format (OOK) and without the use of complex WDM devices.

Other recent work [10] describes an indoor optical bidirectional wireless link with an aggregate capacity over 100 Gbit/s where the link operates over ~3 m range at 224 Gbit/s (6 × 37.4 Gbit/s) and 112 Gbit/s (3 × 37.4 Gbit/s) with a wide FOV of 60 and 36 degrees, respectively. They reported their technique, which combines a narrow beam (±0.15 degrees) link with a wide-angle holographic beam steering at both transmitter and receiver using liquid crystal SLMs.

The steering at the transmitter end directs light to the receiver while at the receiver end the light is steered back onto the optical axis of the collection system to allow it to be coupled into a fiber. A coverage area of 3.4 m in diameter in the receiver plane was reported for operation at 3 m. The important conclusion from their demonstration is that this technique can help to achieve a practical extension of optical fiber capacities to free-space systems, which can eventually lead to the terabit communications system.

5.3.3 Li-Fi: Toward All-Optical Networks for Internet Connectivity

5.3.3.1 Li-Fi Basics

Li-Fi (light fidelity) is a wireless optical networking technology that uses LEDs or LDs for transmission of data using optical wavelength. Li-Fi serves both illumination and communication simultaneously. Li-Fi bulbs are outfitted with a chip capable of modulating light for optical transmission. The term Li-Fi, coined by Professor Harald Haas at the University of Edinburgh, is the pioneer in developing this technology and referred to the idea as "data through illumination." He prompted his research work at TED Global talk in July 2011 [14]. Li-Fi can be thought of as an optical solution to replace Wi-Fi and is based on VLC using as optical carrier for both illumination and data transmission at a very high data rate. Both LEDs and LDs can be used as VLC source, which can be switched on and off faster than the human eye can detect, and this flickering invisible on-off activity enables a kind of data transmission using binary codes: switching on an LED or LD is a logical "1," switching it off is a logical "0."

A data stream can be generated by encoding in the light by varying the flickering rate of the LED or LD, which acts as a transmitting source modulating the light with the data signal. The data rate using LDs can be achieved beyond 3 Gbit/s or even 100s of Gbit/s by researchers as mentioned earlier. Various multiplexing techniques and by parallel data transmission using an array of transmitters each transmitting different data streams can also be used to increase VLC data rate. The Li-Fi emitter source system can consist of bulb (visible light) and a chip, which may have power supply, all in an enclosure. The working principle of Li-Fi technology is as follows: The transmitter is the LED/LD source and a photo detector is the light sensor (this can be a smart phone, laptop, tablet, or similar items equipped with optical devices). Li-Fi technology works by sending over the light and an LED/LD bulb is then flicked from OFF to ON in order to generate light signals. For communication purpose the data needs to be first encoded and then modulated, depending on the particular modulation scheme. For processing the data, the LED/LD bulb is connected with a microchip so that the LED/LD can be turned ON and OFF at very high speed so that the intensity of light can be modulated in high speeds with varying amplitude.

In order to retrieve data from a visible light source we need to have a card reader or dongle to function as a wireless modem to be plugged into the laptop or tablet perhaps via USB. A sensor receives the downlink light from the visible light source whereas an IR component can send a signal back via uplink toward the visible light source. Multiple users can be connected with any single light source already equipped with a networking component without a loss of connectivity because of moving. In principle, every light in the global world can connect to the Internet using this Li-Fi technology. The Li-Fi thus provides a very effective high-speed bidirectional networked and mobile communication of data using visible light sources. Multiple light bulbs form a wireless network. Li-Fi is a potential solution for satisfying the tremendous demand for achieving connectivity from using lights that illuminate homes, offices, cars, and streets also connecting to us to data.

For data communication purpose Li-Fi can be effectively used for the receiving platform where an ultrasmall camera in a mobile phone can replace the photodetector for scanning and retrieving data. Similarly, HDTV can be equipped with Li-Fi, which then serves as a platform for connecting smart phone, gamepad, tablet, and other peripherals in a room can all exchange data interactively. Li-Fi can be integrated with the street lights, which can then provide Internet access to a mobile car and all other devices on-board. Li-Fi–based VLC communication systems use an efficient modulating format such as optical orthogonal frequency division (O-OFDM) modulation suitable for very high data rates with multiple-access connectivity for both indoor and outdoor scenarios.

All the communication components must be directly in light where the bulbs are integrated with a microchip, turning them into a *hotspot* for Internet connection—in short, a complete communication. Internet connectivity with Li-Fi technology requires bulbs outfitted with microchips and the devices equipped with dongles for uplink and downlink data exchange without losing connectivity. By the year 2020, the number of devices including mobile users to be connected for Internet of Things (IoT) applications will exceed 20 billion, which evidently shows the tremendous demands of connectivity of tomorrow. Only optical wireless networks can provide the only solution where Li-Fi plays a major role.

Fig. 5.4 shows a general scheme for establishing Internet connectivity from a Li-Fi technology–based access point. The idea of using Li-Fi in the most cost-effective way is to reuse existing infrastructure with their original purpose without installing a costly new cable or wiring, or to use DC power from one Ethernet pluggable device to another without the need for extra power source. For example, use of power line communication (PLC) and power over Ethernet (POE) are currently considered to make Li-Fi technology very versatile for broadband Internet connections and to use existing infrastructure. An example of using Ethernet over PLC in automotive networks has been reported [15].

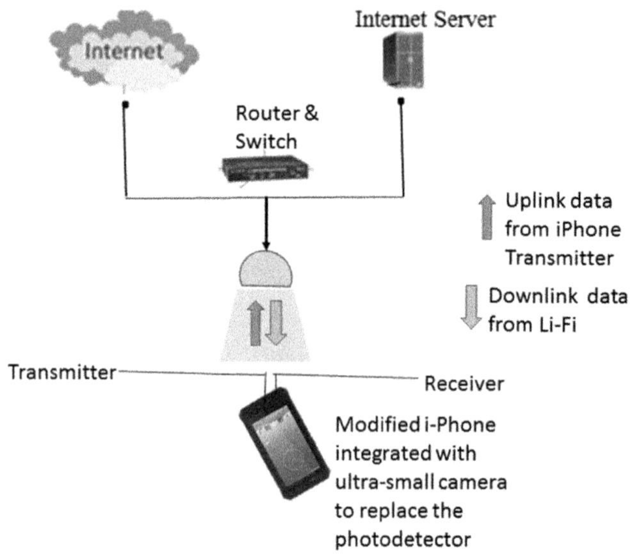

FIGURE 5.4

A general architecture of connecting to the Internet with Li-Fi technology. The Li-Fi access point can be simply created at the Li-Fi light sources and a number of Li-Fi sources can be connected via optical wireless communication network.

A visible light source (LED/LD) with a microchip containing signal processing and optical communication capabilities streams data embedded in the beam at high data rate to the photodetector. A receiver equipped with a dongle converts the changes in amplitude into an electrical signal and demodulates back into a binary data stream, which is then transmitted to a computer and mobile device that are Internet-enabled devices to recognize binary data stream as web, video, and audio applications. Fig. 5.5 shows the basic principles of a Li-Fi providing Internet connectivity.

From the working principle of a Li-Fi technology it is evident that each LED/LD light bulb equipped with a small microchip could be converted into a wireless router, thereby using a *light bulb to access the Internet*. Very fast Internet connectivity at tens and hundreds of Gbits/s will make VLC the future of Internet connectivity where data for PCs, laptops, smart phones, and tablets are transmitted through light.

5.3.3.2 Li-Fi Internet Connectivity and Network Architectures: How Li-Fi Sources Create Hotspots for Internet

Li-Fi uses light for data transmission and can be easily used in dense environments because of less interference. Typically, Li-Fi covers a distance of about 10 m and can provide data rates in excess of multiple Gbit/s. Fig. 5.6 shows

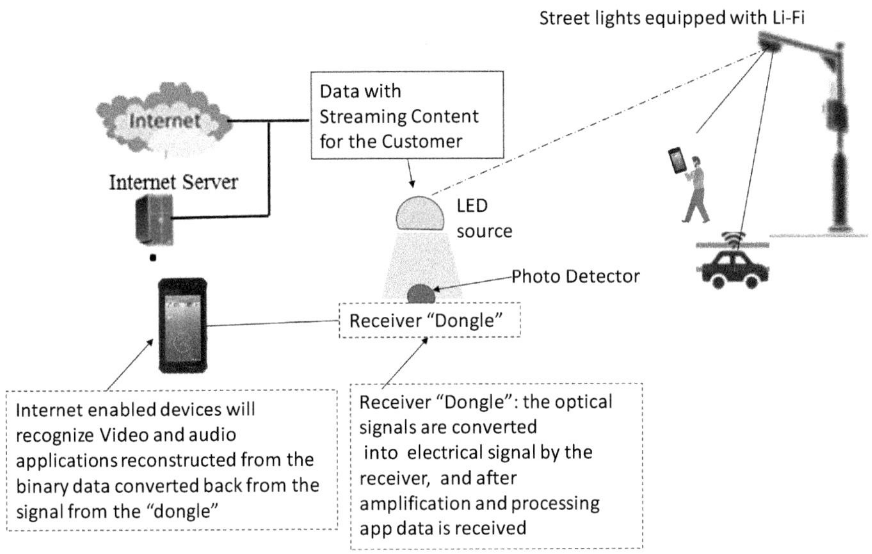

FIGURE 5.5
Basic principle of Li-Fi connecting street lights to iPhones and moving cars, providing Internet connectivity.

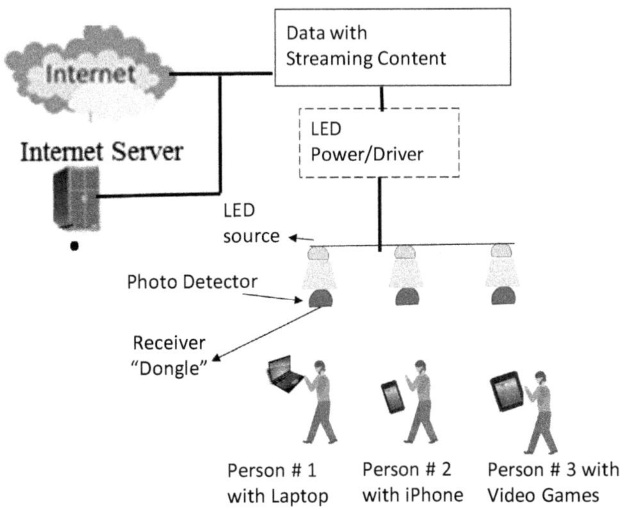

FIGURE 5.6
Li-Fi Internet architecture for different types of data applications and for multiple users.

a Li-Fi Internet architecture. Three major components accomplish Internet connectivity: (1) LED/LD lamp driver equipped with driver software push the streaming content (generated by the server and the Internet) to the LED/LD lamps, (2) LED/LD lamp, and (3) Li-Fi dongle composed of a photodetector, amplification/processing, and applications for different types of data to use Li-Fi Internet services such as browsing Internet on a laptop, tablet, or smart phone by various users. LED/LD lamps can be placed in any location in a home or an office building where multiple users can connect to the Internet from anywhere and simultaneously browse different applications of data as shown in Fig. 5.6.

5.3.4 Optical Wireless for a Home Access Network

Indoor small-cell cellular networks such as optical attocell based on VLC can be designed to provide efficient indoor optical communications in terms of data rate. Some of the following concepts can be practical and feasible to establish FSO and laser communications from indoor all the way to global locations:

- Indoor optical communications: LED or LDs together with Li-Fi to direct to an optical tower (either on a rooftop of a house or to a central optical cell tower with optical antenna)
- Concept of small cells to connect the optical links
- Small cells to LAN or metropolitan area network (MAN) with optical towers (OTs) equipped with omnidirectional antennas
- Continue the optical link to the OGS (solving last-mile problem and using fiber backhaul) for directing the optical beam to a laser FSOC-based transceiver satellites; other platforms can be planes, HAPs, balloons
- Intersatellite optical links for reversing the process of using another satellite and back to remote places, if needed

There is a missing puzzle piece for connecting Li-Fi (wherever available or installed even if in an extremely *remote* location) with an optical tower and then to an optical gateway system center and finally to a laser-equipped satellite for global Internet connectivity. A potential solution is to use both FSO communication transceivers with usual IR or LD establishing a link between a fixed location and another fixed location (cities between buildings, for example) and then fiber optic backhaul and FSO to reach OGSs to send the link to a low earth orbit (LEO) satellite that has Internet from a server and Internet provider. The fixed FSO locations can be equipped with Li-Fi's with LED/LD lamps, which then are finally connected with anybody walking in the street, car, or any mobile platform. In short, the solution is a combination of Li-Fi links and FSO fixed systems to cover the worldwide OWC using Internet connectivity. Therefore, it is extremely important to understand and analyze the propagation effect of LED/LD visible lights, optical/laser propagation for horizontal and slant paths for a propagation length of a few meters to few hundreds of meters for evaluating communication parameters like SNR and BER. This was discussed in detail with the necessary mathematical description in the previous chapter.

5.3.4.1 Broadband Li-Fi—Based Wireless Access to Streetlights, Moving Platforms

Li-Fi in homes or business buildings can provide simple, and reliable and secure wireless communications. Users can figure out the best coverage locations by simply seeing the lights from the Li-Fi lamps. Streetlights, building lights, and transportation lights can serve as Li-Fi hotspots for the Internet and therefore are able to communicate wirelessly between static or mobile platforms using smart phones, tablets, or portable laptops. Wherever there is LED/LD lighting infrastructure there can be a wireless Li-Fi communication network. A Li-Fi access point (LED/LD light) in a Li-Fi network has a unique IP address that will utilize the technology of small cell network, which can be very flexible in coverage and mobility. *In short, Li-Fi has the potential to provide high speed, dense, and reliable networks for enterprise environments, which can extend to cover smart buildings, cities, and the nation, and eventually can be a part of the broadband global Internet connectivity.* The 14 billion light bulbs existing worldwide today can be replaced with inexpensive LED/LD lamps that can transmit high speed data. *With this Li-Fi technology, when implemented worldwide, every streetlight would be a free access point.* The 14 billion lamps in the world will gradually become Internet masts to satisfy the increasing demand of mobile connectivity. There is thus the potential of only needing to hover under a streetlamp to get public Internet access to download a movie from the lamp or surf the Internet will provide cheap high-speed Web access to every street corner. For example, the traffic control lights, streetlights, and cars with headlights and backlights can be modified to have LED/LD-based lamps so that users walking in the streets as well as in cars can all communicate with each other while moving without any loss of connectivity. For *smart transports* Li-Fi can provide not only high-speed, secure, and reliable wireless communication for users of transports. For *smart cities*, Li-Fi can play a major role in making truly smart cities by enhancing communications wirelessly between streetlights, building lights, and transportation lights. For transforming into *smart lights*, anywhere where there is already existing LED light infrastructure, there can be a potential wireless Li-Fi communication network. Thus, each Li-Fi access point (LED light) in a Li-Fi network can have a unique IP address that will allow utilization of the power of a small cell network. Public Li-Fi hotspots can therefore be used for establishing communication between the users. Fig. 5.5 shows the public Li-Fi hotspot concepts and the typical users connecting with each other by Internet. Li-Fi is the latest technology that can provide fast speed Internet access almost anywhere there are visible lamps. Li-Fi can be used for data transmission, and can easily provide high speed Internet via every light source such as overhead reading bulb inside an airplane. By implementing this technology worldwide every street lamp would be a free access point. In another example, the ceiling bulbs inside a mall can create their own constellation of navigation beacons. Smart phones equipped with a camera automatically receive these signals, which switch the navigation software to use the information to connect with nearest spot the smart phone is searching.

From a network point of view, each Li-Fi access point (LED light) in a Li-Fi network can have a unique IPS address that can utilize the power of a small network. Users can establish communications by using Li-Fi hotspots.

5.4 INTEGRATION ARCHITECTURES FOR IN-BUILDING OPTICAL WIRELESS ACCESS NETWORKS: A BIG PICTURE

The tremendous growth in the traffic of multimedia services demands to have some hot spots in places such as conference centers, hotels, airports, businesses, and shopping malls, which therefore must have wireless access. For all-optical network concept, wired access can be replaced with high bandwidth fiber optics and then integrate with optical wireless access technologies [16]. In fact, recent advances in transmission and broadband access technologies are promising to bring the information highway containing multimedia services to houses worldwide. This section describes and discusses potential broadband access specifying connectivity mechanisms from the local signal (data) provider to the end user; for example, home, car, or somebody in the street with a smart phone. To establish two-way communications, the end user is equipped with network-enabled devices.

This chapter thus includes proposed architectures for exploring *small cell networks* to connect with the global Internet using all-optical technology and utilizing OTs integrated with omnidirectional optical antenna. The access points are therefore very essential to connect the digitally controlled devices to a network that ultimately leads to the Internet.

5.4.1 Internet Connectivity From Home-Based, Ground-Fixed, Ground-Mobile, Man-Portable Terminals

A broadband home access architecture includes basically the following technological components, which are needed to finally connect with the Internet: broadband local loop, home gateway, and home area network devices (processing tools) in home environments. Fig. 5.7 shows the generic system architecture of the in-building optical-wireless access network for fiber-based and wireless multimedia services using optical technology. The broadband local loop of communication networks supports the connection between the local provider to the consumers by bringing information to the home via fiber optics, optical wireless (FSOC), and laser-equipped satellite communications. The goal is to bring high-speed information to homes, offices, campuses, and so on. The communication architecture system should be able to provide

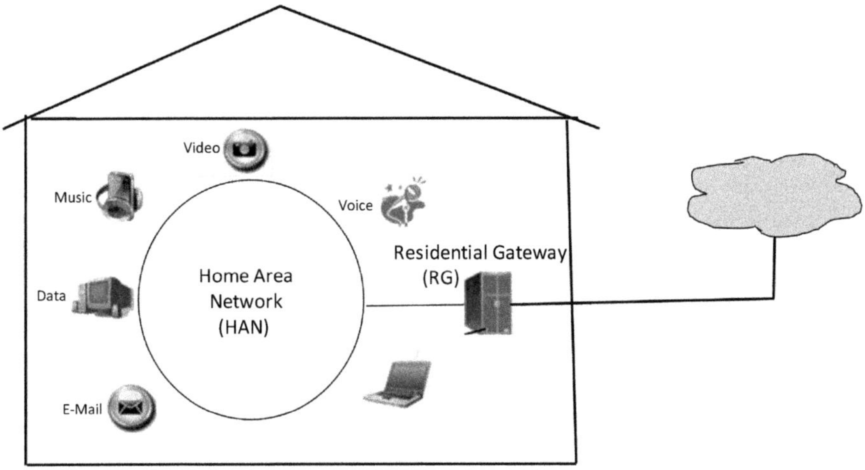

FIGURE 5.7
Broadband home access architecture based on wireless optical technologies.

complete coverage of both fixed and mobile users and applications. A network integration among Internet-equipped devices and the Internet is therefore required to provide seamless two-way communications. In Fig. 5.7 home gateway and optical networks are combined to realize the traffic providing and the resources control. The residential gateway is the interface device, which offers an effective bidirectional communication channel to every networked device in the home with broadband local loop. It can also be considered as the centralized access point between the home and the outside world with different broadband and local area network technologies as well as personal area networks. Various wireless data services and also the multimedia services in Ethernet passive optical network are then distributed to the terminal rooms via indoor communications, which can use LED or LD sources and the communications can be accomplished either by LOS or by NLOS by reflection or scattering in the rooms. The object is to integrate home wired and wireless network with the external Internet with the existing passive optical network facilities. The broadband home access architecture therefore serves the purpose of connecting the next generation of Internet-based vendors with consumers located in the comfort of their own homes.

5.4.2 Basic Architecture of Broadband Optical Wireless Access

The basic purpose of the wireless networks is to transmit and receive data over the air, which is the propagation media for communication. Since this book addresses mainly the optical communications for data transfer, this section will be applicable to the FSO communications through the atmosphere.

It is also possible to have optical wireless technology for a home network backbone solution, but will be used in any case to interconnect a class of devices forming a *subnetwork* for mobile communications such as a person carrying a smart phone in the street or somebody wanting Internet connection from the car. This way, these mobility networks will interface with other subnetworks and with the Internet by connecting to any type of home network backbone. The propagation technology for optical wireless networks can be IR technology with LED/LD as transmitters as well as VLC technology as described earlier in the Li-Fi technology.

5.4.3 Infrared LED/LD and Visible Light Communications Technology

Indoor point-to-point systems are not too different from outdoor systems design in operation except atmospheric loss has no effect for all indoor systems and the communication power link budget is therefore determined almost entirely by the transmitter launch power, free space loss, and the receiver sensitivity. IR transmission techniques for LOS and NLOS for indoor link configurations rely on the existence of a direct path between the transmitter and the receiver as well as the degree of directionality, that is, source beam-angle and nondirected NLOS systems, which are generally referred as diffuse systems. The indoor communication system consists of a light source (transmitter), either an LED or LD, free space as the propagation medium, and a detector (using APD or PIN diodes). Information, typically in the form of digital or analog signals, is input to electronic circuitry that modulates through an optical system into free-space propagation medium. The received information is collected by an optical system (an optical filter to reject optical noise, a lens or concentrator that focuses light on the optical signal detectors and thereafter to signal processing electronics). The wavelength band from 780 to 950 nm is the best choice for an indoor optical wireless system and low cost LEDs and LDs are readily available that can be used. LEDs can be more favorable because of relaxed safety regulations for eye safety, low cost, and high reliability compared to LDs. An array of LEDs can be used to design higher power transmitters than a single LED transmitter. The system usually uses intensity modulation with direct detection (IM/DD) scheme for modulation and detection. Direct detection is performed by PIN photo-diodes or APDs, which produces an electric current proportional to the incident optical power. FSO is that part of the transmitted power that is lost or not captured by the receiver's aperture (due to diffraction property of the initial beam). For a point-to-point system with a slightly divergent beam, the free-space loss would be about 20 dB, whereas an indoor system using a wide-angle beam could have a free loss of about 40 dB or more. The signal fading can be observed in both indoor and outdoor optical wireless systems. The fading is due to the reception of signals via different paths by the receiver.

Some of these interfere destructively (out of phase) so that the received signal power effectively decreases. This type of degradation is also known as multipath signal fading. Depending on the applications and system requirement, there are several possible transmission techniques available by appropriately choosing the transmitter, the receiver, and the optics to direct the beams: (1) directed radiation, (2) diffuse radiation, and (3) quasi-diffuse radiation. However, compared to the direct transmission, this diffuse technique requires higher transmitter power, large FOV for the receiver, and suffers multipath dispersion. Due to this multipath dispersion effect the original transmitted pulses are broadened when they reach the receiver, which causes ISI at higher data rates or in larger cell system [17]. In quasi-diffuse technique, a BS with relatively broad coverage and made of either passive or active reflector is mounted on the ceiling. A quasi-diffuse link based on multispot diffusing can also be established. Multiple narrow-beam transmitters and an angle diversity receiver with several narrow FOV detectors aimed in different directions can be used. This system is exposed to fewer multipaths and achieves lower path losses requiring lower power transmitter than a wide-beam diffuse system at the expense of increased complexity.

Among several techniques, the most common modulation schemes for indoor systems considered in the past are OOK and pulse-position modulation (PPM). OFDM can be used for a multiple-subcarrier modulation scheme. In OFDM applied to parallel data transmission, high data rate can be achieved by transmitting orthogonal subcarriers. Combining OFDM with any multiple access scheme makes it a powerful tool for indoor optical wireless applications. For OFDM IM/DD optical systems applications, two schemes have been used: DC-biased OFDM [18] and asymmetrically clipped optical OFDM [19]. Multiple access techniques define the way several users get access simultaneously to the available network services. This way, different user's signals can occupy the same time slot, code, or carrier frequency. A single cell topology using a single optical access point (OAP) per user or per room and a cellular topology with spatial reuse using multiple OAPs are examples of topologies that can be used for indoor systems. Electrical multiplexing (such as time domain multiple access, frequency domain multiple access, or code-division multiple access) is also possible to realize multiple access in a single cell per room or cellular topology. For optical multiplexing techniques wavelength-division multiple access and space-division multiple access can be used.

5.4.4 Architectures of Networks for Local Area Network, Metropolitan Area Network, and Wide Area Network With All-Optical Technologies

Note that a LAN is a group of computers and network devices connected together, usually within the same building, and is a high-speed connection of hundreds of Mb/s to hundreds or more of Gb/s. LANs are designed to cover a small area, usually an office or home. The larger network is covered by a MAN, which can include several buildings in the same city or town. MANs can connect multiple LANs across a city or multiple cities to form a cohesive whole. A wide area network (WAN) compared to MAN is not restricted to a geographical location confined within the bounds of a state or country. A WAN connects several LANs, and may be limited to an enterprise or accessible to the public. The network technology is high speed and the *Internet is an example of a worldwide public WAN*. Different architectures of FSO networks are needed to address all of these network types. The basic architecture starts with a point-to-point and a point-to-multipoint configuration with some examples of optical wireless configurations below. Fig. 5.8 shows an optical wireless network in ring, star, and mesh architectures, showing the concepts of point-to-point and point-to-multipoint scenarios connecting different buildings.

For an optical wireless FSO network in ring architecture, a pair of transmitter and receiver units (transceiver) are installed on top of each building and an

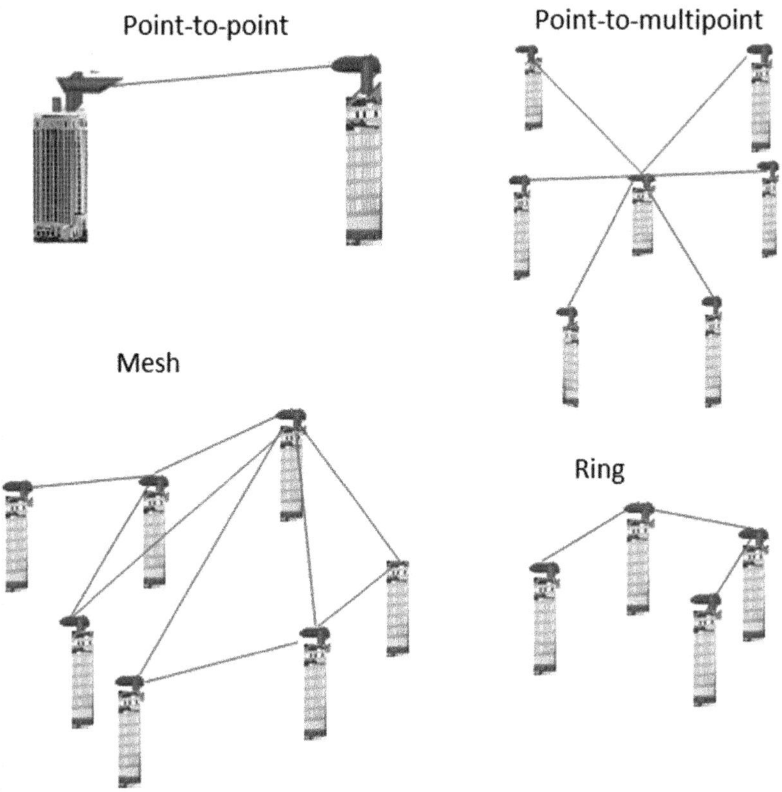

FIGURE 5.8
Optical wireless in ring, star, and mesh architectures with free-space optical networks.

LOS optical link is maintained between adjacent locations. If LOS is not possible, optical repeaters can be used to divert the links so that LOS is maintained. Buildings can be up to a few hundred meters apart and still suffer from atmospheric turbulence effects, although with a wider transmitting angle the effects can be minimized to achieve an acceptable BER. In case there is a broken communication link, the indirect connection can be used. To increase the reliability and the security against failure some additional redundant links can be installed. For an FSO network in a star architecture, an optical multipoint unit can be selected at the roof of a building and a number of user terminals with the optical transceivers of each of the users are connected to the optical multipoint hub station via FSO links. The optical multipoint unit is interconnected with the high bandwidth backbone network; for example, a broadband fiber backbone. In order to provide additional reliability in case the optical multipoint has a failure, a redundant multipoint unit needs to be installed. For a point-to-multipoint architecture, the optical multipoint unit is connected with a switch or router to the backbone network. Meshed architecture has the benefits of both architectures mentioned and enjoys the flexibility also. All these architectures finally need to be connected with the access network, which is usually located between the individual subscriber and the network provider's backbone network in order to finally have the Internet connection. Fibers do not always reach to the customer's location and there can be a gap of about a mile from the fiber. This is the same as the last-mile problem where FSO links can bridge the gap. The same networking technologies used in fiber link can be used for FSO link also, and thus it is possible to couple from FSO systems directly to the network fiber. Other advanced optical devices such as WDM can be easily used for data transfers using many wavelengths simultaneously for many channels at the same time to increase the data rates.

5.4.5 Free Space Optical/Laser Communications: All-Optical Network

It is therefore possible to design an architecture for concurrently providing all optical broadband wireless and fiber-optics guided services for home, and for public places such as a mall, conference center, and airport. The content provider sends the data to a central office equipped with an omnidirectional optical antenna and optical networks where the data can be can be transmitted by wireless optical technique (FSO) and via fiber backhaul to a platform that includes shopping mall, conference center, or airport. Visible light sources can provide the optical wireless connectivity from various hotspots as mentioned in the Li-Fi technology. *Internet connectivity and two-way communications between users with Li-Fi devices anywhere in the street, near buildings in metro city need access points.* This hybrid optical network architecture system can allow fiber-guided and wireless transmissions of the same content such as HDTV, data, and voice up to much faster than the current available networks. The customers can enjoy the same services either by plugging into the fiber-based connection in the wall or access the same information through a wireless system, via all-optical network technology. The customer premises can be conference centers, airports, hotels, shopping malls, or from their own homes and small offices.

5.4.6 Broadband Optical Wireless Access

There are a number of optical systems competing for the last-mile where free-space optics is a viable method to communicate between two locations or platforms. This also addresses the issues of broadband optical wireless access. The basic requirement for establishing the connectivity is to develop FSO communication links and network architectures to provide very high data rate for both downlink (from the hub to the terminal) and uplink (from the terminal to the hub). This asymmetry is basic characteristics of the uplink and downlink for Web traffic. FSO connectivity includes point-to-point, point-to-multipoint, and multipoint-to-multipoint for bringing broadband communication information and providing networking capabilities among the end users. Broadband wireless access services provide Internet access, voice, audio, image, and streaming video capabilities.

It is obvious that OWC systems will play a vital role as alternative access technology in the global Internet connectivity for future next-generation networks based on full-OWC links [20]. The full OWC system using all-optical technology will operate by directly propagating an optical beam from a broadband fiber termination point through free-space using an optical antenna for FSO communication and using an optical transceiver. If the optical signal's transmitted power is not sufficient for transmission through the atmosphere, an optical erbium-doped fiber amplifier (EDFA) can be used at the receiver end of the signal. Therefore, a combination of EDFA and dense wavelength division multiplexing (DWDM) technology can make the all-OWCs possible for accomplishing extremely high data rates in the ranges of multiple Gbit/s through the atmosphere [21]. The system will be capable of offering error-free transmission at multi-Gbit/s data rates. The main requirement is to develop and design seamless coupling a free-space propagated optical beam to a fiber (which is usually the end point or another point in the broadband fiber-backhaul). The design and development of various innovative wireless communication systems will be the only possible solution to provide the users with access to broadband communication (Internet connectivity) services at *anytime* from *any location*.

The goal for next-generation all-optical networks is the spread of broadband technologies to any remote and underserved areas where the provisions of high bandwidth fiber or other broadband technologies are not practically feasible. This book addresses this issue of developing innovative concept technologies to accomplish full OWC systems. Fig. 5.9 depicts a scenario and a general global network architecture that uses a full-OWC system in urban or rural areas. Later in this chapter, other types of platforms such as satellite, UAVs, HAPs, balloons, and airplanes will be discussed so that complete two-way communication links can be achieved in any remote place worldwide, anytime.

An optimum choice of an operating wavelength of 1550 nm should help in accomplishing this goal. The transceivers can operate at this wavelength where the communication channels are the atmospheric propagation paths. The problems of turbulence and scattering effects of the optical propagation to achieve multigigabit data rate can be easily solved with the recent development of adaptive optics (AO)-based transceivers. Another advantage of choosing the 1550 nm is that the design concept of the full OWC system includes the recent developments of efficient coupling of the received optical beam to a fiber. High bandwidth optical fiber is one option for access technology and the research and development of telecommunication network developed on fiber technology for the lasercom for the last many years can now be integrated with an FSO communication system to complete the two-way communication at very high speed data rate. The different architectures of broadband fiber optics technology for access technology can be used with different architectures to

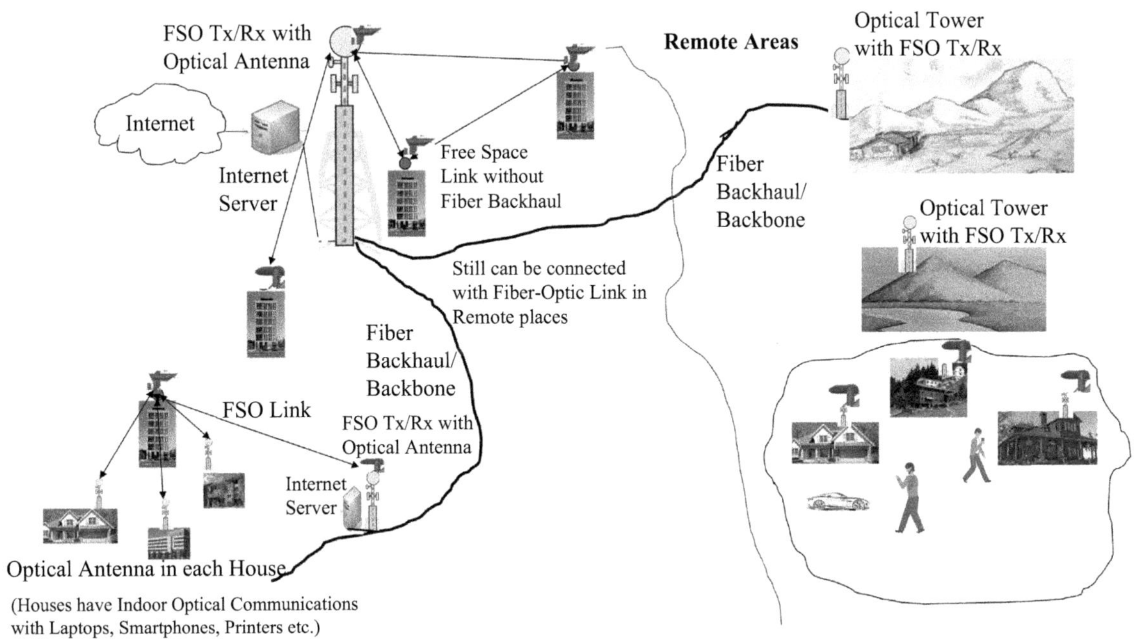

FIGURE 5.9
A general scenario of optical wireless communication system deployment to provide Internet connectivity.

combine with free-space wireless optical links and networks for point-to-point or point-to-multipoint architectures ranging from fiber to the home through fiber to buildings/offices.

The OWCs have the advantages of ease of deployment, license-free operation, high transmission security, high data rate communication, and two-way communication (duplex) systems. The wireless optical communications are therefore suitable for terrestrial and space based communications offering high data rate information exchange which are absolutely necessary in many modern Internet-based applications. OWC systems are deployed for the so-called last-mile access, enterprise connectivity, and fiber backup in long haul communication links. For the operating wavelength, operating near 1550 nm wavelength is the most practical because of the availability of the EDFA and LDs for DWDM. For the essential architecture for achieving a full OWC system, an optical beam is emitted from a fiber termination to free space using an *optical antenna* (to be discussed later in this chapter), which can be located in the optical cell tower. The optical antenna can be of omnidirectional type so that the antenna can accept and transmit the optical beam from a wide FOV to connect the beam to another point in a wide angle.

FSO together with optical fiber networks can provide the broadband wireless solution for closing the last-mile connectivity gap throughout MANs, LANs, and WANs. Terrestrial short distance and long distance links, optical links between HAPs and UAVs, airborne terminal (airplanes), satellite platform to ground, and intersatellite links will be part of the global all-optical Internet connectivity to anywhere including remote locations and at anytime. FSO thus offers the last-mile bottleneck solution for the access network and in space networks for remote broadband global connectivity. Free-space laser communications in various atmospheric communication channels, which includes atmospheric turbulence and scattering medium, are already discussed in detail [15,22]. The propagation channel includes indoor communications (inside a home or building), terrestrial FSO links for short ranges such as between buildings on campus or different buildings of a campus, hospital, or office. This type of terrestrial link can also include the path between various optical cell towers within a city or on a mountain (to achieve both downlink and uplink communications).

To establish longer range optical links three types of optical propagation paths need to be considered. One link is up- or downlink between ground systems to the aircraft or HAP/UAV platforms through the troposphere; other links can be between airborne platform and a satellite platform (both up- and downlink), between ground and satellites, or optical ISLs. Note that each of these links have different propagation channels where atmospheric turbulence and scattering effects are more pronounced at the altitudes close to the ground (up to a few hundred meters or a few Km). Atmospheric effects for the links through the troposphere are obviously less severe than terrestrial links. Satellite orbits propagation links are above the atmosphere where there are practically no atmospheric turbulence effects. This is true also for ISLs where the optical beam suffers just absorption (path loss) due to geometric loss only between two points.

For establishing efficient networks from the point of view of connectivity performance, the channel capacity, reliability, and availability of all these different types of links are important considerations. Different FSO technology and systems will be required to design different network architecture for the links mentioned and therefore different network architectures have different availabilities and reliabilities. Basically, an FSO unit for data transmission requires transceiver units designed at both locations, the transmitting and receiving ends. The details of the subsystems and the modulation techniques of an FSO system is described elsewhere [22].

In order to quantify the FSO communication system performance, a detailed link analysis is needed with requirement of subsystems like transmitter laser/LD, modulation techniques, encoder for the information, receiver including photodiodes, and amplifier are discussed [22] and is not repeated here. Some of the link parameters needed to design a proper network system are data rates and BERs [22] and cost effectiveness. For shorter ranges from a few hundred to about 300 m a wide beam angle (of the order of a few tens of mrad to about 60 mrad) should provide a high reliability link operation under various atmospheric conditions. In this type of FSO system a telescope with tracking and aligning capability are not required. The other types of systems may require a collimated beam leading to a smaller beam divergence of about 2–6 mrad and a direct coupling from fiber-optic technology for coupling directly from fiber to FSO provides a very efficient system with much optical loss at the interface. The transceivers can be mounted at two fixed platforms on stable mounts and the system has very useful short range terrestrial applications including last-mile access. Multiple transmitters and receivers can be employed to mitigate atmospheric turbulence effects [23,24]. For much more longer ranges, the required beam divergence is less than 1 mrad and an accurate ATP system with AO compensation for atmospheric turbulence is required. In addition to direct detection scheme for the received signal, coherent detection is also used. The FSO system can be applied to satellite-to-ground applications for achieving the best network performance. The choice of wavelength of 1550 nm is due to the fact that this is the best transmission line through the atmosphere as well as eye safe compared to other shorter wavelength LDs available; for example, at 850 nm. The scattering losses such as Mie-scattering in haze or light fog are smaller at longer wavelength because of inverse fourth power of wavelength criteria. Atmospheric channel effects on free-space laser communications can be found in details in the chapter by Jennifer C. Ricklin et al. in

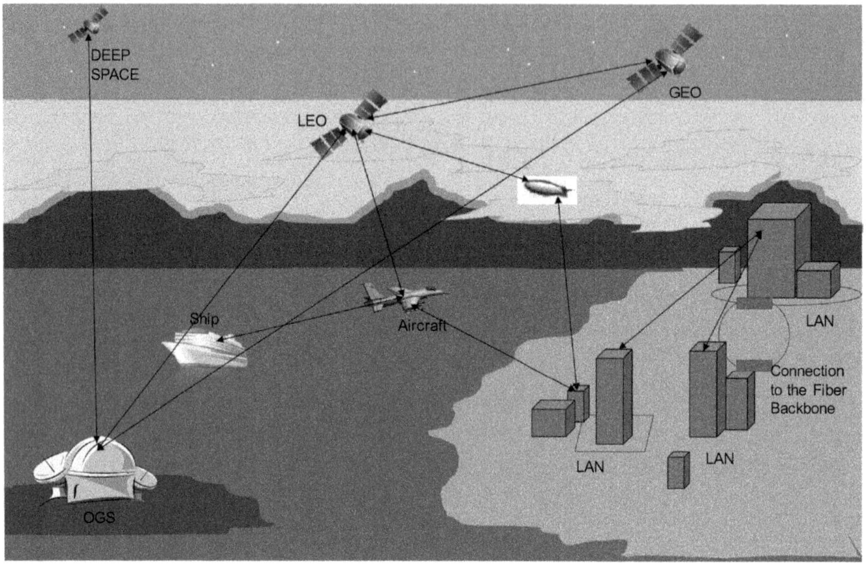

FIGURE 5.10
Global all-optical network connectivity using both FSO and fiber optics backbone.

Ref. [22]. Fig. 5.10 depicts an overview of a global all-optical network around the world, which includes all possible FSO links and fiber optics backbone.

Fig. 5.10 considers all possible scenarios of data transfer via various optical links such as terrestrial FSO between buildings, offices, hospitals, campus, horizontal and slant paths, airborne platforms, and space communications links between aircraft, UAV, HAP, and satellites. Solving the last-mile problem by FSO, the optical fibers providing the backbone for the networks integrated with FSO networks and links can thus be integrated for developing a seamless global network, and therefore will serve for connecting with the Internet all over the world. All the networks components and devices are now available, and are all-optical technology offering extremely fast response, and are compact and portable. These may include optical routers, optical modems, and similar components necessary to build a complete network system for connection and interface to the network.

5.4.6.1 *Satellite Architecture for Data From Anywhere to Everywhere: Optical Communications—Based for Global Internet Connectivity*

Today's world is becoming more and more Internet-connected, cloud-based, and data-driven, which requires instant infrastructure anywhere and anytime and at high speed broadband capacity. Satellites receive Internet at the most remote places in the world, and telephone, faxes, videos, and telecommunications via satellite signals. In order to satisfy the requirements of high bandwidth, satellite signals must be optical, where two-way communications must be via optical links. Therefore, data networks must be taken into space and satellites can be platforms for completing the Internet connectivity. Satellites can be routers in the optical networks and can be connected with optical/laser links to form optical backbones in space. Using these optical links (consisting of both FSO wireless and high bandwidth fiber optics) can provide broadband connectivity from a customer to the final destination, anywhere around the globe including any remote locations. The global broadband Internet connectivity thus combines high bandwidth fibers and flexibility of satellites at various heights with potential to provide 5G satellite backhaul to the cellular industry for creating high data rate secure networks.

It is clear from the previous discussion that there are opportunities for satellite communication in 5G especially for laser-based satellites for broadband communications worldwide. Satellite communication is becoming an important element in the big picture of the 5G system, complementing fixed and wireless terrestrial communication. This section will discuss scenarios for integration of satellite components in future global broadband networks based on all-optical technology. Users will want one network that meets their requirements, whereas application developers will not design their solutions to accommodate different network characteristics such as propagation delay to address the latency issues. From a network architecture point of view, suitable for different scenarios, seamless compatibility of satellite and terrestrial 5G networks is essential for global connectivity. Data delivery to remote areas will be possible by satellite networks integrated with terrestrial networks. Various communication links and networks architectures can be designed and developed using FSO-systems that can include (1) terrestrial FSO (between buildings, offices, campus, hospitals, and meeting and conference centers), horizontal and slant paths (e.g., from a ground to top of the mountain location), airborne and space communications (between aircraft, UAVs/drones), HAPs, balloons, and satellites (both LEO and geosynchronous orbit [GEO]). Fig. 5.11 shows the concept of all optical networks around the world connecting FSO links and networks to the backbone provided by the high bandwidth fibers. Optical wireless global connectivity using FSO networks and FSO links can be achieved *anytime, anywhere* in the world including any remote locations.

Optical wireless can thus play an important role as a broadband *access technology* in the future net generation global network. As Fig. 5.10 showed, a full-OWC system can be accomplished by directly propagating an optical beam from a fiber termination point through free space using a full optical wireless transceiver. FSO communication systems can support the needs for increased bandwidth for information collection systems. FSO communication systems offer space-based and airborne users the same and even much more bandwidth

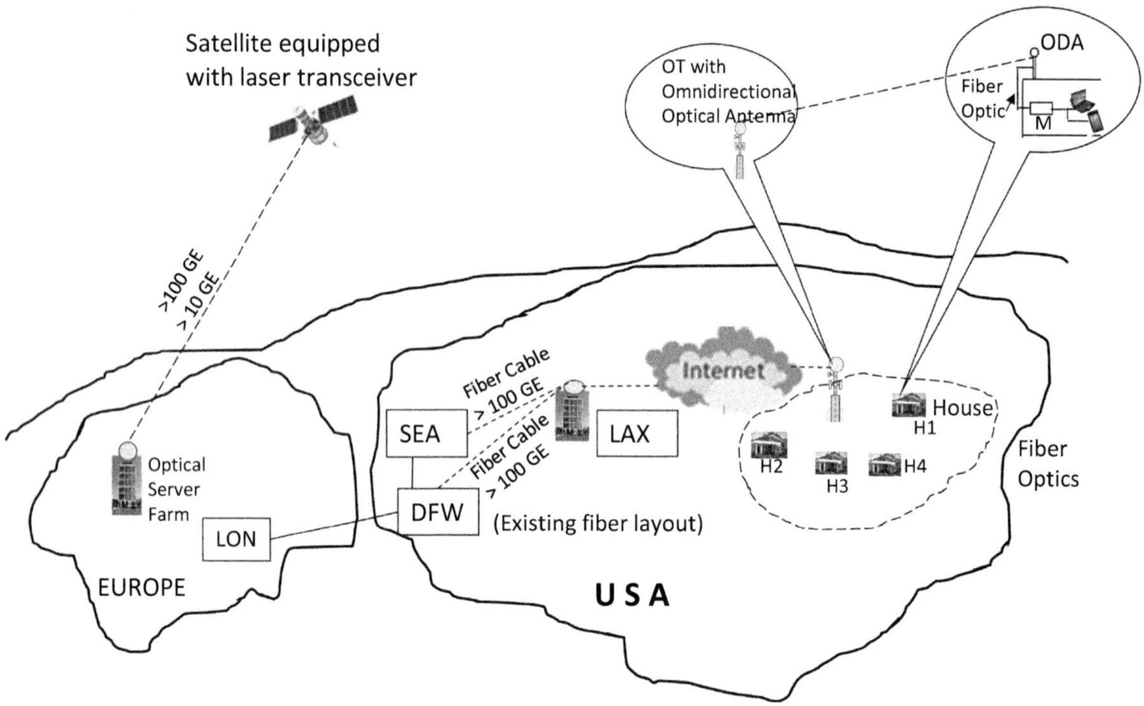

FIGURE 5.11
Concept of all optical networks around the world connecting FSO links and networks to the backbone provided by the high bandwidth fibers: Optical towers equipped with omnidirectional optical antennas.

benefits that fiber optic systems provide for applications on land. Proven technologies exist now with capabilities to develop a family of lasercom terminals to meet the next-generation communication systems.

Optical communications architecture can consist of the following segments to develop the family of lasercom terminals:

- *GEO terminal*: High data-rate GEO backbone or GEO-user applications. Optical telescopes designed for this terminal can be located on a gimbal to provide the required aperture size as determined by the communication link budget between GEO and earth or between GEO and LEO satellite. A large elevation and azimuth field will be incorporated in the telescope/gimbal design.
- *LEO terminal*: This terminal will be needed for high data-rate LEO or shorter range GEO user applications. The optical design involves the communications performance parameters required to achieve acceptable BER and SNR from LEO to fixed terminals on the ground such as OGS at a very high data rate.

- *Airborne terminal*: This type of terminal includes high data rate airborne applications between aircraft and GEO, between aircrafts, between other airborne platforms such as HAP, UAV, balloons, and fixed ground-based optical terminals. Optical network designs should include from airborne up to and including relay to GEO satellites. Optical communication system design needs to address pointing and tracking solutions. A seamless handover operation of communications is a key factor so that there is no interruption of connectivity.

It is important to note that the reliability and availability of FSO links and FSO networks will depend on reliability of each single component for all electrical and optical components built into the terminal, including the connection and interface to the network. Depending on the range and location of FSO links, atmospheric effects such as turbulence and scattering on optical communication performances need to be considered. For up- and downlinks between Earth and satellite through the troposphere the optical links are affected by the atmosphere but much less than the terrestrial links close to the ground. Optical ISLs are not affected by atmosphere because the links are above the atmosphere. The optimum choice of the wavelength for atmospheric and space transmission is usually 1550 nm (which is the telecom choice wavelength with optical devices for communications already available today) where the WDM components are used to transmit and receive multichannel optical wavelength using the same laser for tremendously increasing data rate in the multi-Gbit/s ranges. For choosing the appropriate detectors at this wavelength they are less sensitive and have a smaller receive surface area compared to the Si-APD detectors operating at 850 nm.

For optical amplifiers, EDFA are available today to increase the received signal optically and is essential to achieve high data rate FSO communication systems, which can find the importance also whenever a relay link is necessary. The Doppler shift (e.g., for interorbit, intersatellite links the satellite in the lower orbit travelling at a higher speed than the satellite at the higher one causes a relative speed between them, which is translated to a frequency shift of the received frequency) is lower also at this frequency than others. Both types of detection like direct and coherent detection schemes can be applied, however coherent detection requires much more precise alignment requirement. Modulation schemes such as OOK, PPM, binary phase-shift keying (BPSK), and channel capacity and coding for atmospheric FSO communications can be found in Refs. [16,22].

The use of FSO for connecting a satellite-to-Earth station to the backbone or to an optical LAN is described. Optical wireless LANs have the possibility to set up a network cell with several mobile users sharing a broadband connection using

FSO links. The connection to laptops or PCs in remote locations can be established via optical LAN/FSO/satellite link. An optical wireless LAN access point is interconnected by a network switch or a hub with the FSO system. The data stream can be transmitted over the FSO link to the outdoor OGS equipment. Today, the global telecommunications optical network is possible due to the massive expansion over the last few years in the areas of FSO communication links and the tremendous growth of the optical fiber long-haul, WAN, followed by a recent emphasis on MANs. LANs and gigabit Ethernet ports are now being deployed. LANs, MANs, and WANs are the three tiers for the terrestrial network where all-optical technology (FSO-based and fiber-optics) can provide access technology for full OWC systems. The satellite backbone-network node can be properly designed to support all the necessary (LAN/MAN/WAN) functions similar to their terrestrial counterparts. A high-bandwidth last-mile bridge between the LANs and the MANs or WANs is needed for complete high data rate communication. An example where FSO has been very efficient is the success in providing a LAN/campus connectivity market, which can include a link between a newsroom and a broadcasting station, or a dedicated link between two high-rise nodes in a large building complex. FSO systems are currently being deployed in many applications such as last-mile access connecting end users with Internet service providers or other networks, metro area extensions by carriers to extend their MAN fiber rings, enterprise connectivity to interconnect LAN segments in buildings, as well as fiber backup (as redundant link in place of a second fiber link). In conclusion FSO is now a viable and practical choice for connecting LAN, WAN, and MAN carrying voice, video, and data at the speed of light.

A corresponding transceiver at the receiver end directly couples the free-space propagated optical beam into an optical fiber connection port. This is a future concept technology for seamless coupling a free-space propagated to a fiber, which is capable of offering stable multigigabit per sec data rates over extended periods of time and still accomplishing global Internet connectivity to access broadband communication system, from any location, *anytime, anywhere*, which is the main goal of this book. Already this section has pointed out the broadband access (fiber-backhaul and wireless), multimedia, voice over Internet, video streaming, and IPTV will benefit from the technology of all-optical global network connectivity. This concept is becoming more practical because of the interest in the use of optical wireless links in digital transmission of signal through the atmosphere brought about by the increased demand of wireless links, which can be faster and easily deployable anywhere to offer high-speed, broad bandwidth communication links. Fig. 5.12 shows a general scenario depicting full-optical communication system deployment as a universal platform to effectively provide optical wireless service in urban and rural areas.

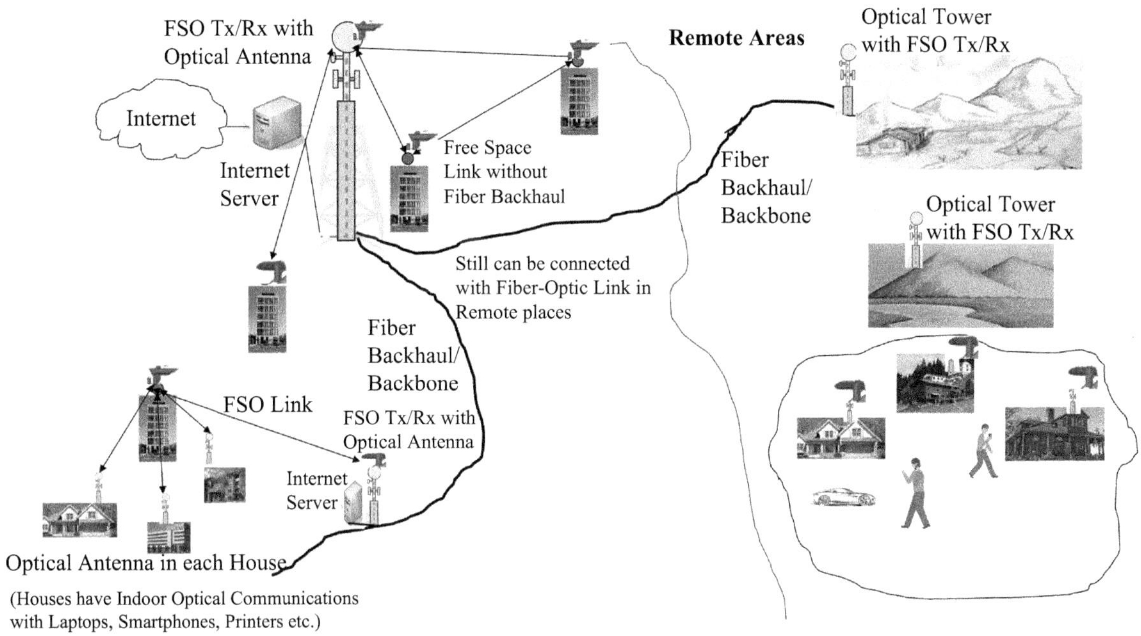

FIGURE 5.12
All-optical wireless deployment scenario and a generic architecture.

The role of LEO and GEO satellites is to establish FSO links with GEO and LEO communications networks to provide satellite backhaul for terrestrial small cells and mobile devices in order to bring high-performance broadband Internet access to people all over the world *anytime, anywhere*.

5.4.6.2 Satellite-Based Free-Space Optical Communication System Architecture and Network for All-Optical High Data Rate Global Connectivity

This section will describe the LEO satellite-to-ground system architecture. The concept technology proposed here for the FSO communication system architecture has the following main elements *inside* the satellite: (1) multigigabit per second laser communication terminal in LEO links in continuous-wave mode to the optical ground station, (2) beacon receiver optics to sense and detect a laser beacon from the ground system by a wide-field-view CCD camera mounted on a two-axis gimbal, and (3) capability of high bandwidth fibers for data transfer for either transmitting and/or receiving to/from various nodes.

Using existing fiber cables to make fast communication forming a network of underground landlines it is almost impossible to achieve global coverage. LEO satellites with capabilities of high bandwidth fibers (e.g., single-mode fibers) offer an extremely high data rate in hundreds of Gbit/s or much more in the Tbit/s range. The fiber optics layout in the LEO satellite will provide a solution for all data communication challenges. These fiber connections will help to connect the all-optical communication devices such as optical routers, switches, and optical microdata chips, which are necessary to manipulate the data stream at very high speeds both for transmitting from the laser communication terminal down to ground-based terminals as well as receiving communication data from the ground-based terminals. Since light travels through space faster than the fibers, there is definitely a potential possibility of creating global base data networks and the Internet connectivity with space-to-ground links and even without underground fiber cables, or perhaps using them whenever it is practical from a logistic point of view. In addition, each LEO satellite can have a number of transmitting lasers interconnected into the others whose transmitting beams can create a mesh data network covering the entire globe. The optical ground systems can be equipped with an AO system, which can incorporate wavefront sensor and a deformable mirror and a real-time control system typically needed for an AO system. Fig. 5.13 shows satellite-based FSO communication system architecture.

In order to complete the space-to-ground link, a laser beam is transmitted from the sender, which is a laser communication terminal flying on the LEO spacecraft. An AO ground station (OGS) is used for pointing, acquisition, and tracking of the laser beam from space. The AO system corrects for the forphase front distortions due to the turbulence near the ground up to a few km and then launches the corrected beam into a single mode fiber. The laser

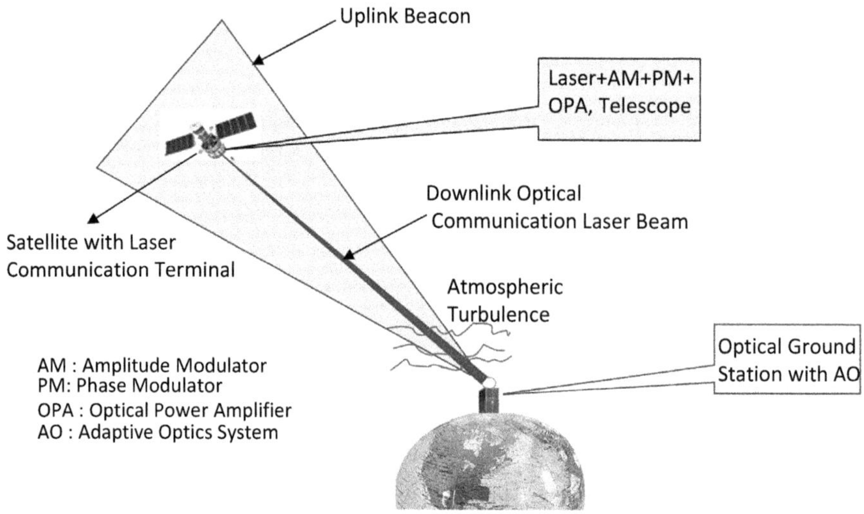

FIGURE 5.13

Satellite-based free-space optical communication system architecture and network for all-optical high data rate global connectivity.

beam tracking for FSO communication on an LEO satellite system architecture consists of a 1550-nm downlink beam as primary downlink method and an uplink beacon around 976-nm for acquisition and tracking. Laser transceivers at both satellite and ground terminals establish two-way high data rate communication. In order to accomplish global Internet connectivity, the architecture will include an optical communication system for ground-to-satellite and inter-satellite links. Various techniques, both at the physical layer as well as at other layers (link, network, or transport layer), will be applied to combat the adverse effects of the atmosphere. The major challenge for ISLs is pointing to and from a moving platform, which therefore requires a very tight ATP system for the optical beam to reliably reach the receiver. Therefore, due to increasing demands for high data rate and large communication capacity it is important to build all-optical communication architecture that includes ground-to-satellite optical communication links that are connected to satellite optical networks and satellite-to-ground optical links as already shown previously in Fig. 5.11 for space FSO links.

5.5 FREE-SPACE OPTICAL LINKS FROM VARIOUS PLATFORMS: HIGH-ALTITUDE PLATFORMS, UNMANNED AIR VEHICLES/DRONES, AND BALLOONS

5.5.1 High-Altitude Platform-Based Optical Wireless Communications

FSO technology can be used very effectively for establishing high capacity backhaul links from HAPs or satellite-based networks. FSO communication has been shown to provide gigabit capacity backhaul links due to its low cost and rapid deployment speed in comparison with conventional backhaul technologies like RF or optical fibers. This section focuses on optical backhaul links between HAPs, satellite, and ground, which can serve as future broadband backhaul communication channels. HAPs are quasi-stationary vehicles like helium-filled airships or aircrafts located at an altitude of 17–25 km in the stratosphere. HAPs are placed far from the atmospheric region so they can provide better channel conditions than satellites and better LOS conditions in almost all coverage areas, thus offering much more network flexibility. At this altitude, the effects of atmosphere on optical beam propagation are less severe than the turbulence effects close to the ground. HAP altitudes are cloud free so that more reliable communication links are possible between HAPs and satellites. Besides broadband capability, HAPs offer large coverage area (3–7 km) and flexible capacity increase through spot beam resizing and quick deployment. HAPs can be used to relay the high capacity *optical* data through the atmosphere to the ground locations and therefore their backhaul *optical* link can easily be connected to the core through terrestrial gateway stations. In other words, HAP-based optical networks and communication architectures can be developed for a complete combined optical backhaul for connecting very high data rate connectivity for both terrestrial and space systems simultaneously.

Using HAPs for completing the satellite-to-ground optical link on-board regenerative HAPs payload can perform the task in two parts: (1) optical link from satellite-to-HAP the atmosphere can be assumed to behave like almost free-space with not much atmospheric effects, and (2) HAP-to-ground (terrestrial) optical links, which will be somewhat affected by the lower atmospheric turbulence and scattering effects, but can be effectively mitigated for establishing high data rate optical communications. On-board satellite data processing time can also be reduced to make it more efficient, and with efficient optical networks design to channel and distribute enormous data at extremely high data rate. Fig. 5.14 depicts a concept architecture where HAPs can stand alone as well as integrated with both satellite and terrestrial systems. Broadcasting and broadband services over a large coverage area including suburban and remote areas can be provided by the integrated ground-HAP-satellite system at a low cost of deployment. Large solid-state in terabyte-size memories can be used as multiple *optical payloads* to store data from the satellite and transmit to the ground terminals, optimizing the satellite visibility time [26].

Fig. 5.15 depicts an integrated system architecture for satellite-HAP-UAV-terrestrial terminals for FSO communications emphasizing to handle site diversity and redundancy.

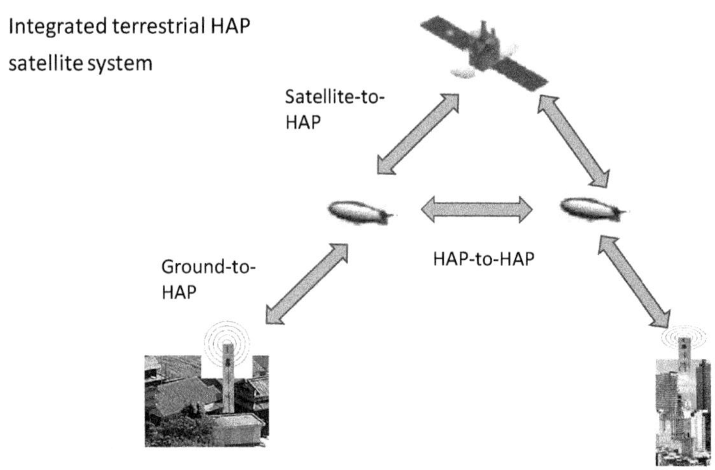

FIGURE 5.14
Concept architecture showing high-altitude platforms integrated with satellite and terrestrial systems.

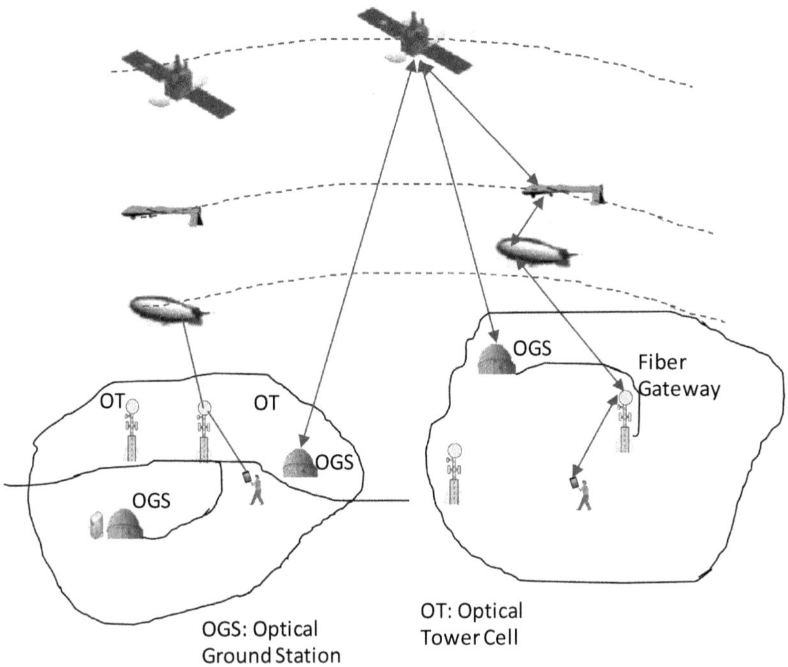

FIGURE 5.15

Integrated system architecture for satellite-UAV-HAP-terrestrial terminals showing site diversity and redundancy.

5.5.2 Unmanned Air Vehicle—Based Optical Wireless Communications

The use of UAVs for delivering wireless connectivity is considered by the telecommunication operators and over-the-top service providers in commercial networks. A network of UAVs operating at a certain height above ground can provide wireless service within coverage areas shaped by their optical antennas (integrated in the UAVs) with the UAVs using the existing terrestrial base station network for FSO wireless backhaul to connect to a dedicated optical ground station. Therefore, the UAVs must optimize their connectivity to deliver reliable service to the end user while simultaneously meeting their own wireless backhaul requirements. A low-altitude UAV network above the built-up urban area may be used. Deploying FSO technology for mobile links between UAVs and fixed OGSs also involve several challenges to establish the ability of a mobile FSO system to operate in different atmospheric conditions. UAV FSO communications and different scenarios for free space communication links such as ground-to-UAV mobile FSO channel, UAV-to-ground FSO link, UAV swarms with different architectures such as ring, star, and meshed types and alignment and tracking of an FSO link to a UAV are already discussed [16]. A demonstration of the FSO communication link at 2.5 Gbit/s is presented [27] for a UAV altitude of 15.8–18.3 km using a 200-mW downlink laser at 1550 nm. They reported BER of 10^{-9}, which needed the pointing requirement on the flight terminal of 19.5 μrad and a bias error of 14.5 μrad with a probability of pointing-induced fades (due to turbulence) of 0.1%. UAV links can therefore be effectively used for high-speed connections where terrestrial links are not available [28–31]. Today's network and associated techniques require gigabit capacity data rates with low-cost backhaul solutions. Next-generation mobile networks (5G) driving the need for high capacity backhaul links involve integration of terrestrial and space networks where UAVs can provide last-mile connectivity, which eventually can establish Internet access to remote areas using flexible FSO links.

UAVs will play a major role in today's connected society by bringing advantages to a wide range of industries, commercial, healthcare, public safety, logistics, and utilities. Current mobile networks need to optimize for flying objects. A UAV traffic management system would help pilots avoid collisions between UAVs and aircraft. All FSO communications can be done using existing optical networks and via secure interfaces on mobile networks. UAV's potential for establishing ultrabroadband connectivity will be able to address the unprecedented growth in demand for mobile and fixed broadband, which will be ready for 5G. Therefore, UAV in combination with FSO will be needed everywhere that large amounts of data have to be delivered in real time. Optical solutions with potential data rates of tens and hundreds of Gbit/s and using OFDM-based multicarrier transmissions have the potential for accomplishing future communication means for UAV scenarios.

For the upcoming 5G wireless network and beyond a reliable and efficient backhaul/fronthaul framework is required. Recently much interest in the unmanned flying platforms of various types including UAVs, drones, balloons, and HAPs can be used to provide wireless communication services based on FSO. Network flying platforms are capable of transporting the backhaul/fronthaul traffic between the access and core network via high data rate point-to-point FSO links. Different mitigation techniques exist for compensating FSO propagation effects distorted by the atmosphere. Emerging backhaul/fronthaul requirements for the 5G+ wireless networks can therefore be met even in the presence of ultradense heterogeneous small cells that are very often necessary to accomplish broadband connectivity without loss. Normally the traffic of small cells is delivered to a central LOS hub. A communication system with UAVs as platforms and the flying altitude can range from a few hundred meters to several kilometers (typically 20 km) depending on the coverage area, weather condition, and the UAV's communication capability, which should be able to provide a potential solution. Some excellent papers are available in the literature discussing UAV platforms for FSO links for communication networks.

5.5.3 Balloon-Based Optical Wireless Communications Connectivity

5.5.3.1 Recent Story in Puerto Rico

This section discusses the developments in establishing global Internet connectivity in remote places using balloon platforms. Recently balloon platform–based communication links in remote places have been demonstrated by Alphabet's Project Loon [32] by providing connectivity to hard-hit areas in Puerto Rico to satisfy the emergency need for communication with the outside world. In Project Loon, the telecommunications company would transmit Internet to a network of solar-powered high-flying balloons. Leon balloons are actually made up of two balloons, one inside the other, where the outer balloon is filled with helium to help the balloon reach the right altitude, while the inner balloon takes on air to descend or vents it to rise [33]. The balloons equipped with appropriate networks and transceivers will then transmit that Internet to users on any rural and remote locations worldwide. The communication technology involves sending a fleet of balloons to serve as cell phone towers in the sky and thus develop balloon-based efforts. The balloons can also link up with base stations and pass signals between themselves, which can act as relays for communications to accomplish much longer distances on the earth.

Balloons are much cheaper to operate compared to the costs of building and launching communication satellites on a daily basis or even for a global network. The residents using the Internet will not be required to pay for that last-mile of fiber-optic connectivity. A balloon can stay as much as 20 km above the ground for as long as 6 months at a time. This height obviously helps for a balloon to provide a clear LOS link to a large area of the ground, including remote areas. Balloons can thus serve as cell phone towers in the sky for thousands of customers simultaneously. Carrier partners around the world can build their services on top of the balloon backbones. Using sophisticated models for predicting the wind pattern at various altitudes it gives the best chance of keeping the balloons together close to the desired areas. If the balloons are equipped with optical devices, then FSO communication links can be established, which can offer Internet connectivity and high data rates using all-optical technology. FSO communication has the potential to build a high bandwidth backhaul network with wireless optical communication systems using a balloon mesh network.

In a recent paper [34], an algorithm to track the drifted balloon and evaluate the tracking system by state estimation of the drifting balloon's position with respect to another nearest balloon has been proposed. A backup optical link to keep the network intact in case of a failure due to a balloon's misalignment with an adjacent balloon is also discussed. Significant improvements have been made in the project with connected devices to provide 4G-type speeds.

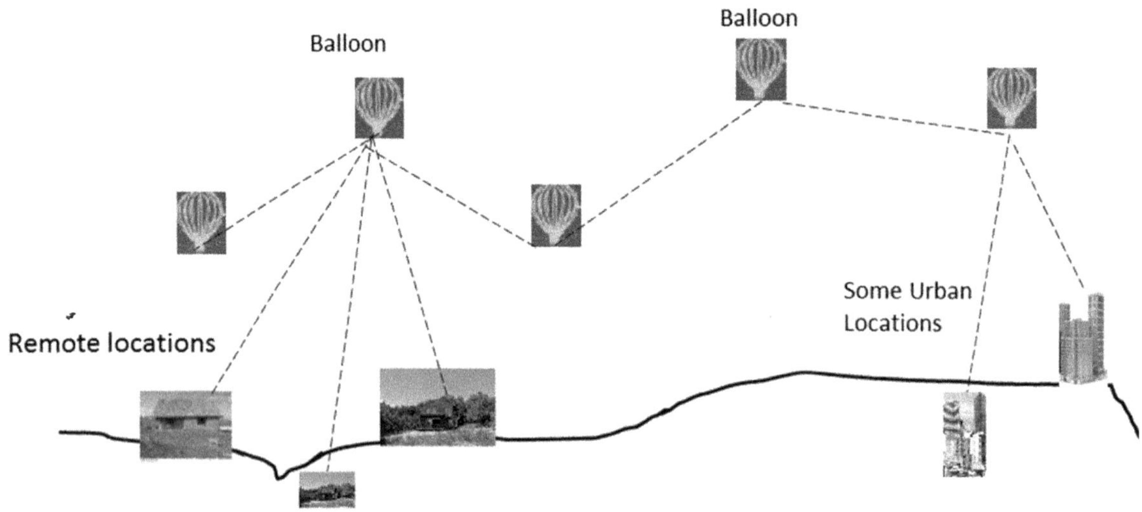

FIGURE 5.16
A schematic of a basic concept similar to Project Loon to build ring of connectivity around the world.

With all-optical technology–based connectivity the users will be able to enjoy high data rate 5G-type speeds in the future. Fig. 5.16 shows a concept architecture similar to Project Loon in order to build a ring of connectivity around the world.

5.5.4 Internet Access in an Airplane

More and more airplanes are offering Internet connection to a passenger. Internet connection on an airplane can be established by satellite and air-to-ground methods. High data rate broadband Internet connection on an airplane can be established with optical transceiver-equipped satellite and air-to-ground optical wireless link. Fig. 5.17 shows an optical-satellite–based architecture for Internet access in an airplane.

In order to obtain an Internet connection in an airplane from air-to-ground the cell towers on the ground can be used. However, over the water there are no cell towers and therefore the connection is blocked; the only option for the airplane to get connectivity is via satellite directly. Signals from the airplane go into space to an orbiting satellite such as a GEO satellite 22,300 miles up. The signal can be directed from the satellite to ground via ground stations. A constellation of LEO satellites in future development will provide lower high bandwidth connections with FSO communication links in airplanes, satellites, and ground optical cells (as depicted in Fig. 5.17) to achieve high data rate connectivity for emails, fast Internet search, and watching videos and movies in an airplane.

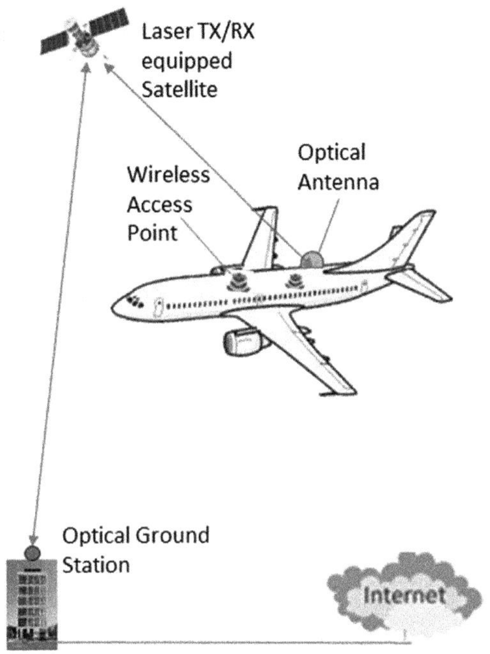

FIGURE 5.17
A satellite-based architecture of Internet access in an airplane.

5.5.5 Constellation of Satellites for Broadband Global Internet Connectivity

It is possible to provide global telecommunications service through satellite constellations with satellite beams for cheap, ubiquitous broadband service using FSO links. Using constellations of several hundred LEO satellites a global broadband Internet can be accessed anywhere in the world. This will also allow local service providers to extend their networks in terms of both scope of services and geographic reach so that local service can be provided through a global network. Launching the smaller satellites can be much less expensive than larger satellites, largely mass-produced to satisfy the demands of consumers for high-speed Internet connectivity almost everywhere—airplanes, cruise ships, and remote places worldwide. Some of the companies involved in proposing and developing networks of satellites in LEO to provide high-speed broadband access around the globe are SpaceX, OneWeb, and Boing.

Integrating all-optical technologies discussed in this section with the small satellites constellation can even make the global Internet connectivity provide much higher data rates than exist today. The technology concept of small satellite networks will then not require miles of expensive high-speed fiber-optic cables. OneWeb plans to develop 700 satellite constellations in the future

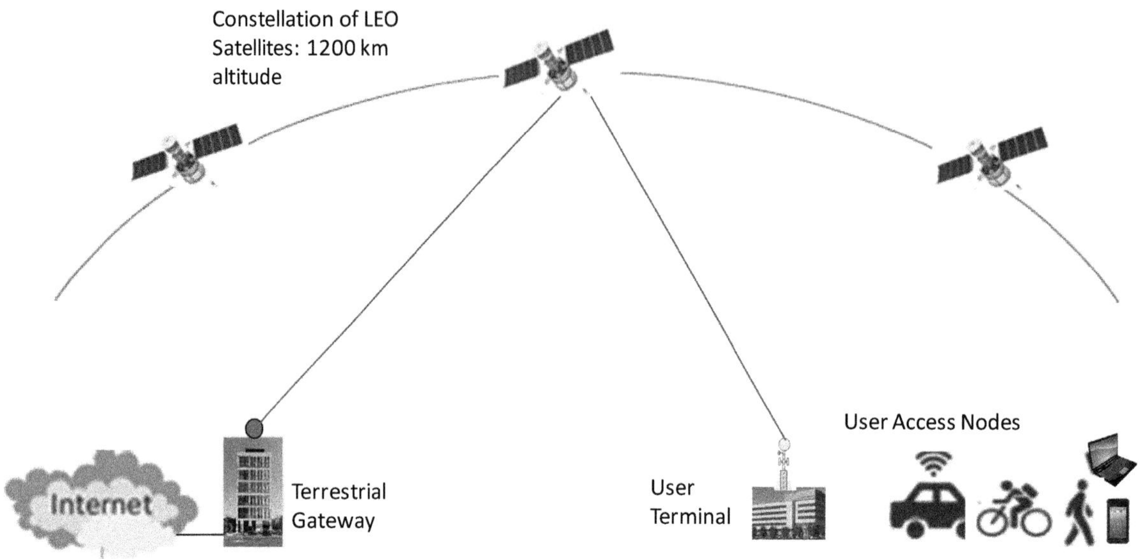

FIGURE 5.18
Global communication networks with a constellation of satellites.

that will connect every unconnected school to the Internet. The OneWeb network will integrate with terrestrial networks to extend 4G and Wi-Fi services to include all kinds of mobile devices. Because of advances in miniaturized parts, devices, and extremely fast microprocessors, the concept of satellite swarms to provide high data rate everywhere will be possible. Fig. 5.18 shows a global communication network with a constellation of satellites. The communication network will provide satellite backhaul for terrestrial small cells and high-performance Internet access to users and places, such as homes, health centers, businesses, and other end users. Putting satellites in LEO will also help in fast data transfer between the satellites for designing broadband communication networks to establish high speed connectivity and capacity on the satellites to telecommunications operators.

High-speed network is essential in designing the communication technology for the constellation of satellites as well as high-performance wireless air interface for end-to-end satellite communications with implementation of wireless coding, modulation, and protocols necessary to deliver connectivity. The efficient wireless air interface will be needed to enable intrasatellite, intersatellite and intergateway seamless handoffs for uninterrupted connectivity.

A facilitation of a new age of entrepreneurship in the developed world has created a great need and benefit of the *ubiquity* of the Internet. Today's business is increasingly data-driven, cloud-based, and covers international and intercontinental locations. Satellite systems equipped with laser beams are the only viable solution to meet these tremendous broadband connectivity requirements responding to specific customer needs in the enterprise covering things such as emails, HDTV, financial, commercial, entertainment, education, telemedicine, and almost everything customers may need today. LEO satellites with laser transceivers are closer to Earth (LEO is an orbit with an altitude above Earth's surface of 2000 km [1243 mi]) than MEO orbit, which is above LEO and below GEO (altitude of 35,786 km [22,236 mi]).

With optical/laser ISLs, LEO satellite has many advantages when it comes to throughput, latency, and true global coverage for providing Internet connectivity worldwide. LEO-based optical satellite can be used to develop a new type of satellite constellation and offer very high-speed secure data network for point-to-point to multipoint data solutions covering remote locations worldwide. Optical satellites can provide faster high-throughput links than existing satellite systems as well as terrestrial fiber links on long distance network routes. Furthermore, the LEO constellation using polar orbits can provide full global coverage and global point-to-point for Internet connectivity. When a user performs data uplinks to the constellation of satellites, which themselves are interconnected through optical/laser links, data will travel from satellite to satellite until it reaches its downlink destination. Depending on the types and specifications of optical/laser transceivers used, the reliability of terrestrial/space propagation paths chosen and optical network routing and modulation/demodulation techniques used, the potential data rates for full-duplex connectivity per link of a few Gbit/s to 100s of Gbit/s can be accomplished with already developed high speed compact optical devices available today. As mentioned earlier optical wireless access with star or mesh architecture can be used to suit multiple enterprise connectivity scenarios and applications, making the network configuration very flexible. Because of shorter distance between Earth and the LEO satellite compared to the distances for MEO or GEO satellites, the latency encountered for connections between Earth and any of the satellites of the constellation will be less than about 20 ms. When combined with LEO satellite's backbone in space even for larger distances it can be below 120 ms and is less than the current latency on the London-to-Singapore route, which can be well in excess of 180 ms [25]. By proper choice of efficient encryption, a very high secure connectivity can be developed between any two locations on the Earth.

In summary, LEO satellites create a new concept in satellite telecommunications, offering communications services far beyond the existing services with much higher speed, reliability, and security required for a critical data network. Because of a tremendous increasing need to transport cellular signals at long distances at high speeds, optical/laser—based LEO satellites can play a very important role in the newer 5G systems. All optical global Internet connectivity is a potential technological solution for providing seamless, global data coverage including any remote locations to satisfy the continued growth in modern Internet use, streaming media, smart phone use, mobile apps, and the IoT.

5.5.6 Intersatellite Links for Global Optical Connectivity

A constellation of LEO satellite systems can provide backhaul connectivity to terrestrial 5G relay nodes that create an on-ground 5G network. The important communication parameters such as large delays and Doppler shifts as a result of long propagation path between ground and satellites and/or path between intersatellites can pose severe challenges to terrestrial-based systems impacting high data rate Internet connectivity. The efficient integration of satellite and terrestrial networks is therefore absolutely necessary in order for a heterogenous global communication system. Although the integration of terrestrial systems with GEO satellites would be useful for large-capacity coverage, as mentioned before the large delays and Doppler shifts in GEO will cause significant challenges in terms of designing networks, interfaces and seamless hands-off. Because of the lower altitudes, LEO satellites can be used to avoid these issues, but in order to fully provide global coverage with less delays and Doppler shifts hundreds of satellites must be deployed. The optical networks and innovative solutions for FSO-based links are needed. The performance of optical intersatellite communications based on laser communication terminals have been demonstrated in LEO-to-LEO ISLs as well as investigating beam propagation effects for LEO-to-ground links.

A full duplex data rate of 5.625 Gbps at a BER of less than 10^{-9} for the link was reported [35] using homodyne BPSK modulation scheme. LEO-to-GEO laser communications were also demonstrated and reported [36,37]. The work reported in [35] described the detail performance characteristics of the following links: LEO-to-LEO, LEO-to-ground, LEO-to-GEO, GEO-to-ground, and UAV-to-LEO/GEO links. The results of these developments for ISLs are important to eventually establish FSO-based Internet connectivity using laser-based satellites.

Various challenges faced by FSO communication systems for ground-to-satellite, satellite-to-ground, and ISLs are recently reported [26] with performance mitigation techniques in order to have high link availability and reliability. Reference [26] also provided various techniques both at the physical layer and at the other layers (link, network, or transport) for mitigating atmospheric effects as well as a recently developed technique using angular momentum for utilizing high capacity advantage of optical carrier for space-based and near-Earth optical communication links. The use of space-based backlinks from HAPs or satellite-based network for backhaul solutions using FSO technology has also been discussed. Orbital angular momentum for improving the quality and data returns is mentioned in their paper. FSO communication technology provides an efficient solution for cellular carriers using 4G technology by providing a backhaul connection between cell towers. Using ultrashort pulse laser, a high data rate up to 10 Gbps backhaul connection without deploying fiber cables with a potential for the wireless backhaul capacity can provide even beyond 100 Gbps using advanced modulation schemes to satisfy the requirement of future 5G cellular network.

The FSO-based 5G Internet system will integrate and interconnect satellites with airborne and ground-based transceivers via optical wireless links, which are supported by terabit side laser backbone wide, located at synchronized altitude. All-optical networks will allow connectivity to remote locations worldwide by integrating terrestrial and space networks with the help of HAPs, UAVs, and balloons by providing last-mile connectivity as already discussed earlier in this chapter.

LEO and GEO satellites can provide Internet access to the ground. With suitable protocols and sophisticated algorithms, it will be possible to modify the other TCP layers like transport or network layers to satisfy the optical network requirement. Optical intersatellite communications with the Alphasat and Sentinel-1A in orbit is also reported in recent literature [38] discussing the feasibility, reliability, and simplicity of an optical GEO-relay to provide high speed, near-real-time access for LEO satellites.

More advanced laser communications for next-generation information networks will be established in the future using the proven technologies and capabilities being developed for a family of lasercom terminals to meet the needs of next-generation communication systems. This is bridging all the segments of an optical communication systems architecture and is providing laser communication hardware solutions for each of these segments: (1) GEO terminal for high data rate GEO-backbone or GEO-user applications, (2) LEO terminal for high data rate LEO or-shorter range GEO-user applications for optical ISLs, and (3) airborne terminal (aircraft-ground, ground-aircraft, and aircraft-aircraft) developed for high data rate airborne applications up to and including relay to GEO satellites, high bandwidth optical communication between aircraft and aircraft to a satellite requiring a unique approach to provide the pointing and tracking solution for the terminal.

A communication relay platform interconnects intelligence, surveillance, and reconnaissance aircraft; ground-mobile, ground-fixed, and man-portable assets via FSO communication links. The communication relay aircraft in turn can be linked back to the global information grid via optical link to the cross-linked GEO platforms.

5.6 SOME RELATED TECHNOLOGY CONCEPTS TO ENHANCE OPTICAL WIRELESS LINK CONNECTIVITY

Chapter 5 identifies a number of recent technological developments that have shown promise for enhancing and developing innovative OWC/FSO systems. These include (1) MRR-based FSO duplex communications (includes fiber-based amplified retromodulators and MEMS deformable mirror retroreflector modulator) and (2) Wi-Fi/Li-Fi FSO optics via LED lights. The chapter explains the physics of atmospheric reciprocity leading to *reciprocity-enhanced optical communication* through atmospheric turbulence. This section describes the reciprocity theorem, which holds for bidirectional two-way optical propagation through atmospheric turbulence and thus provides mitigation (or overcomes) turbulence for high data rate communication and low delay requirement in achieving bidirectional optical communication links. Fiber-array—based (multiple channel) amplified retromodulator is discussed by a recent patent by the author and his colleague to provide a pixelated fiber array system for both incoming and outgoing optical beams, eliminating the need for complex aligning and pointing in FSO communication at multigigabit data rate. The link budget analysis for a satellite-based system laser interrogator and low-power consuming Gb/s amplified fiber retromodulator shows the feasibility of space-to-ground communication. Recently newly developed technology Wi-Fi/Li-Fi to increase the bandwidth of conventional Wi-Fi systems several times using FSO transmission via LED lights is described, where up to 100 Mb/s rate data can be sent to *each user* for an indoor system. The increased bandwidth should eliminate problems like videostreaming that stalls and buffers in home applications; for example, gaming systems or watching movies with tablets/computers. There is a potential of achieving up to 1000× higher capacity per area than traditional Wi-Fi systems. The chapter discusses the latest advancement of achieving Li-Fi Internet breakthrough for 224 Gb/s connection broadcast with an LED bulb for ~3 m wireless link. The chapter discusses how a transceiver unit be integrated into every smart phone, tablet, and laptop and form part of the 4G and 5G mobile network infrastructure to use this Li-Fi technology. OWC/FSO features create this huge technological opportunity.

5.7 SUMMARY AND CONCLUSIONS

An all-optical network concept for establishing global Internet connectivity including remote locations is presented and discussed relating to the trends in FSO communications technology development. Based on the recent successful demonstrations of achieving an optical link between a satellite terminal and the ground station using a laser satellite, the potential of future high data rate communication *anytime, anywhere* is becoming more clear and convincing.

In addition, plans for communication based on integrating the constellation of satellites, small and nanosatellites, HAPs, UAVs, and balloons show a clear pathway for the real possibility of establishing multigigabit level Internet connectivity all over the world in almost any totally remote location. On the home-based and even in the streets levels the newest Li-Fi technology will play an extremely important role for providing Internet hotspots for both fixed and mobile terminals which will be all connected eventually to the Global Internet gateways and optical nodes. Finally, all of these links are based on all-optical technology, which requires extremely fast and compact devices being continuously developed today. This global Internet connectivity will be possible only if the airborne satellites and the fixed terrestrial terminals can all be integrated with optical devices and laser transceivers.

In conclusion, the concepts presented in this chapter clearly is convincing for establishing global connectivity using innovative technologies including advanced fiber optics, laser transceiver, adaptive optics, ATP devices, and optical networks configurations. One advanced optical technology for creating multigigabit duplex laser communications is the MRR, which has tremendous potential use for all airborne, space, and fixed platforms, and is discussed in the following appendix.

APPENDIX A: MODULATING RETROREFLECTOR APPLICATIONS FOR AIRBORNE AND SATELLITE-BASED FREE-SPACE OPTICAL COMMUNICATION LINKS

Direct FSO links such as between satellite and ground; intersatellite, between UAV, HAP, balloon, and ground; as well as between them with active terminals on both ends are required for establishing high bandwidth global Internet connectivity. Optical MRRs couple passive retroreflectors with electrooptic modulators to allow long-range FSO communication with a laser, and ATP required on only one end of the FSO communication link. Retromodulators require very little power draw and offer extremely small form factors and mass. There is an increasing demand for high data transfer rate, light weight, power, and size for communication terminals, establishing multiple communications nodes and mobility. Retrocommunication (i.e., communications with retromodulation) is attractive in these cases where semipassive optical nodes operating by retromodulation are more suitable than conventional transceivers implementations. A retroreflective communications system comprises a laser transmitter/receiver station and a remote retroreflector that can be switched to on and off states. The interrogating laser illuminates the remote station where the laser light is modulated and reflected (in the same direction) back to the transmitting unit equipped with an associated detector (receiver) recovering the data signal originating from the remote retromodulator.

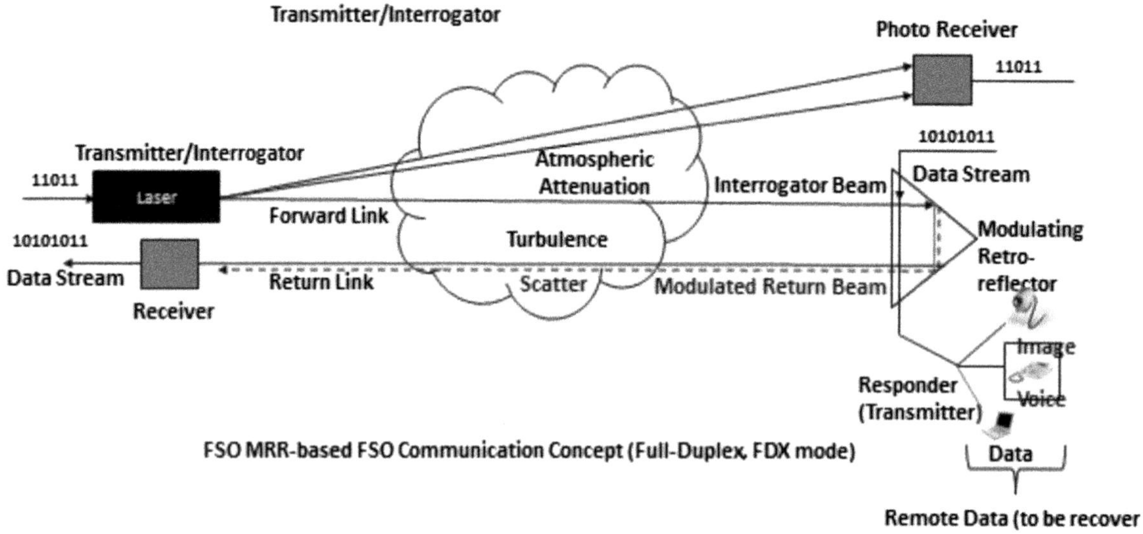

FIGURE A.1
MRR-based FSO communication system concept (full duplex).

The remote terminal can only communicate when it is interrogated by a laser transmitter. The incoming light is first modulated in accordance with the input data stream, and then retroreflected directly back to the remote receiver (located at the interrogating laser site). The modulated reflected light can then be interpreted as a stream of bits recovering the data information. Some of the advantages are potential for achieving high-capacity secure communications with low weight ($\sim$10–100 g) and small volume, low energy/power consumption (less than 100 mW), a large FOV resulting in reduced pointing requirements for the interrogating laser transceiver, and no active laser transmitter required by the MRR. These features make MRR a very attractive optical communication device to install in a satellite, UAV, HAP, and small/nanosatellites. Fig. A.1 depicts an MRR-based FSO full-duplex communication system concept.

Various MRR technologies including electrooptic modulators, multiple quantum well, and MEMS are already explained and described in detail [16] and therefore are not repeated. Only the essential highlights of this subject from Ref. [16] specifically relevant to this chapter such as satellite, UAV, HAP, and small satellites are mentioned next. This section emphasizes a unique type of MRR technology, which is fiber-optics based.

Fiber-Based Amplified Retromodulator Using Single-Mode Optical Fiber: A New Concept in Modulating Retroreflector Technology

A new concept of MRR is described in Ref. [39], which uses an amplified retromodulator (ARM). The amplifier increases the effective area of the retromodulator more than 300 times to make the system as effective as a larger aperture passive retromodulator without the increased weight and power consumption of a larger retromodulator. It was pointed out in Ref. [39] that a high-efficiency optical coupling of FSO signals into a single-mode fiber (SMF) was possible by combining high efficiency FSO-SMF optical couplers with high-speed modulators and very low-power consumption EDFAs. It was possible to develop an ARM with a return signal 2000 times the return signal from an identical aperture conventional MRR that can simultaneously be operated at several Gbps modulation rates. The total power consumption of an amplified 2.5-Gbps ARM used only 120 mW of electrical power.

The intensity incident upon a typical MRR is given by:

$$I_{\text{inc}} = \frac{P_T \cdot \eta_T \cdot \eta_{\text{Atm}}}{\Omega_T \cdot R^2} \tag{A.1}$$

where $P_T$is the transmitted power, η_T represents the efficiency of the transmit optics, η_{Atm} is the atmospheric transmission efficiency, Ω_T represents the divergence of the transmitted beam, and R represents the link range. The return signal for a conventional MRR is given by [40]:

$$\begin{aligned} P_S &= I_{\text{inc}} \cdot A_{\text{eff\_retro}} \cdot \eta_{\text{receiver}} \cdot \eta_{\text{Atm}} \cdot \frac{A_{\text{receiver}}}{\Omega_r \cdot R^2} \\ &= \frac{P_T \cdot \eta_T \cdot \eta_{\text{Atm}}^2 \cdot \eta_{\text{receiver}} \cdot A_{\text{receiver}}}{\Omega_T \cdot \Omega_r \cdot R^4} \cdot A_{\text{eff\_retro}} \end{aligned} \tag{A.2}$$

where $A_{\text{eff\_retro}}$ is the effective area of the retromodulator, A_{receiver} is the receiver area, Ω_r is the divergence of the retroreflected beam, and λ is the interrogator laser wavelength. If the retromodulator has a gain of G, then the effective area of the retromodulator is given by

$$A_{\text{eff\_retro}} = G \cdot A_{\text{retro}} \tag{A.3}$$

where A_{retro}represents the physical area of the retromodulator. From these equations the return signal from an ARM with a gain G is obtained as follows [39]:

$$P_s = \frac{P_T \cdot \eta_T \cdot \eta_{\text{Atm}}^2 \cdot \eta_{\text{receiver}} \cdot A_{\text{receiver}}}{\Omega_T \cdot \Omega_r \cdot R^4} \cdot A_{\text{retro}} \cdot G \tag{A.4}$$

Eq. (A.4) clearly shows that the return signal from an ARM with a gain G increases the received signal by a factor G compared with an MRR with the same aperture. With commercially available EDFA systems with a 40-dB small signal gain, it is possible to increase the effective area of the retromodulator by nearly 4 orders of magnitude [39]. The effect of the spontaneous emission noise on the receiver SNR can be neglected.

Fiber Array—Based (Multiple Channel) Amplified Retromodulator

The ARM described earlier is limited to its extremely small FOV of about ±0.004 degrees only. In order to overcome this limitation, a recent patent by the author and his colleague [41] (Patent No. US 8,301,032 B2, Oct. 30, 2012) describes a wide FOV amplified fiber retrosystem. The concept is to provide a pixelated fiber array system for both incoming and outgoing optical beams to maintaining one-to-one correlation between each set of lenslet/fiber array, which can also determine the exact location of the source. The patent describes a means of achieving a wide FOV fiber retrosystem where the remote device can accept a wide angle of interrogating signal and is explained and described in detail [16].

Effects of Atmospheric Turbulence on the Amplified Fiber Retromodulator

The effect of atmospheric turbulence on the amplified fiber retromodulator system using an array of fiber couplers needs to be evaluated. This can be best understood by estimating the variance of angle-of-arrival fluctuations caused by the presence of atmospheric turbulence, which can be written as

$$\sigma_a^2 = 2.914 D^{-1/3} H^{-5/3} \int_0^H z^{5/3} C_n^2(z) dz \tag{A.5}$$

where D = aperture diameter, C_n^2 is the turbulence strength, and H is the altitude. If the communication link is along a slant path, then $C_n^2(z)$ should be replaced by $\sec(\theta).C_n^2(z)$ where θ is the zenith angle (away from the vertical) and the limit of integration should be taken as the slant range. This is exactly how this ARM technology can be applied and analyzed for satellite, airborne, HAP, balloon, and small (including nanosatellite) satellite constellation to provide the potential concept of establishing worldwide Internet connectivity with *all-optical technology*.

An Example of Link Analysis for a Satellite-Based System Laser Interrogator and Low-Power Gbps Amplified Fiber Retromodulator [16]

Fig. A.2 shows the simulation result for a satellite-based laser interrogator and ground-based ARM. The range of the satellite was assumed to be 370 km, the atmospheric transmission efficiency = 0.5. The retroreflected beam is received by the satellite receiver of 6-inch diameter. The transmitter efficiency was assumed to be 0.5, the required BER = 10^{-9}, the SNR = 144, and the gain of the retromodulator system was taken to be 4×10^5. The simulation result shows the received power at satellite as a function of required laser transmitter power on the satellite for different values of the divergent angles of the

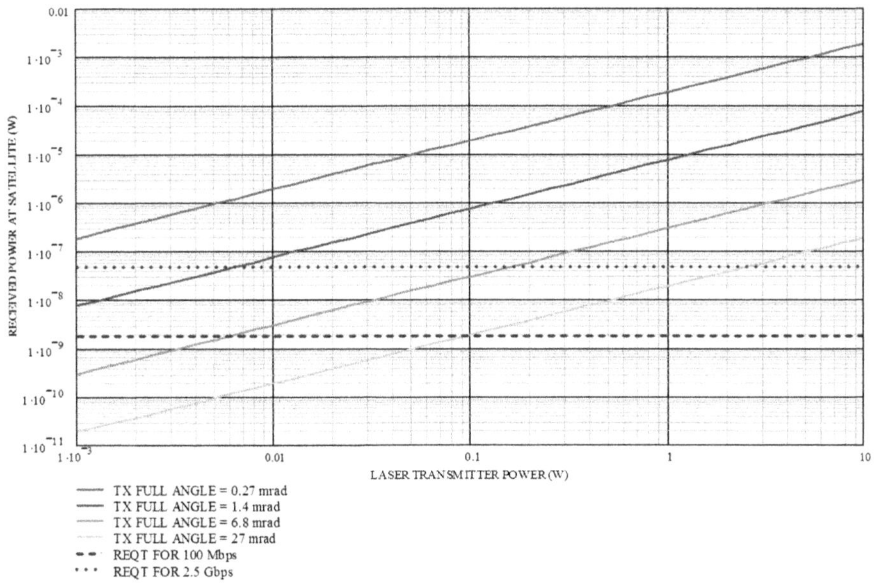

FIGURE A.2
Plot of the received power versus required laser transmitter power: an example of a satellite-based laser interrogator and a ground-based fiber retrosystem using fiber array.

transmitter. The two horizontal dashed lines represent the needed received power at the satellite for 100 Mbps and 2.5 Gbps data rates. To achieve a data rate of 2.5 Gbps for a transmitter divergent beam of 6.8 mrad requires 160 mW of laser power, whereas for a transmitter divergent beam of 27 mrad, it requires about 2.5 W of laser power. These numbers are very practical and realistic in order to design a satellite-based FSO communication system.

Other Related Research and Development With Modulating Retroreflector and Beacon Placed on Unmanned Air Vehicle, High-Altitude Platform, Satellite, and Small Satellite

MEMS MRR for an air-to-ground FSO communication system between the airborne and ground station enabling ATP has been reported [42,43]. A new technique for the pointing, based on a liquid crystal device, was used in order to optimize the ground-station performance. The MEMS device located at the UAV is capable of returning the light from a distant interrogating laser without any additional pointing requirement onboard, thus fully eliminating the ATP on one end of the link, resulting in a considerable reduction of power, size, and weight onboard the UAV. The development and implementation of ATP systems for a stratospheric testbed (HAP/balloon/aircraft) for use with optical free-space terminal on a high platform was reported [44]. Two techniques were considered: (1) to place a beacon laser at the receiver and point toward a HAP where the ATP system could find the beacon, and (2) to place a retroreflector at the receiver. Jet Propulsion Laboratory researchers reported [45] a robust ATP subsystem for 2.5 Gbit/s UAV at altitude of 18 km with a 200 mW downlink laser at 1550 nm for a BER of 1E-9 to ground FSO communication link. A beacon laser from the ground was used. Recently a development of a CubeSat-sized laser beacon tracking capable of achieving a submilliradian altitude knowledge accuracy with low fade probability during various sky conditions is reported [46]. The work is useful for establishing a high data rate FSO communication link on a CubeSat platform on LEO, which is directly relevant in establishing potential global Internet connectivity.

References

[1] E. Leitgeb, M. Gebhart, U. Birnbacher, Optical networks, last mile access and applications, in: Arun K. Majumdar, Jennifer C. Ricklin (Eds.), Chapter in Free-space Laser Communicationss: Principles and Advances, Springer, New York, 2008, pp. 273–302.

[2] M. Gebhart, P. Schrotter, U. Birnbacher, E. Leitgeb, Satellite communications, free space optics and wireless LAN combined: worldwide broadband wireless access independent of terrestrial infrastructure, in: IEEE MELECON 2004, May 12–15, 2004, pp. 449–452. Dubrovnik, Croatia.

[3] Space: The Ultimate Network Edge, Data Storage Knowledge, October 17, 2016 http://www.datacenterknowledge.com/archives/2016/10/17/space-the-ultimate-network-edge.

[4] T. Bui, S. Kiravittaya, K. Sripimanwat, N. Nguyen, A comprehensive lighting configuration for efficient indoor visible light communication networks, International Journal of Optics 2016 (2016) 9. Article ID 8969514.

[5] C. Chen, D.A. Basnayaka, H. Haas, Downlink performance of optical attocell networks, Journal of Lightwave Technology 34 (1) (January 1, 2016) 137–156.

[6] D.C. O'Brien, G. Faulkner, H. Le-Minh, O. Bouchet, M.E. Tabach, M. Wolf, J.W. Walewski, S. Randel, S. Nerreter, M. Franke, K.-D. Langer, J. Grubor, T. Kamalakis, Gigabit optical wireless for a home access network, in: Proc. PIMRC, vol. 2009, 2009, pp. 1–5.

[7] H. Le-Minh, D. O'Brien, G. Faulkner, A Gigabit/s indoor optical wireless system for home access networks, in: CSNDSP, 2010, p. 2010.

[8] H. Le Minh, Z. Ghassemlooy, D. O'Brien, G. Faulkner, Indoor gigabit optical wireless communications: challenges and possibilities, in: The 12th International Conference on Transparent Optical Networks (ICTON), 27 June- 1 July 2010, Bavarian Academy of Sciences and Humanities, Germany, Munich, 2010.

[9] C. Lee, C. Shen, C. Cozzan, R.M. Farrell, J.S. Speck, S. Nakamura, B.S. Ool, S.P. DenBaars, Gigabit-per-second white light-based visible light communication using near-ultraviolet laser diode and red-, green-, and blue-emitting phosphors, Optics Express 25 (15) (July 24, 2017).

[10] A. Gomez, K. Shi, C. Quintana, M. Sato, G. Faulkner, B.C. Thomsen, D. O'Brien, Beyond 100-Gb/s indoor wide field-of-view optical wireless communications, IEEE Photonics Technology Letters 27 (4) (February 15, 2015).

[11] S. Watson, S.P. Najda, P. Perlin, M. Leszczynski, G. Targowski, S. Grzanka, M.A. Watson, H. White, A.E. Kelly, Multi-gigabit Data Transmission Using a Directly Modulated GaN Laser Diode for Visible Light Communication through Plastic Optical Fiber and Water, Summer Topicals Meeting Series (SUM), July 13–15, 2015, https://doi.org/10.1109/PHOSST.2015.7248278. Nassau, Bahamas, WD1.2 (Contributed), IEEE.

[12] D. Tsonev, S. Videv, H. Haas, Towards a 100 Gb/s visible light wireless access network, Optics Express 23 (2) (January 26, 2015), https://doi.org/10.1364/OE.23.001627.

[13] A.T. Hussein, M.T. Alresheedi, J.M.H. Elmirghani, 20 Gb/s mobile indoor visible light communication system employing beam steering and computer generated holograms, Journal of Lightwave Technology 33 (24) (December 15, 2015) 5242–5260.

[14] H. Haas, Wireless Data from Every Light Bulb, TED Global, Edinburgh, July 2011.

[15] M. Strobl, T. Waas, S. Moehne, Using ethernet over powerline communication in automotive networks, in: 2012 Proceedings of the Tenth Workshop on Intelligent Solutions in Embedded Systems (WISE), Klagenfurt, Austria, 5–6 July 2012.

[16] A.K. Majumdar, Advanced Free Space Optics (FSO): A Systems Approach, Springer, New York, 2015.

[17] D.J.T. Heatley, D.R. Wisely, I. Neild, P. Cochrane, Optical wireless: the story so far, in: IEEE Communications Magazine, December 1998, pp. 72–82.

[18] H. Elgala, R. Mesleh, H. Haas, Indoor broadcasting via white LEDs and OFDM, IEEE Transactions on Consumer Electronics 55 (3) (August 2009) 1127–1134.

[19] J. Armstrong, A. Lowery, Power efficient optical OFDM, Electronics Letters 42 (6) (16 Mar. 2006) 370–372.

[20] E. Leitgeb, S. Sheikh Muhammad, C. Chlestil, M. Gebhart, U. Birnbacher, N. Perlot, H. Henniger, D. Giggenbach, J. Horwath, E. Duca, V. Carrozzo, S. Betti, Book Chapter 4 for SJA-2413 "Clear Sky Optics", of the e-Book "Influence of the variability of the propagation channel on mobile, fixed multimedia and optical satellite communications", Satellite Communications Network of Excellence, SATNEX, IST Network of Excellence No 507052, January 2008.

[21] M. Matsumoto, K. Kazaura, P. Dat, A. Shah, K. Omae, T. Suzuki, K. Wakamori, T. Higashino, K. Tsukamoto, S. Komaki, An Alternate Access Technology for Next Generation Networks Based on Full-Optical Wireless Communication Links, ITU, 2008. Kaleidoscope 92-61-12441-0/CFP0838E.

[22] A.K. Majumdar, J.C. Ricklin, Free-space Laser Communications: Principles and Advances, Springer, New York, 2008.

[23] I.I. Kim, H. Hakakha, P. Adhikari, E. Korevaar, A.K. Majumdar, "Scintillation redusction using multiple transmitters," in free-Sapace laser communications technologies IX, Proceedings of SPIE 2990 (1997).

[24] A.K. Majumdar, G.H. Fortescue, Wide- beam atmospheric optical communication for aircraft application using semiconductor diodes, Applied Optics 22 (16) (1983) 2495–2504.

[25] http://leosat.com/solution/ (LeoSat Enterprises).

[26] H. Kaushal, G. Kaddoum, Optical communication in space: challenges and mitigation techniques, IEEE arXiv 1705.10630v1 ([cs.IT] 28 May 2017), https://doi.org/10.1109/COMST.2016.2603518.

[27] G.G. Ortiz, S. Lee, S. Monacos, M. Wright, A Biswas, Design and development of a robust ATP subsystem for the Atair UAV-to-Ground Lasercom 2.5 Gbps demonstration, Proceedings of SPIE 4975 (2003) 103–114.

[28] C. Chlestil, E. Leitgeb, S. Sheikh Muhammad, A. Friedl, K. Zettl, N. P. Schmitt, W. Rehm, N. Perlot, Optical Wireless on Swarm UAVs for High Bit Rate Applications.

[29] K. Zettl, S. S. Muhammad, C. Chlestil, E. Leitgeb, A. Friedl, N. P. Schmitt, W. Rehm, HIGH BIT Rate Optical Wireless Systems for SWARM Unmanned Aerial Vehicles: A Feasibility Study.

[30] M. Alzenad, M.Z. Shakir, H. Yanikomeroglu, M.-S. Alouini, FSO-based Vertical Backhaul/Fronthaul Framework for 5G+ Wireless Networks.

[31] K. Antonis Hatziefremidis, E. Zarganis, H.C. Leligou, N. Pleros, K.E. Zoiros, Performance evaluation of optical channel transmission between UAVs and Ground Stations, MATEC Web Conference 41 (01005) (2006), https://doi.org/10.1051/matecconf/20164101005 published by EDP Sciences, 2016.

[32] Arstechnica in Google: https://arstechnica.com/tech-policy/2017/10/how-google-is-using-balloons-to-extend-cellular-service-in-puerto-rico/.

[33] Google Website: The Verge, "Inside Project Loon: Google's internet in the sky is almost open for business," by Ben Popper, March 2, 2015.

[34] M. B. Awan, S. Mohan, Balloon Mesh Free Space Optical Communication with Tracking and Link Switching, IEEE PAPER, Member, IEEE University of Arkansas at Little Rock, USA. University of Arkansas at Little Rock, USA.

[35] F. Heine, H. Kampfner, R. Lange, R. Czichy, R. Meyer, M. Lutzer, Optical inter-satellite communication operational, in: The 2010 Military Communications Conference-IEEE, 2010, 978-1-4244-8180-4/10.

[36] T. Nielsen, G. Oppenhäuser, In-orbit test result of an operational inter-satellite link between ARTEMIS and SPOT-4, SILEX, Proceedings of SPIE 4635 (2002).

[37] M. Toyoshima, et al., Ground-to-satellite optical link tests between the Japanese laser communication terminal and the European geostationary satellite ARTEMIS, Proceedings of SPIE 5338 (2004).

[38] E. Benzi, I. Shurmer, N. Policella, D. Troendle, M. Lutzer, S. Kuhlmann, M. James, Optical inter-satellite communication: the alphasat and sentinel-1A in-orbit experience, in: Space Ops Conferences, May 16–20, 2016. Daejeon, Korea, 14th International Conference on Space Operations.

[39] T.M. Shay, R. Kumar, 2.5-Gbps amplified retro-modulator for free-space optical communications, Proceedings of SPIE 5550 (2004) 122–129.

[40] G.C. Gilbreath, W.S. Rabinovich, T.J. Meehan, M.J. Vilcheck, R. Mahon, R. Burris, M. Ferraro, I. Sokolsky, J.A. Vasquez, C.S. Bovais, K. Cochrell, K.C. Goins, R. Barbehenn, D.S. Katzer, K. Ikossi-Anastasiou, M.J. Montes, Compact, lightweight payload for covert data link using a multiple quantum well modulating retro-reflector on a small rotary-wing unmanned airborne vehicle, Proceedings of SPIE 4127 (July 2000) 57–67.

[41] A.K. Majumdar, T.M. Shay, Wide Field-of-View Amplified Fiber-Retro for Secure High Data Rate Communications and Remote Data Transfer, October 30, 2012. US Patent, No. US 8,301,032 B2, date of patent.

[42] A. Carrasco-Casado, E. Oton, Low-impact air-to-ground free-space optical communication system design and first results, in: IEEE ICSOS Conference, May 2011, pp. 109–112.

[43] A. Carrasco-Casado, R. Vergaz, J.M. Sanchez-Pena, Design and early development of a UAV terminal and a ground station for laser communications, Proceedings of SPIE 8184 (29) (September 2011).

[44] B. Epple (Thesis at the Institut Fur Informatik), Development and Implementation of a Pointing, Acquisition and Tracking System for Optical Free-space Communication Systems on High Altitude Platform, vol. 20, Der Ludwig-Maximilians-Universtitat Munchen, May 2005.

[45] Design and Development of a Robust ATP Subsystem for the Altair UAV-to-ground Laser-comm 2.5 Gbps Demonstration, Free-Space Laser Communications Technologies XV, San Jose, CA, USA, 01-28-2003. http://hdl.handle.net/2014/7008.

[46] T. Nguyen, Laser Beacon Tracking for Free-space Optical Communication on Small-satellite Platforms in Low-earth Orbit (M.S. thesis in Aerospace Engineering), MIT, August 2015.

CHAPTER 6

Demonstration of Successful LEO/GO/Small Satellites/UAV/Airplanes to Ground Optical Communication Experiments: Opening Future Potential for All-Optical Networks Around the World for Global Internet Connectivity

6.1 INTRODUCTION

Fixed cell towers to establish high speed Internet connectivity cannot be practically installed and connected throughout the world because of the presence of oceans and seas in between all the countries. Laying fiber cables underseas is not a solution because of cost, reliability, and many hard-to reach islands. The only viable solution is to utilize space for connecting between two points worldwide. Space-to-ground optical communication link with integration into ground terrestrial networks has the only potential to establish multi-gigabit/second bidirectional optical communication link worldwide *anytime, anywhere*. The space and airborne platforms include satellites with orbits at various altitudes and unmanned air vehicles (UAVs), high-altitude platforms (HAP), and most recently, stratospheric balloons to reach the Internet connectivity even in remote locations. Because of the recent advances in research and development of compact, fast, and portable optical components such as Microelectromechanical Systems (MEMS)-based technology and optical chips, commercially available off-the-shelf items are available that are light, and that consume very low power, size, and weight. These compact optical devices together with advanced research and development of optical networks (both hardware and software) make it now feasible to establish potential all-optical global Internet connectivity as described in this book.

Satellite broadband service has been the only choice for people to connect to the web in remote places where cable and other telephone company's broadband cannot reach. But there are some issues with this system: cost, a big dish on the roof that must be aimed at just the right point in the sky, high

Optical Wireless Communications for Broadband Global Internet Connectivity.
https://doi.org/10.1016/B978-0-12-813365-1.00006-0

latency of the available service, and a poor value on a price per megabit basis. To obtain gigabit-level Internet is almost impossible with this satellite. But mass-produced, low-flying satellites (e.g., a constellation of low earth orbit [LEO] satellites) may bring high-value gigabit data rate Internet service globally *anytime, anywhere* with the potential all-optical technology as described in this book. This is the future and next wave of satellite (including small nanosatellites) technology entering orbit over the Earth over the next few years. In addition to satellites located at various altitudes (such as geosynchronous orbit [GEO] and LEO), other platforms such as HAPs, stratospheric balloons, UAVs, and aerial platforms equipped with free-space optical (FSO) transceivers will have the potential of providing high data rate Internet connectivity in remote locations to serve unserved and underserved populations. This chapter addresses first the basic logistics for using these low-orbit satellites and then the recent advanced technological breakthroughs and developments made by any experimental demonstration. The demonstrations of any space platform or aerial platform-based experiments or tests will help to back the proposed concepts technology described in this book for establishing optical wireless communications for broadband global Internet connectivity.

6.2 RECENT RESEARCH AND DEVELOPMENT CONCEPTS AND TECHNOLOGY IN SUPPORT OF THE SPACE, AIRBORNE (INCLUDING AIR-TO-AIR), MOVING, AND GROUND-BASED PLATFORMS WITH SECURE, REAL-TIME DRIVES

These supporting concepts are relevant to space terminal and airborne platforms for connectivity. Optimal location of optical ground stations (OGSs) to serve the LEO spacecraft needs the estimation of availability of optical ground networks, architecting the ground segment of an optical space communication network.

6.2.1 Selecting Optical Ground Stations Optimum for the Availability of Satellite-Based Free-Space Optical Communication Networks

For space-to-ground optical links, one of the propagation issues is the difficulty of penetrating through atmospheric clouds due to larger scattering effects due to clouds at optical wavelength. It is therefore necessary to analyze possible architectures for the ground segment of a fully optical space relay-communications network that can be developed to serve LEO missions. In order to develop reliable space-to-ground optical communications using optical systems, one of the main issues is the atmospheric propagation effects of optical communication that causes the difficulty of penetrating through atmospheric clouds. To mitigate these effects, geographical diversity has been proposed [1] as an alternative solution. Because of recent demonstrations of LEO laser-equipped satellite-to-ground stations for optical communication, it is important to consider various factors such as correlated cloud coverage among all OGSs and trade-offs between minimal cloud probability, minimal latency, and proximity to supporting infrastructure when selecting optimum locations for networks of ground stations. Architectures for the possible OGSs for LEO satellites have been reported [1–3], analyzing the number and locations of OSGs including GEO satellite relay satellite, providing a new cloud model for determining network availability. Weather data from the National Oceanic and Atmospheric Administration and the cloud fraction dataset from Aqua's and Terra's MODIS instruments to characterize weather conditions across the globe were used for his study. Results are presented [1–3] to show a maximum availability of 95.5% with 12 additional OGSs. Intersatellite links (ISLs) to combat the effects of cloud coverage were shown to increase the availability from 80% with six ground stations, to 98.7% with the same number of ground stations. The results from this research also provide a potential for establishing any point on Earth to be selected as a candidate location for an OGS.

To compute the network availability the link outage probability (LOP) is defined as [2] the probability that all the OGSs that are in line-of-sight (LOS) with a satellite fail due to cloud coverage at the same time

$$LOP_P = \prod_{g \in G} \theta_{gL,gl} \tag{6.1}$$

where G is the set of ground stations in the LOS from point P, and gL, and gl denote the latitude and longitude of ground station, respectively. The availability of the network is expressed in terms of monthly network availability (NA_m) and is the complement of the monthly LOP (LOP_m) as follows [2].

$$NA_m = 1 - LOP_m \tag{6.2}$$

The cost model is defined as the estimation of the lifecycle cost of a ground segment architecture composed of nonrecurring and recurring operational costs.

Fig. 6.1 shows locations of the OGSs considered in the research reported [1–3] for different candidate location sets.

Fig. 6.2 shows comparison of the investment cost (in $M) from the simulation of Pareto Fronts (the set of Pareto optimal outcomes that involve solutions in an optimization problem with multiple objective functions) of the different candidate locations for different location sets, as defined in the previous figure plus one set of unconstrained optimization where candidate locations contains any latitude and longitude.

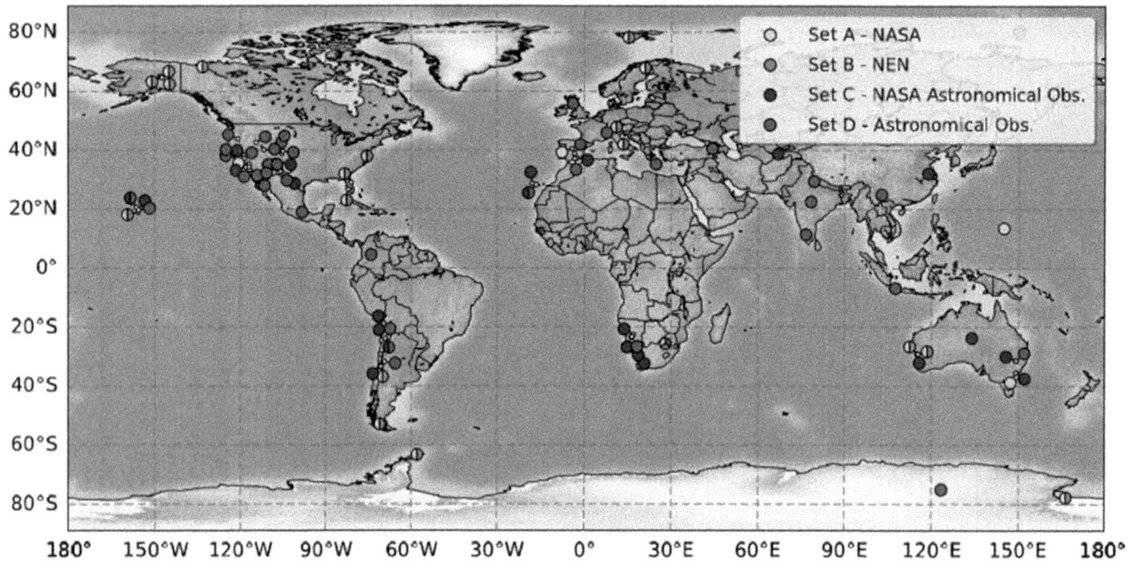

FIGURE 6.1

Locations of the optical ground stations as reported in [1]. Set A: NASA near earth network (NEN) assets; Set B: All NASA assets, NASA owned assets in the NEN, deep space network, and space network); Set C: NASA observatories; Set D: Astronomical observatories. *Courtesy of, and permission from, Iñigo del Portillo, from his M.S. thesis in Aeronautics and Astronautics, MIT, June 2016.*

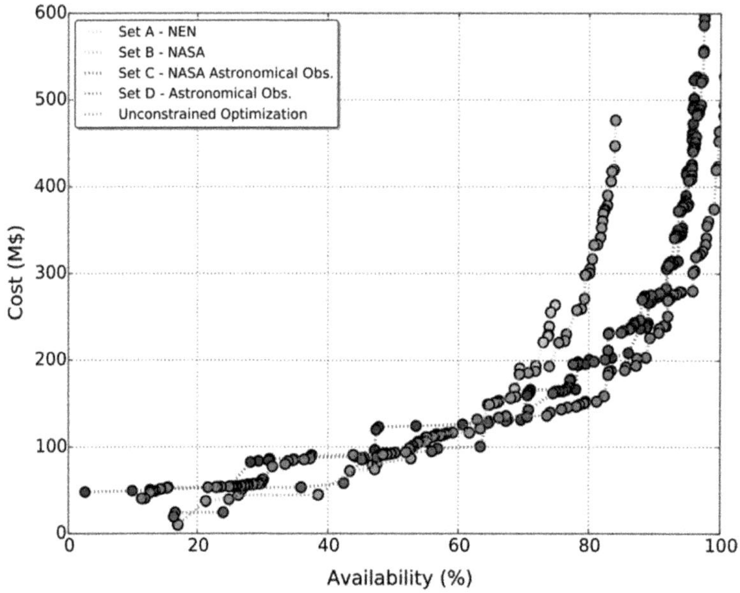

FIGURE 6.2

Investment cost comparison for the different candidate location sets. *Courtesy of, and permission from, Iñigo del Portillo, from his M.S. thesis in Aeronautics and Astronautics, MIT, June 2016.*

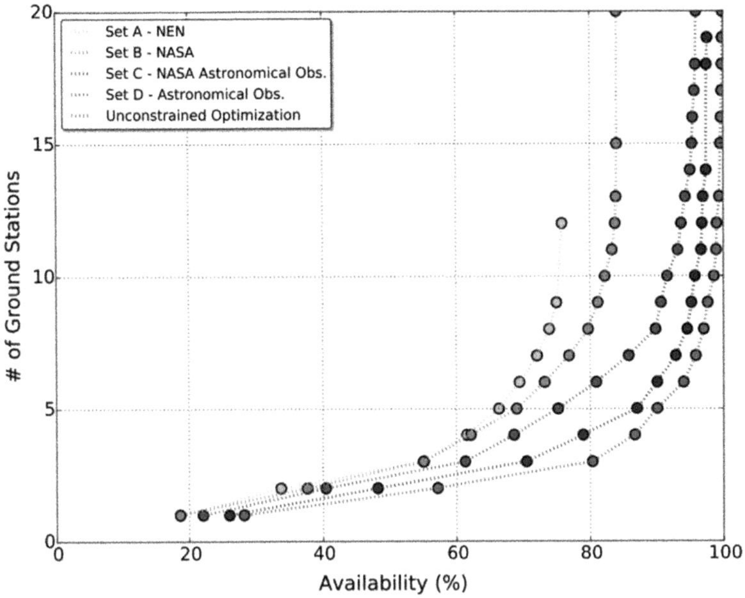

FIGURE 6.3
Availability versus the number of optical ground stations for different candidate location sets. *Courtesy of, and permission from, Iñigo del Portillo, from his M.S. thesis in Aeronautics and Astronautics, MIT, June 2016.*

Fig. 6.3 illustrates availability for different sets including the set with unconstrained optimization as a function of the number of ground stations. The availability attainable is always higher in the unconstrained set.

6.2.2 A Note on Seamless Handoff and Handover Between Terrestrial Terminals and Satellites

Future global communication networks require fast and integrated service to ubiquitous users on-demand, *anytime, anywhere*. In order to provide complete coverage to any location on Earth, satellites will play important role in global communication infrastructure. High-speed worldwide Internet connectivity needs the global information network to offer broadband multimedia data services. Broadband satellite networks must provide a large array of services such as video on demand, multimedia traffic, fast Internet access, interactive video, and other existing Internet-based applications. High data rate in gigabits per second can be achieved with LEO satellite–equipped laser transceivers with FSO link. These satellite systems can extend the existing terrestrial networks to provide complete global coverage. GEO satellites can also be integrated with LEO satellites in the global communication network where ISLs can connect satellites both within the same orbit as well as satellites in adjacent orbits.

The footprints of a satellite are a circular area on the earth's surface. For laser-based communication links with satellite systems such as LEO, for example, the area is much smaller than the area using radio frequency (RF)-based technology. Thus the area of cells within a footprint is also considerably smaller in an optical link than RF link. *Handover* [4], which is basically changing the point of connection while communicating, therefore needs to address transferring of an ongoing connection to a new spotbeam or satellite. For Internet protocol (IP)-based data communication using satellites as IP nodes, network-layer handovers are also required so that end terminals (satellites or user) that have IP connectivity may need to change their IP addresses during moving. Therefore, both the *link-layer handover* as well as *network-layer handover* are necessary.

6.2.2.1 Low Earth Orbit Satellite Systems Handover

LEO satellites are not stationary for a fixed user on the earth's surface, and supporting continuous communication over an LEO satellite system requires both changing one or more links and IP addresses of the communication endpoints. Handovers in satellite networks can be the following: link-layer handover, spotbeam handover, satellite handover, gateway handover, and ISL handover where the performances in the handovers can be evaluated in terms of connection-level quality of service criteria such as *call blocking* probability (P_b) and *forced termination* probability (P_f) [5]. There has been much research and development available in the literature for many categories of handover schemes in classical communications systems that are not repeated here and are out of scope of this book. Also, note that integration among different segments (terrestrial indoor, terrestrial outdoor, and satellite) is required to provide real global and contiguous coverage without interruption. Efficient algorithms must be implemented to perform intersegment handover so that switching among different segments can occur without dropping the call. Seamless handover in software-defined satellite networking (SDSN) architecture is recently reported [6], which simplifies networking among versatile satellites and enables new protocols to be easily tested and deployed. Routing can be possible with ISLs, which offer the following advantages: reduced number of gateways needed, forward connections of data packets within the satellite network as long as possible, and only one uplink and one downlink per direction needed for the connection of two users. A constellation of satellites with ISL can also provide routing.

Average handover latency is given by [7,8].

$$T_{\text{latency}} = (1 - PH)(T_{\text{retry}} + T_{\text{reconnect}}) + P_H T_H \tag{6.3}$$

where T_{retry} and T_H are the times required to re-setup the handover process if a handover fails and the handover time, respectively [5,7]; $T_{\text{reconnect}}$ is the waiting time between the occurrence of a handover failure and the start of r-connection; and P_H is the handover success probability, which can be a constant for the particular handover scheme that contains both the successful signaling probability, P_S, and the nonblocking probability $(1 - P_b)$. The handover success probability in Eq. (6.3) for SDSN schemes can be modified as follows [8]:

$$P_H^* = 1 - \left[1 - P_S\left(1 - P_b^*\right)\right](1 - P_O) \tag{6.4}$$

where P_O is the outage probability of the current main data link. The paper [8] shows that the average latency with the SDSN scheme is less than one-third of that by hard handover scheme. For an LEO system, when the height of the satellite increases, the reduction in the latency is an important parameter in the communication network design.

The other quantity of interest is the quality of experience (QoE), which is affected by latency and the handover failure rate. The user satisfaction score is given by [8].

$$S_u = r \cdot P_H - \exp\left(-1/T_{\text{latency}}\right) - p \cdot (1 - P_H) \tag{6.5}$$

where r and p are the handover success reward score and the handover failure penalty score, and $T_{latency}$ is as defined earlier. From the analysis it was reported [7] that the SDSN handover provides the best user QoE where there is an improvement of QoE even by increasing the height of the LEO satellite.

Telecommunication satellite platforms are the only viable solutions to establish the two-way communications between any two locations on Earth. Optical communication is the technology that could be used in the future for fast Internet connections from space by connecting satellite downlink signals to various ground-based stations connected by fiber optic networks. Typically, the satellites can be placed at the following altitudes:

- LEO: Altitudes from 160 to 1400 km, one-way delay of 50 ms
- Medium earth orbit (MEO): Altitudes ranging from 10 to 15,000 km, one-way delay of 100 ms
- GEO: At 35,786 km, and one-way delay of 250 ms

The two-way delays determine the latency of the communication system, which is an important factor in designing a satellite-based FSO communication system.

One of the main issues of using optical systems is the space-to-ground link (SGL), as it is practically impossible for the laser beam to penetrate atmospheric clouds. As a possible solution, geographic diversity of OGSs has been proposed as an alternative to mitigate these effects. A different systems architecture approach is needed to analyze the ground segment of an optical space communication network to serve LEO missions. This includes optimized number and location of OGSs, use of geostationary relay satellites combined with direct-to-earth links and cross-links among relay satellites. A report [1] shows that the maximum availability of 95.5% can be achieved using an architecture similar to the actual system (the tracking and data relay satellite system) and 12 additional OGSs for establishing optimum global optical communication links. In addition, incorporating ISLs can add an extra mitigation layer to combat the effects of cloud coverage; for example, using ISL can increase the availability from 80% with six ground stations to 98.7% with the same number of ground stations [1]. It is possible to design optical links with SGLs with both LEO and MEO orbit satellites and the ISL links with GEO orbit satellite.

6.3 DEMONSTRATIONS AND EXPERIMENTS OF OPTICAL COMMUNICATIONS FROM LOW EARTH ORBIT-TO-GROUND AND AIRBORNE PLATFORM-TO-GROUND

This section describes some of the recent demonstrations and experiments reported by various laboratories/organizations with the goal of proving the feasibility of communicating optically from space to Earth. Initially a one-way communication link from the LEO platform to ground-based station demonstrations have been reported by the researchers for the downlink scenario with the ultimate goal to incorporate optical transceivers at both terminal ends to prove possible two-way optical (duplex) communications. Finally, the successful demonstrations will provide designs of high data rate (tens and hundreds of Gb/s) optical communications systems for the future applications of establishing broadband Internet connectivity worldwide, *anytime, anywhere*.

Brief descriptions of the optical transmitting and receiving systems the researchers used for their experiments will be described with recent results and the maximum data rates achieved. It is to be noted that atmospheric effects such as turbulence and scattering put a limitation on ultimately the achievable data rate. The effects are more important near the ground, a few hundred meters to kilometers, but effects are less at higher altitudes. The LEO satellite's communication path from the platform to the ground station suffers most atmospheric effects near the ground. To achieve very high data rates, transmitter (and receiver for a transceiver system) on the platform as well as the receiver (and transmitter) optical systems must be optimized in terms of power, wavelength, modulation and detection, mitigation techniques used such as adaptive optics or selecting different locations of OGSs, and the acquisition, tracking, and pointing (ATP) systems. The size, weight, and power requirement at the space platform and the low cost also need to be considered. The performance of the optical communications overall systems will be evaluated by comparing the parameters like signal-to-noise ratio (SNR) and bit error rate (BER). Based on these criteria different systems have been demonstrated.

6.3.1 Jet Propulsion Laboratory /NASA Demonstrations

6.3.1.1 Jet Propulsion Laboratory /NASA Low Earth Orbit-to-Ground Communication Link

Experiment for the optical payload for lasercom science recently demonstrated [9] a successful achievement of 50 Mb/s data rate between the international space station (ISS) platform and an OGS in diverse day and night conditions. This is an important step in exploring the potential of future global broadband Internet connectivity by entirely optical wave technology paving the way for high data rate, multi-Gb/s optical links with LEO satellites, and seamless connection to terrestrial fiber networks.

6.3.1.1.1 Technological Considerations for the Jet Propulsion Laboratory /NASA Low Earth Orbit-to-Ground Communication Link

6.3.1.1.1.1 Optical Ground Station The OGS performs two basic functions: the optical communication receiver detects the modulated optical communication signal transmitted from the flight system located on the ISS, and it maintains an accurate LOS between the ISS and the ground station. The mitigation of atmospheric turbulence is necessary because the optical wave from the transmitter to the OGS suffers random phase and amplitude fluctuations that are more pronounced for the first few kilometers of the earth's atmosphere. This is because strength of turbulence parameter $C_n^2(h)$ varies with altitude and is typically within the range of 10^{-13} to 10^{-15} $m^{-2/3}$. The atmospheric effects limit the optical communication performance in terms of available maximum data rate capacity due to the received low SNR that increases the BER. Adaptive optic (AO) correction of the atmospheric turbulence is a practical and viable solution to mitigate this effect. An AO system requires a reference signal using a beacon so that distortions due to atmospheric turbulence can be measured, which then reconstruct the undistorted wavefront.

6.3.1.1.2 Creating Beacon for Pointing the Downlink Transmitting Beam Accurately Toward the Receiver in Optical Ground Stations

The optical downlink signal from the flight terminal on the ISS requires accurate aiming and pointing toward the ground-based receiver, which is accomplished by creating a beacon that consists of a multibeam beacon transmitter with four 976 nm 2.5 W multimode fiber-coupled continuous wave (cw) laser diodes on the OGS. The beam divergence transmitted to the ISS was 1.6 mrad. A low-cost silicon camera was used at the transmitting terminal on the ISS to help pointing toward the receiver at OGS. Note that the atmospheric transmission properties are slightly different for the transmitting (1550 nm laser) and the probe receiving beam (at 976 nm) to accurately point toward the OGS. A common aperture on the OGS is designed for transmitting the beacon and receiving the downlink. Optical isolation was taken into consideration in the design of the coaligned transmitting/receiving aperture optics to minimize the backscattered radiation from optics, telescope, surfaces, and the atmospheric aerosol scattering contributing unacceptable receiver signal noise.

6.3.1.1.3 Transmitter System at the International Space Station Location for Downlink Beam

The laser transmitter and the associated electronics were enclosed in an atmospheric pressure-sealed container in the flight system at the ISS platform. The transmitter is a fiber-based 2.5 W average power laser transmitter emitting at 1550 nm where the transmitting terminal includes the two-axis azimuth/elevation gimbal and a bistatic optical head with a charge-coupled device sensor. The laser output fed through a fiber patch cord to the optical collimator aligned with a tracking camera. The transmitting beam divergence used was approximately 1.0 mrad beam divergence. The range from the ISS location to the OGS was 800 km. A spectral filtering on the flight terminal on-board ISS was used to reject background noise. The transmitting signal from the space-to-ground used on-off-keying (OOK) modulation scheme with a Reed-Solomon forward error correcting code at 50 Mb/s nominal data rate. The power received at the OGS receiver from the transmitting terminal at the ISS was of the order of tens of nanowatts of downlink power that can be detected using commercial detectors and receivers with required sensitivity.

The telescope system at the ground station has a 1m diameter mirror mounted on a fast-tracking elevation over azimuth gimbal with a coudé path for the received light signal and a field-of-view (FOV) of 2.6 mrad. The telescope is connected with fiber optic and electrical cables to the control room [9]. The details of the optical train for transmitting and receiving lasers are discussed in [9–11].

6.3.1.1.4 Adaptive Optic System for Atmospheric Corrections of Downlink Laser Beam

The AO system incorporates a self-referenced interferometer (SRI) wavefront sensor, a Boston Micro Machines 1000-actuator MEMS deformable mirror (DM) with a direct-drive DM controller, and a 19 kHz wavefront sensor (WFS) [14].

6.3.1.1.4.1 Self-Referenced Interferometer for Adaptive Optics Application SRI is a self-referenced point-diffraction interferometer with a capability of a new technique for unwrapping phase in an optical system for laser beam projection over long atmospheric paths to compensate effectively for strong turbulence in order to reconstruct the branch–point wavefront contributions for AO. The basic principle of SRI technique is the following: The received signal in an SRI is split between two paths. The signal in one path couples into a single-mode fiber to spatially filter the beam. When the light from the two paths are recombined, an interference pattern is formed. The light traversing the fiber is further split to form four interferograms, formed from phase delays of 0, $\pi/2$, π, and $3\pi/2$. The phase surface Φ can be estimated from the measured interferograms, I_1, I_2, I_3, and I_4 at phase shifts of 0, $\pi/2$, π, and $3\pi/2$, respectively, using the equation

$$\Phi = \tan^{-1}(I_1 - I_3, I_4 - I_2) \tag{6.6}$$

A four-quadrant inverse tangent operator is used to estimate Φ [12–15]. The branch-point component can be recovered by subtracting the least-square phase from the wrapped phase computed in Eq. (6.6), and then finally the full reconstructed phase can be obtained in principle [12–15]. The reconstruction algorithm is required to eliminate the sharp unwrapping cuts and also to minimize the impact of any branch cuts in the field. The branch–point component can be recovered by subtracting the least-squares phase from the wrapped phase computed in Eq. (6.6). Finally, the full reconstructed phase can be evaluated from Eq. (6.7) [12–15]:

$$\Phi_R = \Phi_{LS} + W\{\Phi_{BP}\} = \Phi LS + W\{\Phi_{SRI} - \Phi_{LS}\} \tag{6.7}$$

where Φ_{LS} = the irrotational, least-squares component, Φ_{BP} = the rotational, branch-point component, and W = an operator that wraps its argument into the interval $(-\pi, \pi)$.

There are a number of excellent papers and recent developments in the subject of adaptive optics capabilities of handling branch points applicable to strong turbulence reported by a number of researchers for the last few years that are relevant to free space laser communications. Interested readers may read their extensive works in [15,16].

Referencing point-diffraction interferometer and using the previously mentioned SRI technique for unwrapping phase can improve the AO system for laser beam projection over long atmospheric paths to compensate well for strong turbulence such as Jet Propulsion Laboratory (JPL)/LEO to ground station. The laser communication signal received from the LEO satellite to the OGS can be effectively corrected to increase the SNR of the communication system. The SRI wavefront sensor at the JPL OGS has the capability of correcting high levels of turbulence at fast frame rates since the atmospherically distorted phase is directly measured and applied to MEMS-DM. The technique does not need wavefront reconstruction as explained in [15,16]. The optical path to the receiver passed through relay optics coupling OPALS downlink to the AO system. The receiver has a 200 μm diameter InGaAs Avalanche Photo Diode (APD) detector coupled to a single digitizer for recording the captured high-definition video or text transmitted files. AO corrected beam was split between a tracking camera and SMF-28 fiber for recording power and communication signal. The AO corrected output signal to fiber used a low noise amplifier, and fiber Bragg grating filter to improve SNR of the detected communication signal. The output signal was digitally stored for 2 min of data at 50 Mb/s data rate.

6.3.1.1.5 Essential Results From the Jet Propulsion Laboratory OPALS Demonstration Relevant to Optical Communications Performance

The essential results from the JPL LEO-to-ground optical communications tests include: (1) validation of SRI-based AO correction on the communication downlink: the metrics for the effectiveness are measured Strehl ratios; (2) power-in-fiber (PIF) measurements, (3) accomplished data rate, and (4) BER as a function of time. Strehl ratios were computed by the AO system using the residual error between the instantaneous residual wavefront measured by the SRI WFS and a flat-wavefront-calibration phase value. The observed Strehl value was about 0.6 (closed loop) and <0.2 (open loop). The PIF was about a few nW when the AO loop is closed. Both video and bit patterns were at 50 Mb/s data rate. The BER was between 10^{-3} and 10^{-4} during the tests, a measure of communication system performance. The demonstrated results clearly pave the way for future high data rates at tens and hundreds of Gb/s optical communication and seamless space to ground connectivity for future all-optical high data-rate networks. Fig. 6.4 shows a schematic of OPALS ground system with an AO system.

Fig. 6.5 shows the tracking camera output of an AO system for open loop and closed loop cases. The results clearly show the effectiveness of the AO system in free-space laser communications from a satellite-to-ground terminal.

Fig. 6.6 shows the preliminary results for BER for transmission link closure and shows the BER in the range of 10^{-3} to 10^{-4} for the initial test of the WFS performance for the AO system for correcting the waveform.

NASA's number of missions include Earth's Regime Network Evolution Study (ERNESt), where the architectural concept involves the Space Mobile Network (SMN) envisioned as critical for creating an environment for future challenges. The SMN incorporates a number of technologies including FSO communications with other of today's terrestrial mobile network users. Laser Communications Relay Demonstration (LCRD) in 2019 will demonstrate the architecture to include FSO communications relay capabilities in orbit. The first SMN relay node around the mid-2020s will incorporate some of the new technologies and services onboard.

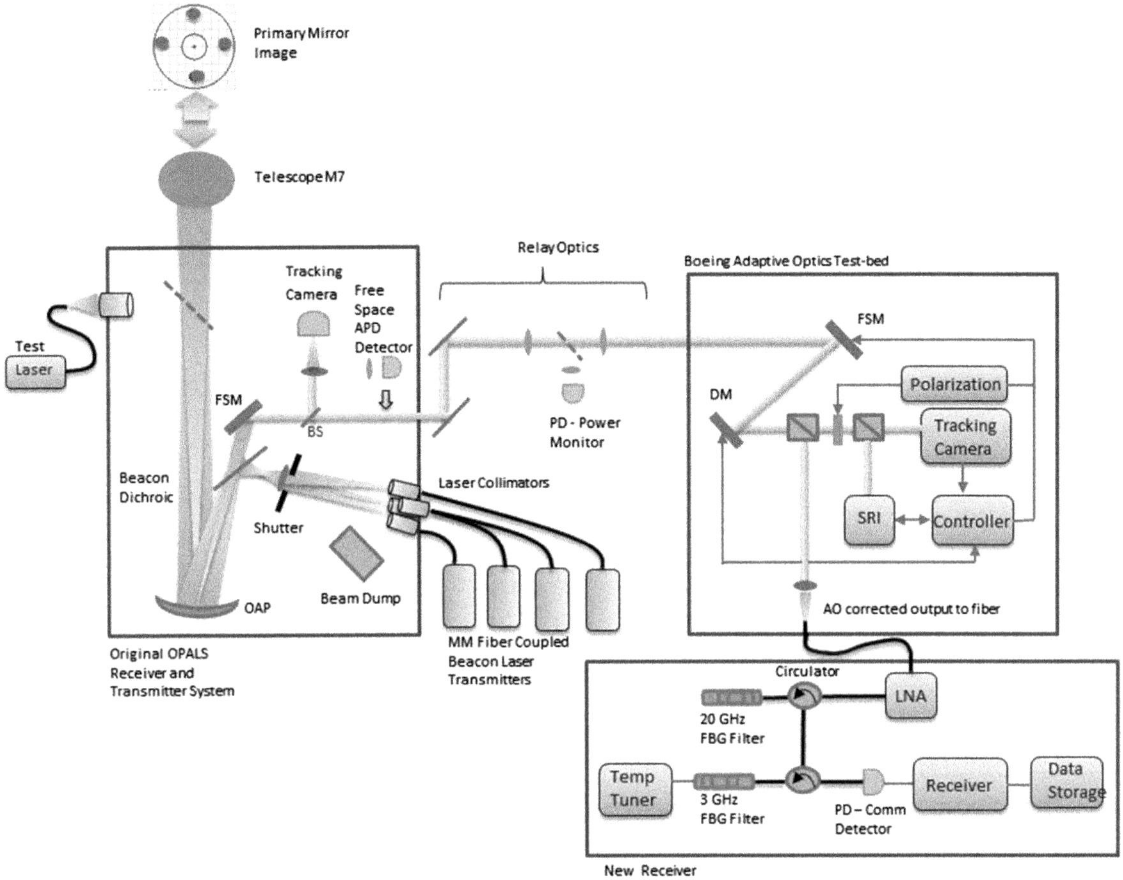

FIGURE 6.4

Schematic of OPALS ground system showing addition of Boeing-provided adaptive optic system. *FBG*, fiber Bragg grating; *FSM*, fast steering mirror; *LNA*, low noise amplifier; *OAP*, off-axis parabola; *PD*, photo detector; *SRI*, self-referencing interferometer. *Courtesy of, and permission from, Dr. Malcolm W. Wright, JPL, and Reprinted with permission from SPIE, 2018 [9].*

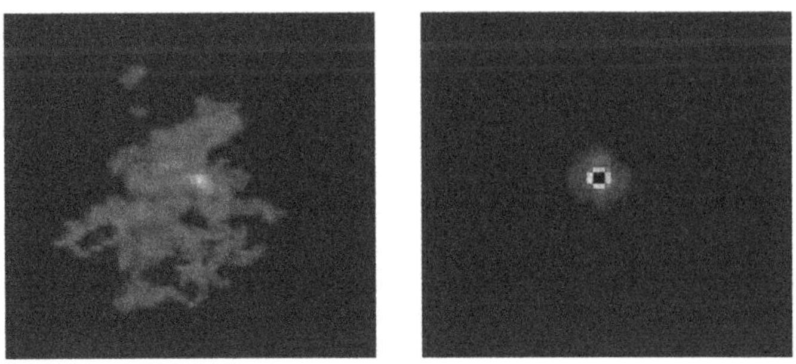

FIGURE 6.5

Tracking camera output of adaptive optic system showing open (left) and closed (right) loop case. *Courtesy of, and permission from, Dr. Malcolm W. Wright, JPL, and Reprinted with permission from SPIE, 2018 [9].*

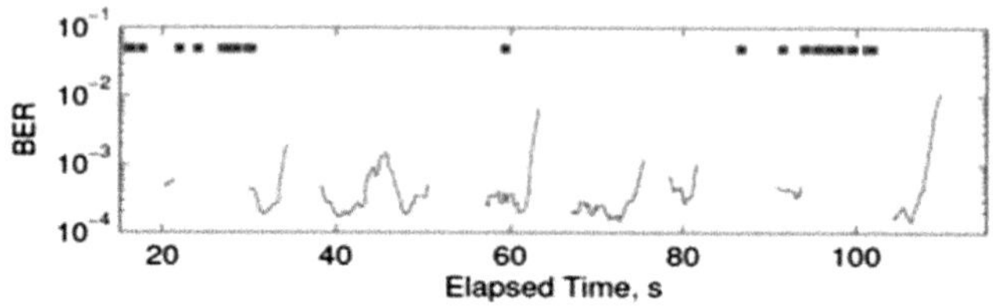

FIGURE 6.6

Bit error rate as a function of time through the second pass showing transmission link closure. Upper points reflect times when there was not enough signal on the wavefront sensor for the adaptive optic system to correct the wavefront. *Reprinted with permission from OSA, 2018 [10].*

6.3.1.2 Jet Propulsion Laboratory /NASA Low Earth Orbit-to-Ground Communication Link

The optical relay demonstrations have been successfully deployed using LEO to GEO and are planned for GEO to ground systems. This will extend the network reach to LEO orbiting assets with either low latency or high bandwidth links. The advantage of a laser communications relay is that the benefits for both deep space and near-earth missions can be integrated to develop a global optical communications network.

This section will discuss the recent NASA project LCRD [17–21], which is a NASA/Goddard Space Flight Center (GSFC)-led technology and is the joint project between NASA/GSFC, NASA/JPL, and Massachusetts Institute of Technology/Lincoln Laboratory (MIT/LL). The mission summary is: high bandwidth geo-sync to ground optical link, downlink at 1.2 Gb/s and uplink at 1.25 Gb/s, two ground stations, and a flight scheduled in 2019. The mission will provide high data rate bidirectional optical communications between Earth and GEO, and will serve as a testbed in space for the network layer protocols. The GEO optical terminal with a mass of 69 kg will transmit at 1550 nm wavelength with optical transmit power 130 W DC, and telescope diameter 108 mm. The differential phase shift keying (DPSK) modem designed by MIT/LL can transmit and receive data at 2.88 Gb/s and beyond 10 Gb/s in the future. The modem will also support 16-ary pulse position modulation (PPM).

There will be two optical communications terminals in space (on a satellite in geostationary orbit) and two ground stations, one in Hawaii and one in Southern California. Optical communications relay will be established from one ground station to another via the communication payload on the GEO terminal. The goal of this mission will also be to achieve high data rate bidirectional optical communications between LEO and GEO as well as between GEO and Earth. In other words, the three communication terminals between LEO, GEO, and Earth will all be connected, which will pave the way to future Internet connectivity. The architecture of the communication links to the ground stations will have the capability of supporting terrestrial, airborne, and LEO users. Modulator for the communication system will consist of a multirate modem designed by MIT/LL with the capability of generating both the PPM and DPSK modulation schemes, described briefly below. PPM will be suitable for power (photon) limited users including deep space links, and the DPSK has the advantage of generating very high data rate (tens of Gb/s).

6.3.1.2.1 Pulse Position Modulation and Differential Phase Shift Keying Modulation and Demodulation Schemes for Atmospheric Path Between Space and Ground Terminals: Bit Error Rate in Presence of Atmospheric Turbulence

When establishing an optical communication between space-based or airborne platform and a ground station on Earth, the communication channel is the atmospheric path. The strength of turbulence is closer to the ground for the first few kilometers and much less at high altitude levels of GEO or LEO satellite platforms. However, the end-to-end communication performance is still limited by the atmospheric turbulence, and scattering medium resulted in limiting the maximum data rate. Depending on the modulation formats, the power received at the receiver determines the available SNR for the communication performance metrics BER.

6.3.1.2.2 Basics of Pulse Position Modulation and Differential Phase Shift Keying

PPM is a modulation technique to increase transmission efficiency in the FSO communication system. PPM is an orthogonal baseband modulation technique and has been studied extensively in optical communication for its superior power efficiency compared to any other baseband modulation techniques and thus makes it well suited for handheld devices that require low consumption. In this scheme, an L-PPM symbol consists of a single pulse of one slot duration within L ($=2^M$ where $M > 0$ is an integer) possible slots with the remaining slots being empty. Information is sent by transmitting a nonzero optical intensity in a single time slot, while other slots remain dark. Each block of $\log_2 M$ data bits is mapped to one of M possible symbols. The notation of M-PPM is commonly used to indicate the order. Each symbol consists of a pulse of constant power P occupying one slot, with the rest of the M-1 slots empty. The position of the pulse corresponds to the decimal value of the $\log_2 M$ data bits. The information is thus encoded by the position of the pulse within the symbol. A PPM signal can be expressed by the following equation

$$x(t) = LP\sum_{R=0}^{L-1} c_k p(t - kT/L)\ldots\ldots\ldots\ldots \quad (6.8)$$

where $[c_0, c_1, \ldots\ldots\ldots\ldots, c_{L-1}]$ is the PPM codeword, and $p(t)$ is a rectangular pulse of duration T/L and unity height. $x(t)$ represents optical power for FSO communication application, and therefore it must satisfy

$$x(t) \geq 0 \text{ and } < x(t) > \geq P \ldots\ldots\ldots\ldots \tag{6.9}$$

where P is the average optical power constraint of the transmitter.

6.3.1.2.3 Bit Error Rate for Pulse Position Modulation in Turbulent Atmosphere

In FSO link, M-ary PPM scheme has the advantage that no thresholds are required for detection. In many power-limited FSO systems including hand-held devices, average transmitted power requirement using M-ary PPM decreases with increasing modulation level M in the absence of atmospheric turbulence. However, in the presence of atmospheric turbulent channel, the increasing modulation level cannot significantly reduce the transmitted power requirement to achieve a desired BER for M-ary PPM. The average BER in the presence of moderate to strong atmospheric turbulence averaged over the random channel state h (assuming to follow a gamma-gamma distribution for $f(h)$) using the M-ary PPM scheme is given by the following equation [22]

$$\langle BER \rangle = \frac{M(\alpha\beta)^{\frac{(\alpha+\beta)}{2}}}{2\Gamma(\alpha)\Gamma(\beta)} \int_0^\infty h^{\frac{(\alpha+\beta)}{2}-1} K_{\alpha-\beta}\left(2\sqrt{\alpha\beta h}\right)$$

$$\mathrm{erfc}\left\{ R P_T \left(\frac{\eta A}{\lambda L}\right)^2 \frac{\sqrt{M \log_2 M}}{2\sigma_n} h \right\} dh \tag{6.10}$$

where α and β are the parameters for characterizing the gamma-gamma distribution for $f(h)$, $K_{\alpha-\beta}(\cdot)$ is the modified Bessel function of the second kind, and order $(\alpha - \beta)$, $\Gamma(\cdot)$ is the gamma function, R is the photodiode responsivity for optical intensity to an electrical signal, $\mathrm{erfc}(\cdot)$ is the complementary error function, P_T is the average transmitted power, η is the efficiency of the transmitter and receiver both, λ is the transmitting wavelength, L is the range or distance between the transmitter and the receiver, and h is the channel state due to atmospheric turbulence, white Gaussian noise with variance, σ_n^2 for the received electrical signal where P_T is the average transmitted power. Eq. (6.3) can be numerically evaluated using the turbulence parameters [22].

6.3.1.2.4 Differential Phase Keying Modulation for Free-Space Optical Communication Link: Bit Error Rate in Atmospheric Turbulence

The DPSK can be viewed as the noncoherent version of the phase-shift keying. It eliminates the need for a coherent reference signal at the receiver by combining two basic operations at the transmitter: (1) *differential encoding* of the input binary wave, and (2) phase-shift keying. DPSK has gained considerable interest for FSO communications due to a ~3 dB sensitivity improvement over the popular OOK, and reduced peak power. Two orthogonal symbols represented by the relative phase between two differentially encoded bits carry the binary information: a 0 represented by no-phase change and a 1 by a π phase difference (or vice versa). The receiver is equipped with a storage capability so that it can measure the relative phase difference between the waveforms received during two successive bit intervals. The phase difference between waveforms received in two successive bit intervals will be independent of the unknown phase θ, if θ contained in the received wave varies slowly (slow enough for it to be considered essentially constant over 2 bit intervals). DPSK is another example of noncoherent orthogonal modulation when it is considered over 2 bit intervals. The average probability of error or BER for DPSK is given by

$$\text{BER} = 1/2\exp\left(-\frac{E_b}{N_0}\right) \quad \text{or}, \ldots = 1/2\exp(-N_{avg}) \tag{6.11}$$

where E_b and N_0 are the signal and noise energy per bit, respectively, and N_{avg} is the SNR, equal to the average number of photons per bit; that is, $N_{avg} = \text{SNR}_{\text{DPSK}}$.

In the case of atmospheric scintillation, the unconditional BER is obtained by averaging Eq. (6.11) over the irradiance fluctuations statistics given by P(I). The BER result in presence of atmospheric turbulence is given by [22]

$$\text{BER}_{\text{Turbulence}} = \int_0^{\infty} 1/2e^{-\text{SNR}_{\text{DPSK}}} P(I)dI \ldots\ldots\ldots \tag{6.12}$$

where $P(I)$ can be assumed as gamma-gamma distribution discussed earlier in this chapter. The BER can be computed from Eq. (6.12) by a numerical solution where $P(I)$ can be assumed as gamma-gamma distribution discussed earlier in this section.

For the LCRD project, the data rates for PPM will be up to 311 Mb/s uplink and downlink, and DPSK up to 1.244 Gb/s (2.880 Gb/s uncoded) uplink and downlink. The ground segment for the operation of this mission will have two ground stations and the mission operations center to schedule, command, and control the communication payload and the two ground stations. Uplink beacon beams from the ground stations to the GEO terminal will help to point toward correct locations on Earth terminals. One of the ground stations in Southern California will be using the JPL Optical Communications Telescope Laboratory equipped with AO systems as described before for the LEO satellite and OGS communication links. The AO system will be needed to correct for the atmospheric turbulence-induced distortions of the optical signals close to the earth. The other ground station in Hawaii will be developed by MIT/LL. The ultimate goal of this LCRD will be to prove the development of the potential future optical service provider in space. The optical links in the future will even be able to support between any two user remote platforms to establish global Internet connectivity. The ATP operations will include establishing optical links between the payload and the fixed OGSs on the earth and between the payload and a moving user platform such as space-borne LEO or an airborne platform. The latter is much more complex. The acquisition sequence will be coordinated by the mission operation center. When using multiple OGSs on the earth, a seamless handover between ground stations will be necessary. This will be important for establishing future potential connectivity to global remote locations based entirely on optical wave technology.

6.3.1.2.5 Highlights of Laser Communications Relay Demonstration Efforts

The LCRD space flight is planned for 2019. The raw data rate is 2.5 Gb/s for the LCRD demo, which incorporates [21]: space modem (26″ L × 6.3″ H and 15.5″ W), 5 × 114 Gb/s transmitter, 442 elements: AWG mux, lasers, modulators, detectors, control elements, 5 × 114 Gb/s receiver 171 Elements: AWG demux, local laser oscillator, 90 deg hybrid, balanced detectors, integrated LCRD LEO-user modem and amplifier, provides pathway to near-earth, low-cost lasercom terminals and uses integrated electronics/photonics where cost-effective, reduced size, weight, power, and cost, receiver preamplifier photonic integrated circuit: Integrated net gain = 20 dB, noise figure of 3.75 dB small-signal-gain regime.

6.3.1.2.6 Low Earth Orbit–to–Geosynchronous Orbit–to–Earth Optical Communication Links

In order to establish optical communication links between LEO and GEO user terminals to be able to support the bidirectional high speed communication between LEO and GEO as well as between GEO and Earth, NASA is developing a next-generation user terminal. Initially putting the terminal on the ISS, the relay operation of LEO-to-GEO-to-Earth will be demonstrated in the future [18]. The optical module is a 10-cm reflective telescope mounted to a two-axis gimbal with the capability of generating ~15 μrad downlink beam and has a spatial acquisition quadrant detector with an FOV of ~2 mrad to detect a scanned uplink signal.

Fig. 6.7 shows NASA's LCRD mission, which will demonstrate optical communications relay services between GEO and Earth over an extended period in order to design and operate future optical communication systems and relay network. It will demonstrate optical-to-optical data relay and will provide an orbit platform to test new international standards for future

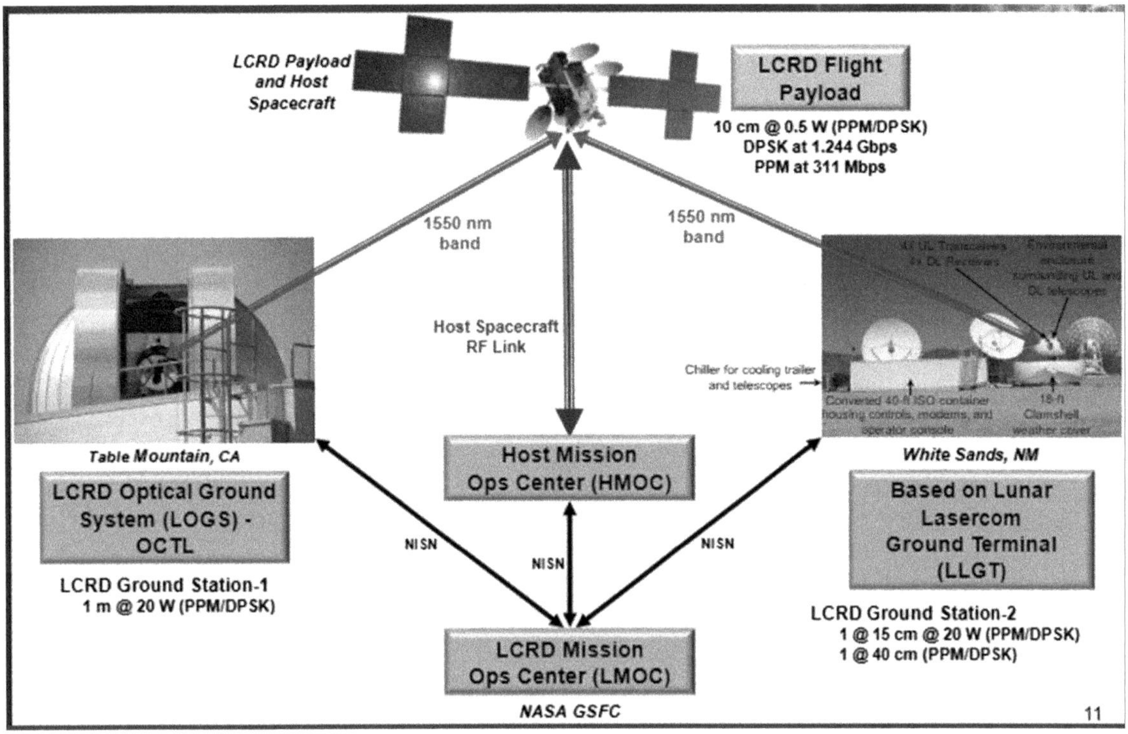

FIGURE 6.7

NASA's Laser Communications Relay Demonstration mission architecture. *Courtesy of, and permission from, Dr. Bernard Edwards, NASA Goddard Space Flight Center, 2018 to reuse.*

interoperability. The mission provides a pathway to the next-generation tracking and data relay satellites.

6.3.2 DLR's Demonstration Experiment

The first high data rate at gigabit laser-based communication called European Data Relay System (EDRS) was achieved by the European Space Agency (ESA) in November 2014. Then in November 2016 researchers at the German Aerospace Center (Deutsches Zentrum für Luft-und Raumfahrt; DLR) reported the world record in free-space laser/optical communications using a satellite at 1.72 terabits per second [23]. This terabit-level high data rate is equivalent to 45 DVDs per second transmission. This has clearly paved the way for developing future global broadband Internet connectivity in a remote area of the world. This section describes briefly the demonstration experiment.

The goal of this demonstration of transmission using laser is to enable Internet access at high data rates outside major cities and to explore the potential of using satellites. A broadband supply via geostationary satellites is possible, opening the pathway for next-generation communications satellites as part this effort where the satellites should be connected to the terrestrial Internet via laser links. The future data rate can be more than 1 terabit per second. The success in transmitting 1.72 terabits per second (Tbit/s) across a distance of 3 km in free space is very encouraging to take the next step. The goal of the DLR THRUST (Terabit-throughput optical system technology) project is the innovative development of next-generation communications satellites to transmit at this high data rate. The successful transmission at 1.72 Tb/s across a distance of 3 km in free-space was reported by the project leader, Dr. Juraj Poliak of the DLR Institute of Communications and Navigation. Another successful test between ground and mountain for a propagation path length of 10.45 km where atmospheric effects like turbulence and scattering media affects the laser communications performance was performed. The atmospheric test results are very useful for the DLR scientists to develop a new free-space optic transmission system that can be a stable optical link for long-term laser communication with the satellite. Fig. 6.8 shows the THRUST transmission terminal and Fig. 6.9 shows the reception terminal developed by DLR.

In order to evaluate the future optical communication system performance with GEO-OGS platforms a laboratory experiment was reported to predict the potential capacity over 1 Tb/s data rate under realistic atmospheric turbulence. Results were reported for expected BER values and statistics for a direct detection of OOK signals at a 43 Gb/s nonreturn-to-zero (NRZ). The optical intensity fluctuations due to turbulence was simulated for both weak turbulence (scintillation index of 0.27) and strong turbulence (scintillation index of 1.80).

FIGURE 6.8

The THRUST transmission terminal developed by the DLR Institute of Communications and Navigation, which was deployed during the long-distance experiment at DLR's site in Weilheim. *Credit: DLR/Bernd Müller © Bernd Müller for DLR.*

FIGURE 6.9
The THRUST reception terminal enables light to be coupled into the 10 μm optical fibers after travelling 10.45 km. *Credit: DLR/Bernd Müller © Bernd Müller for DLR.*

For weak turbulence BERs $<10^{-3}$ were observed and for strong turbulence BER was increased to >30% with a large number of values $<10^{-2}$. The results clearly demonstrated a pathway to achieve over 1 Tb/s with optical communications between GEO-OSG platforms in the future. System design will involve characteristics of advanced dense wavelength division multiplexors (DWDMs), preamplifiers, detector sensitivity, and finally use of efficient turbulence mitigation techniques for a complete communications system.

6.3.2.1 Other Relevant Satellite-Based Optical Communication Research and Development in Germany

In this context, the researchers also published in the 13th Conference on Telecommunications, in 2015 the development of an optical feeder link experimental testbed and presented results [24] to demonstrate more than 1 Tb/s FSO transmission applicable and relevant to a scenario of GEO uplink. The experimental testbed results simulated the worst-case scenario similar to a GEO uplink transmission as encountered due to the severe atmospheric turbulence effects close to the ground. The testbed uses a DWDM-based FSO link over 26 km distance with less than 2 degrees elevation angle. Several DWDM transmitters were multiplexed for transmitting over a single FSO channel to determine the transmission characteristics. Recently the researchers published their work on optical feeder links for very high throughput satellites for the GEO feeder link in the future based on optical communication [25] at 1550 nm wavelength technology. The technology described is called optical feeder link (OGEOFL); using more than 1 Tb/s might be possible via only to one OGS. The concept as a single uplink infrastructure could potentially feed the entire spacecraft. The detail technology for OGEOFLs with DWDM channels for uplink and downlink is already published [25]. The DWDM-based communications (for both uplink and downlink) consists of several data streams separated at individual DWDM channels (around the central transmission laser wavelength of 1550 nm, for example) with central wavelengths, WDM multiplexer, erbium-doped fiber amplifier (EDFA), preamplifier EDFA with automatic gain control, and receiver front ends. A properly designed AO can also improve to correct for the atmospheric turbulence-induced distortions. The number of gateways in the ground optical network can thus be minimized by this DWDM-communication system, which has the potential of providing more than 1 Tb/s for the optical feeder-link in satellite communications.

The design of a near-ground long-range (10 km) atmospheric transmission testbed was also published [26] around the same time in 2015. The testbed provided results for a low elevation 1.8-degree scenario similar to a worst-case GEO uplink case. Like other testbed tests, several DWDM channels were multiplexed to accomplish the total link capacity of 1 Tb/s. The OGEFL technology allowed sending more than 1 Tb/s via only one OGS toward the GEO. This makes the future network architecture of potential OGS-GEO connectivity much simpler with FSO communications technology.

Another project in Germany involves providing data relay services to LEO satellites from GEO orbit by means of an optical band using an EDRS-C satellite operating on a geostationary orbit incorporating a laser communication terminal (LCT) [24]. LCT was appropriately designed to accommodate the small GEO platform to establish the lasercom link. The main payload of the EDRS-C satellite is the data relay payload, which is capable of establishing and maintaining an optical communication link between the LCT located on the EDRS-C and a counterpart LCT on a suitable target LEO satellite to guarantee the relay of the data to the ground in both the advanced mode for which the incoming data rate is 1800 Mb/s and the sentinel mode with incoming data rate of 600 Mb/s. The details of the LCT, platform, and integration are reported in [24].

6.3.3 Satellite Lasercom Demonstration Experiments in Japan

This section describes recent demonstrations of satellite-based optical communications in Japan reported by researchers at the National Institute of Information and Communications Technology (NICT), which has a long history of FSO communications demonstration from the space to the ground. Many of the NICT optical communications missions demonstrated the FSO technology for establishing optical links between microsatellites to OGSs. The recent lasercom mission, Small Optical Transponder (SOTA), started in 2014 and ended in November 2016. The demonstration and the details are already published elsewhere [27]. Some of the essential results and the goals accomplished will be summarized here. Although microsatellite platforms for achieving Gb/s data rate has not yet been possible, it is important to demonstrate the maximum data rate possible with microsatellite platforms that can carry less weight and volume for transceiver hardware with limited space and size. However, the cost to build is much less and will be much cheaper to launch in the orbit. The SOTA lasercom terminal was placed onboard the Space Optical Communications Research Advanced Technology Satellite (SOCRATES), which is inserted in a sun-synchronous near-circular orbit of ~600 km. SOTA mass is only 5.9 kg, and maximum power consumption during operation is below 40 W. The receiving system supports the uplink optical link and also provides ATP toward the OGS beacon. The laser transmitter has four lasers: one at wavelength 976 nm (communication laser), two at 800 nm, and the other one at 1549 nm (communication laser). The communication lasers can transmit at 1 or 10 Mb/s with NRZ-OOK modulation. The communication laser at 1549 nm has divergence angle of 223 μrad for aperture size of ~5 cm, with circular polarization and pointing loss of −5.7 dB whereas the 976 nm communication laser has divergence angle of 500 μrad for aperture size of ~1 cm, with linear polarization and pointing loss of −3.4 dB [27]. Other international OGSs participating in the interoperability tests include ESA, National Center for Space Studies, DLR, and Canadian Space Agency. The SOTA mission also explored the space diversity technique of mitigating atmospheric turbulence to increase the probability of establishing a communication link. For that purpose, the network called INNOVA (In-orbit and Networked Optical Ground Stations Experimental Verification Advanced Testbed) consisting of three different uncorrelated OGS locations was being developed. This network can be operated remotely from control center according to the link availability using the meteorological data predictions such as clear sky or rainy season statistics in each location. Because of the low cost and light weight compact hardware requirement there will be more launching of small satellites to develop satellite constellations to achieve global optical communication networks covering the entire earth.

Fig. 6.10 shows a figure of NICT's small satellite (SOCRATES) and SOTA for satellite optical communication projects.

Fig. 6.11 shows OGS in Japan for NICT's LEO-to-ground project for SOTA.

Fig. 6.12 shows the very-small optical transponder (VSOTA) with onboard components housed and installed in the NICT's RISESAT (Rapid International Scientific Experiment Satellite) for the VSOTA mission.

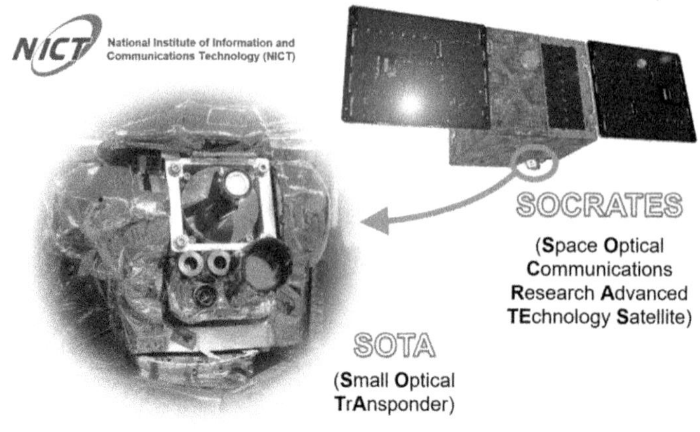

FIGURE 6.10

NICT's satellite optical communications project uses SOTA housed and installed in SOCRATES. *Courtesy of, and permission from, Dr. Morio Toyoshima and Dr. Alberto Carrasco-Casado, NICT, 2018 to reuse.*

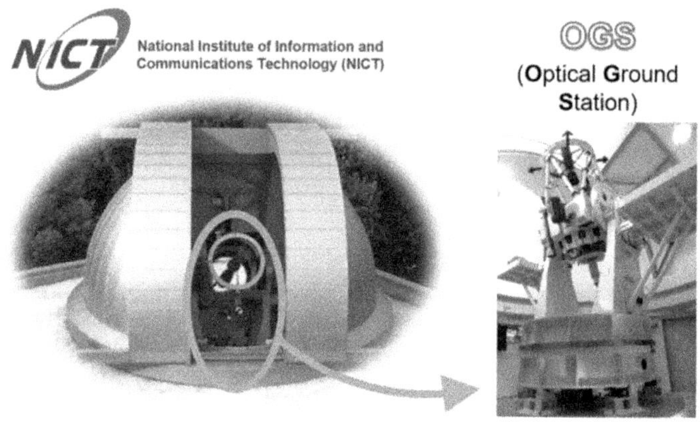

FIGURE 6.11

Optical ground station in Japan for NICT's LEO-to-ground optical communication project. *Courtesy of, and permission from, Dr. Morio Toyoshima and Dr. Alberto Carrasco-Casado, NICT, 2018 to reuse.*

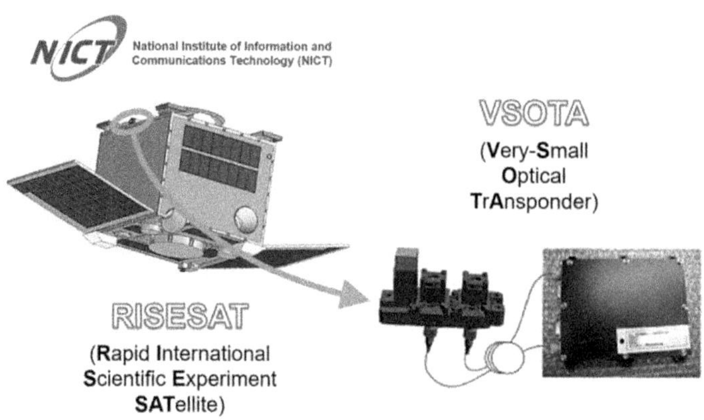

FIGURE 6.12

NICT's VSOTA housed and installed in RISESAT for a satellite communication project in Japan. *Courtesy of, and permission from, Dr. Morio Toyoshima and Dr. Alberto Carrasco-Casado, NICT, 2018 to reuse.*

6.4 FUTURE OPTICAL COMMUNICATIONS DEMONSTRATIONS

Space-based breakthroughs are happening in 2018 with many new missions and experimental demonstrations underway worldwide. These include ESA's EDRS, NASA's LCRD, and Japan's Optical Data Relay Satellite. In addition, a number of companies such as Facebook, SpaceX, Laser Light, OneWeb, and LeoSat are continuously exploring FSO communications for commercial programs and applications for global interconnectivity. One of the goals is to obtain higher data rates and smaller mobile terminals and to develop components suited to a range of architectural requirements. There are constant improvements in flight optics and ATP systems. More and more optical components are becoming available and with many architectural possibilities, the FSO communications industry is becoming a major player in many classes of global and extraglobal telecommunication. New architectures of space and air global connectivity will be possible along with the development of high performance and small terminal size.

6.5 DEMONSTRATION OF FREE-SPACE LASER COMMUNICATION FROM AIRBORNE PLATFORM

6.5.1 Lasercom Airborne Flight Test System Demonstration

Airborne platforms have been used as a pathway for developing space applications of free-space laser communications for many years and are an important element of the evolving space network architecture. Lasercom Airborne Flight Test System (AFTS) was directed toward the development of the design concepts applicable to a high data rate spaceborne laser communication system [20], which has shown a clear pathway to develop concept technology for broadband global Internet connectivity. The successful demonstration of high data rate communications from an airborne terminal established the feasibility of practical operational airborne terminal as a node for either a space-based global communication network or space-to-space or air-to-air data rely. AFTS demonstrated important capabilities including acquisition and tracking from a moving platform and extended free-space communications over an atmospheric channel such as a slant path link. Uplink wavelength was 1.064 μm for 20 Kbit/s tracking beacon, downlink communication link wavelength was 0.53 μm, and the slant path range was 11–100 km [28]. An airborne high data rate transmit terminal was installed in KC-135 aircraft and the ground station receiver was located in New Mexico. The aircraft flew a circular trajectory around the ground station at slant ranges from 11 to 100 km. The system achieved the 1 Gbit/s data rate utilizing 500 Mpps mode-locked, frequency-doubled laser, which generated a stream of 300 ps optical pulses at a fixed rate. The results clearly represented a significant accomplishment in the area of high data rate lasercom from airborne platform and was an important milestone in the demonstration of the maturity of critical technologies for developing both air and space communication applications including global communication networks.

6.5.2 High Data Rate Demonstration Between a Small Aircraft and a Ground Station Free-Space Optical Link

Recently a free-space high data rate optical communication experiment between a Twin Otter aircraft and a ground station [29] was conducted. The airplane was typically flown at an altitude of 12 kft in a semicircular pattern at ranges between 25 and 60 km around the ground station. The capabilities of the system were demonstrated out to 80 km with aircraft at 16 kft. Four ground receivers individually tracked 12 mm aperture and the aircraft used 25 mm aperture with an output beam power of approximately 500 mW for communication downlink laser. Results demonstrated strong evidence of channel reciprocity for single-mode optical transceivers where a priori knowledge of the channel state can be used to design systems that achieve channel capacity with minimal latency. The paper illustrates a schematic of a unidirectional transmitter that implements data frame gating based on measurements of the received optical power and incorporates forward error correction (FEC) scheme and is capable of gigabit-class data rate. Fig. 6.13 shows the pictures of the aircraft and downlink ground station.

6.5.3 Demonstration of High Data Rate Free-Space Laser Communications From a Fast Airborne Platform

Airborne and ground optical communication terminals with smaller size, weight, and power are desirable to establish flexible optical links at high data rate. Recently much attention is on general design aspects of optical terminals for aircraft and UAVs for establishing nodes for Internet connectivity. Experimental demonstration of a high data rate airborne optical downlink from a fast flying platform has been discussed [30]. The aircraft speed was up to 0.7 Mach and included downlink of a huge amount of data during a short ground station flyby. The system described used a single wavelength from the DWDM grid and comprises three main units: the airborne laser terminal, the receiver ground station, and the sensor payload. Images or videos are taken by a payload camera and sent to a computer system that can be seen as a simple streaming device generating a 1.25 Gbit/s data stream at the transmission wavelength in C-band. Direct detection modulation scheme was used. The details of the aircraft terminal and receiver ground station are in [30]. The data communication could achieve up to 50 km.

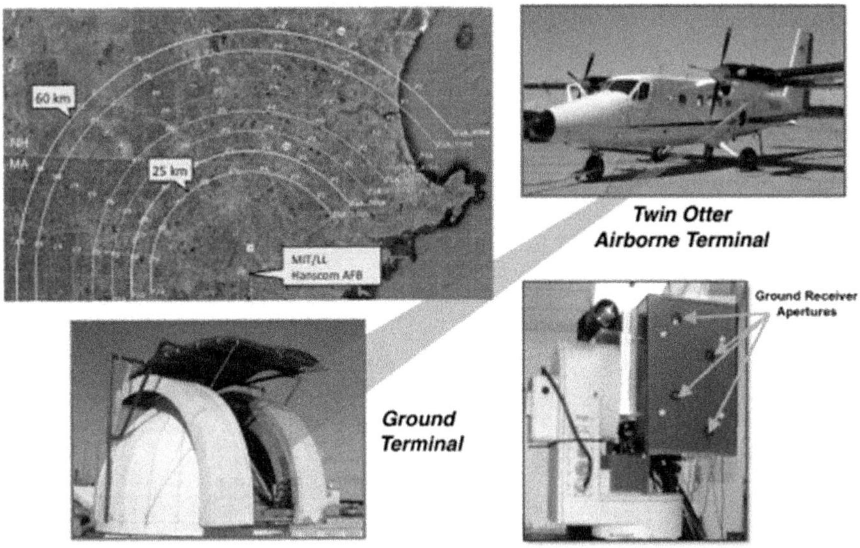

FIGURE 6.13

Experiment shows the 2.7 Gbit/s optical link between a transmitter mounted on a Twin Otter aircraft and a ground-based receiver. The aircraft was flown in a semicircular pattern centered on the location of the ground terminal. Ranges were between 25 and 80 km. *Reprinted with permission from The Optical Society of America, OSA, 2018 [29].*

6.6 DEMONSTRATIONS OF FREE-SPACE LASER COMMUNICATIONS WITH UNMANNED AIR VEHICLES

UAVs have recently emerged as a promising solution for free-space laser communication links to and from ground stations as well as to establish optical links in different scenarios such as between UAV-to-satellite, UAV-to-HAP, UAV-to-UAV. Fig. 6.14 shows an artist concept of the different scenarios how low and high altitude platforms can be connected by optical links to establish an optical network.

6.6.1 Demonstration by Jet Propulsion Laboratory Researchers for Unmanned Air Vehicle-to-Ground

One of the early research and development works involves a 2.5 Gbit/s UAV-to-ground lasercom project at JPL as described in their paper [31]. Design and development of a robust ATP was reported. The demonstration shows the successful bidirectional daytime and nighttime optical communications link from an Altair UAV to stationary ground stations located at the OCT at Wrightwood, California and Maui, Hawaii. The Altair UAV provides GPS data about the

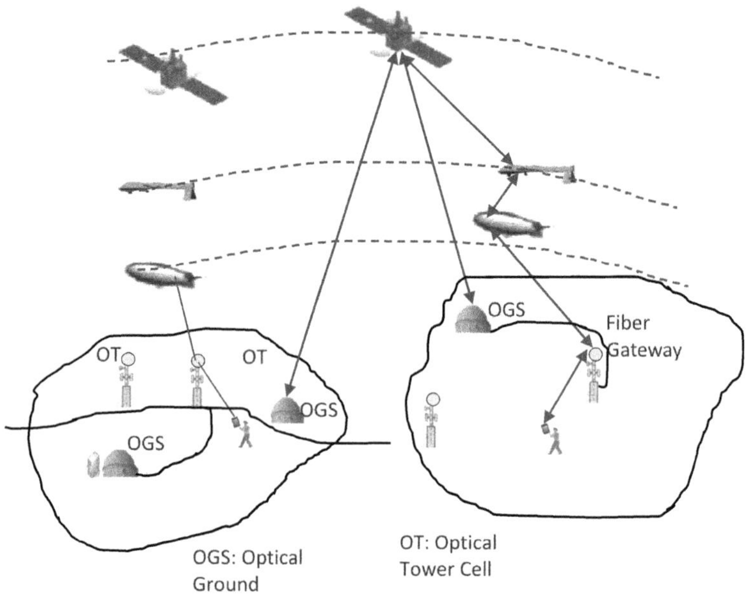

FIGURE 6.14

Artist concept of optical network between unmanned air vehicles and high-altitude platforms connecting to a satellite network with relay functionality.

position knowledge to OCT. The terminal uses this information with a priori knowledge of the ground location to point the gimbal and initially the ground illuminates the aircraft with a beacon signal. The altitude of the UAV is in the range of 15.8–18.3 km and the link range varied between 15 and 100 km depending on the zenith angle. During the experimental demonstration, the UAV used a 200 mW downlink laser at 1550 nm for a BER of 1E-9, the pointing requirements on the flight terminal were a jitter error of 19.5 μrad and a bias error of 14. 5 μrad with a probability of pointing induced fade of 0.1%. The robust ATP subsystem was important to align and communicate with UAV.

6.6.2 Early Demonstration of Unmanned Air Vehicle-to-Ground Using Multiple Quantum Well Retromodulators at Naval Research Laboratory

Other research and development in the area of UAV-to-ground laser communication is presented [32] where the researchers successfully demonstrated real-time streaming video from a small UAV in flight at Chesapeake Bay Detachment in Chesapeake Beach, MD using retromodulators. An array of five Multiple Quantum Well (MQW) modulator operating at 980 nm and a miniature video camera were mounted on a small model helicopter and the signal was compressed. The impedance matching circuitry drove the modulator array at 3 Mbps. The helicopter range varied from 30 to 120 m and the video was returned at 30 frames per second. This was one of the first demonstration experiments showing the utility of retromodulator in free-space laser communication link to simplify the pointing and tracking problems. The concepts will be useful in establishing future UAV-to-ground optical links at higher data rate in the gigabits ranges.

6.6.3 Unmanned Air Vehicle Terminal and a Ground Station Design and Development Program in Spain

This demonstration [33] involved the design and development of a MEMS-based modulating retroreflector as a communication terminal onboard a UAV, which eliminates both the laser transmitter and the ATP subsystems. The retroreflector was made up of three mirrors, one being the MEMS modulator, and the system used OOK modulating a Continuous Wave (CW) laser beam by switching between an energized-diffractive state and an unpowered flat-mirror state. This resulted in reduction of power, size, and weight on the UAV, thus moving the burden to the ground station. In order to determine the position of the UAV in real-time, a beacon-based system was designed with the retromodulator device. Based on a distance-dependent beam-divergence controlling the system and by creating two different optical paths using different states of polarization, the system managed the laser power in an optimal way. The beacon beam used 1550 nm laser. The use of a MEMS-based retromodulator will find a number of useful applications in free-space laser communications with UAV platforms.

6.6.4 Experimental Demonstration of Holographic Tracking System for Unmanned Air Vehicle Free-Space Laser Communication

Recently a design and implementation of a full tracking system for medium-long range UAV communication is described [34]. The demonstration was to track a UAV with built-in MQW based modulated retroreflector (MRR). The base system directs a CW laser interrogator toward a UAV (Cinestar 8's Mikrokopter model) equipped with an MRR. The laser wavelength is 1550 nm. Their results show a real-environment offloading from the UAV to the base station at 2 Mbit/s at a range of 300 m. The modulation scheme used was OOK and the BER achieved is 1.9701×10^{-4}. The demonstration shows a feasibility of achieving data rates of hundreds of Mbit/s at a range of 1–2 km with some improvements in the system. It also showed multitarget tracking and aberration corrections for atmospheric turbulence in future. This experiment is another important one to add to the potential Internet connectivity by optical technology.

6.6.5 Demonstration of Unmanned Air Vehicle Laser Communication Link in the C3PO Project

The system for the C3PO project [34] discusses two experiments, one ground-based and equipped with components for both uplink and downlink communications, and the tracking station for satellite acquisition and pointing and ground-based laser for downlink communication or a laser with modulated mode for uplink communication. The space-based terminal will equip an MRR, which reflects the beam from the ground station and sends back the modulated beam to the ground station where a ground-based optical receiver decodes the data. Laser tracking on a retroreflector-equipped LEO satellite is performed in the first experiment to determine the levels of signal and noise, backscattering effects, tracking accuracy, and tracking acquisition time. In the second experiment, a UAV equipped with MRR prototypes and laser communication is established with a ground station. The UAV experiment evaluates MRR requirements (mass, power consumption), field test operability in atmospheric conditions, and communication concepts (achievability of a high data rate given a certain SNR). A high-power CW laser beam is aimed from the ground station toward a UAV with MRR where UAV position is tracked using both coarse and fine tracking modules working cooperatively. The fine pointing tracking module can accomplish a high pointing accuracy within a limited narrow FOV of ± 0.3 degree. The base station has the optical module in which the system is based on a holographic beam steering module. The UAV targets the implementation of a 1 Gbit/s data link between a UAV and a base station using an MRR for a link range of at least 1 km.

6.6.6 Demonstration of Orbital-Angular-Momentum-Multiplexed Free-Space Optical Communications Link Between a Ground Base and a Moving Unmanned Air Vehicle [35]

We explore the use of orbital-angular-momentum (OAM)-multiplexing to increase the capacity of free.

Recently researchers have demonstrated [35] the performance of orbital-angular-momentum (OAM) beams between a ground station OAM transmitter and an OAM receiver via a moving UAV equipped with a retroreflector hovering in the air up to $\sim$50 m away (i.e., $\sim$100 m round trip). Due to the increased needs of higher-capacity data communication needs between unmanned aerial platforms and their ground stations, advanced research and development efforts are directing toward this effort. At the same time, in addition to high-capacity high data rates, it is also essential that FSO communication provides enhanced privacy and security by minimizing the probability of possible interception. Use of OAM for increasing high-capacity and security has been recently investigated between fixed ground stations but not much in the area of ground station and moving platforms. OAM modes are characterized by a phase front having an angular dependence of the form of $\exp(il\phi)$, where ϕ is the azimuth angle and l is an integer that can be positive, negative, or zero for a clockwise phase helicity, counterclockwise phase helicity, or no helicity, respectively. These helical beams twist as they propagate through the atmosphere. Conventional Gaussian beam is equivalent to OAM with no helicity. A Laguerre-Gaussian beam with nonzero azimuthal order and zero radial order has the intensity of an OAM beam that is circularly symmetric, having a ring shape with little power in the center. Some of the challenges involved in using OAM multiplexing in FSO communications between a ground station and a moving platform are alignment and atmospheric and UAV's propeller-induced turbulence. The demonstration discussed in [35] transmits two multiplexed OAM beams up to ~100m round-trip distance via the UAV, which carries a retroreflector reflecting back the initial transmitting beam to the receiver. The researchers used a 1550 nm laser transmitting beam with a total capacity of 80 Gbit/s capacity round-trip OAM beam carrying a Quadrature Phase Shift Keying (QPSK) signal. Signal distortions due to the turbulence effects and misalignment and the achieved BER were reported in terms of received power. They reported the power fluctuations of 2.1 dB on the desired mode when the UAV hovers in the air, while the cross-talk to the other mode was −19 dB below the power on the desired mode. The power fluctuation on the desired mode is 4.3 dB when the UAV moves in the air while the cross-talk to the other mode in that case was −10 dB, decreasing the channel cross-talk with an increase of OAM mode spacing. The FSO communication link between a ground station with the tracking system uses an OAM transmitter and a receiver.

Fig. 6.15 shows an experimental setup of an FSO communication link between a UAV and a ground station using OAM multiplexing.

Fig. 6.16 shows the photos of the experimental setup.

For their demonstration with the UAV and ground station the experimental setup is shown in Fig. 6.15, including the scenario of ground station (tracking system, OAM transmitter and receiver) and UAV. Fig. 6.16 shows a photograph of the experimental setup with ground station, UAV carrying the gimbal-mounted retroreflector, communication link between ground transmitter and the ground receiver via the UAV and the fast steering mirrors, and position detector of the fine tracking system.

Their results reported [35] the value of atmospheric turbulence parameters of Rytov variance σ^2 of ~0.0028 and turbulence strength Cn^2 of $\sim 9.4 \times 10^{-15}\,m^{-2/3}$ when the propeller was off, and σ^2 was ~0.0080 when the UAV propeller was on. The BER in flight environment was in the range of 1e-1 to 1e-4 for all the measurements. The results reported clearly demonstrate the potential of using OAM format for moving platforms like UAVs to serve

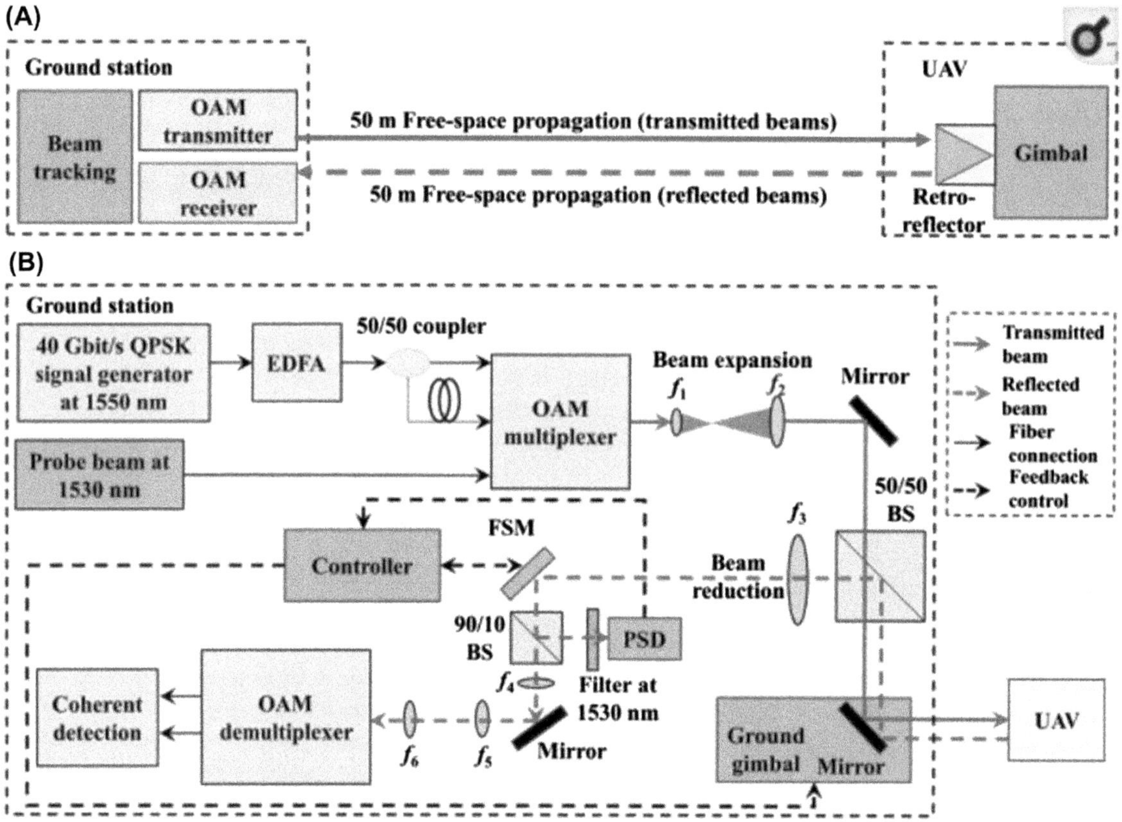

FIGURE 6.15

Experimental setup of an FSO communication link between a UAV and a ground station using OAM multiplexing. (A) Schematic diagram of the ground station and the UAV. (B) Ground station setup with OAM transmitter, the OAM receiver, and beam tracking system. *Reprinted with Permission from SCIENTIFIC REPORTS via Copyright Clearance Center, RightsLink, SPRINGER NATURE, 2018 [35].*

as optical nodes for establishing Internet connectivity with high-capacity data rates and secure communication by providing enhanced privacy and security by minimizing the probability of possible interception.

6.7 DEMONSTRATIONS OF HIGH-ALTITUDE PLATFORM FREE-SPACE LASER COMMUNICATIONS

HAPs are aircraft or airships situated well above the clouds typically at 17–25 km where the atmospheric effects on laser propagation is much less severe than that at or near the ground. Optical links between these HAPs, satellites, and ground stations therefore have the potential to serve as broadband

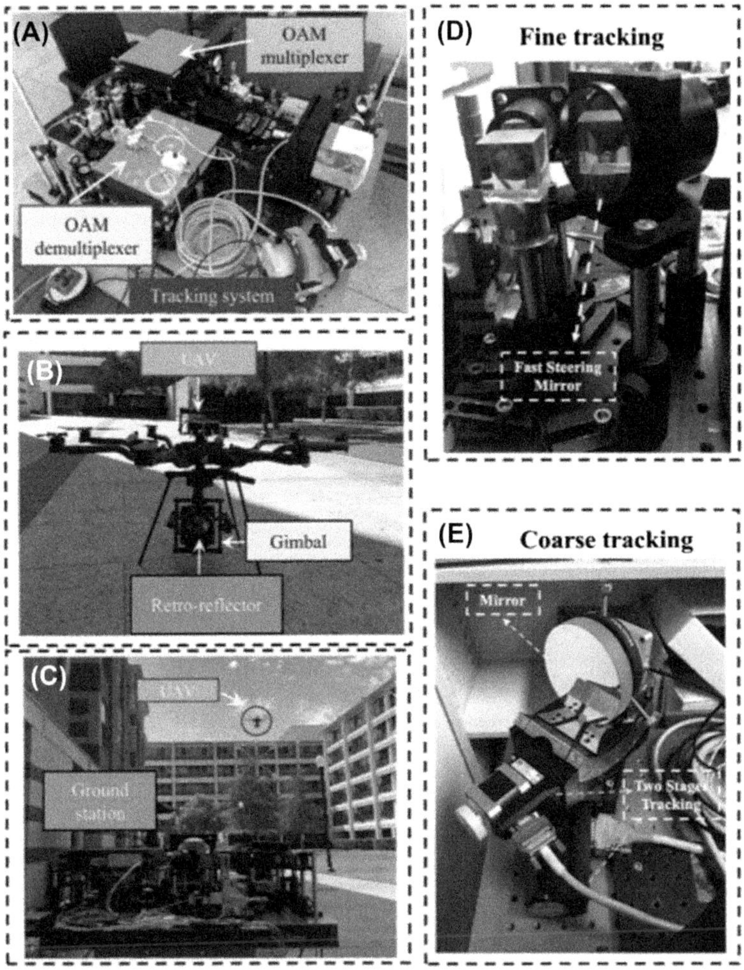

FIGURE 6.16

Photos of the experimental setup. (A) Ground station; (B) UAV, which carries the gimbal-mounted retro-reflector; (C) Communication link between ground transmitter and the ground receiver via UAV; (D) Fast steering mirror and position detector of the tracking system; course tracking system components. *Reprinted with permission from SCIENTIFIC REPORTS via Copyright Clearance Center, RightsLink, SPRINGER NATURE, 2018 [35]*

backhaul channels. Technologies, theoretical studies, and experimental field trials and recent developments in transmitting multi-Gigabits per second data have been discussed [36]. HAPs are quasi-stationary vehicles such as helium-filled airships operating in the stratosphere but below orbiting satellites and are capable of supporting payloads up to 2000 kg. Data rates up to 1.25 Gbit/s have been achieved for optical HAP-to-ground communication trials [37]. A number of theoretical and analyses for the feasibility of HAPs have been made and are described in the following.

6.7.1 Satellite High-Altitude Platform Communications

Some of the published papers in the area of achievable throughput of LEO satellites to the ground discussed several networking scenarios [38], including HAPs as relay stations with data rates of 5.6 Gbit/s, which were assumed for a laser communication terminal on TerraSAR-X [39]. Some of the studies [40] involved the investigation of the feasibility of optical communication links at 1550 nm wavelength through the atmosphere between HAPs and GEO satellites for data rates up to 10.7 Gbit/s with FEC code. The outage probability of better than 10^{-12} could be achieved at BER of 10^{-9}. Other scenarios include HAP-GEO-HAP and HAP-LEO links [40–42].

6.7.2 High-Altitude Platform–to–High-Altitude Platform Communication Links

HAP-to-HAP was analyzed in a theoretical study [41] for 20 km altitude and a distance of 500 km with a communication laser wavelength of 1550 nm, and showed a BER of 10^{-6} at a data rate of 384 Mbit/s and a transmit power of 800 mW. The field trials of the terminals on the ground at wavelength of 1550 nm showed that a bidirectional communication link for a distance up to 2 km achieved a data rate of 2.5 Gbit/s. These results are promising to eventually design and develop HAP-to-HAP optical links to be useful for creating Internet backhaul nodes. The early demonstration performed by JPL for UAV-to-ground link at 18 km altitude for 2.5 Gbit/s at a wavelength 1550 nm for a link distance of 50 km was mentioned earlier.

6.7.3 Ground-to-High-Altitude Platform Optical Communication Links

A unidirectional optical downlink from a stratospheric balloon to a ground station for an EU project Capanina [42] was performed by DLR. The balloon was equipped with a communication transmitter for 1.25 Gbit/s data rate operating at a wavelength of 1550 nm. The experimental and theoretical studies predict the feasibility of HAP-to-HAP and HAP-satellite links to achieve a data rate of 10 Gbit/s with current technology. All these tests and feasibility studies show potential using HAPs to develop global optical broadband Internet connectivity *anytime, anywhere* including remote locations.

References

[1] B. Iñigo del Portillo, Optimal Locations for the Ground Segment of Optical Space Communications Networks (M.S. thesis), Department of Aeronautics and Astronautics, MIT, June 2016.

[2] Inigo del Portillo, M. Sanchez, B. Cameron, E. Crawley, Architecting the ground segment of an optical space communication network, 5–12 March 2016, in: Published in Aerospace Conference, IEEE, Big Sky MT, USA, 2016, https://doi.org/10.1109/AERO.2017.7943631.

[3] Inigo del Portillo, M. Sanchez, B. Cameron, E. Crawley, Optimal location of optical ground stations to serve LEO spacecraft, 4–11 March 2017, in: Published in Aerospace Conference, IEEE, Big Sky MT, USA, 2017, https://doi.org/10.1109/AERO.2016.7500803.

[4] P.K. Chowdhury, M. Atiquzzaman, Handover schemes in a satellite networks, in: IEEE Communications Surveys & Tutorials, 4th Quarter, 2006.

[5] E. Papapetrou, et al., Satellite handover techniques for LEO networks, International Journal of Satellite Communications and Networking 22 (2) (Mar./Apr. 2004) 231–245.

[6] F.L. Gaol, et al. (Eds.), Proc. of the 2011 2nd International Congress CACS, AISC 145, Springer-Verlag, Berlin Heidelberg, 2012, pp. 119–124. springerlink.com.

[7] B. Yang, Y. Wu, X. Chu, et al., Seamless handover in software-defined satellite networking, IEEE Communications Letters 20 (9) (2016) 1768–1771.

[8] C.-W. Lee, M.-C. Chuang, M.C. Chen, Y.S. Sun, Seamless handover for high-speed trains using femtocell-based multiple egress network interfaces, IEEE Transactions on Wireless Communications 13 (12) (Oct. 2014) 6619–6628.

[9] M.W. Wright, J. Kovalik, J. Morris, M. Abrahamson, A. Biswas, LEO-to-ground optical communications link using adaptive optics correction on the OPALS downlink, in: H. Hemmati, D.M. Boroson (Eds.), Free-Space Laser Communication and Atmospheric Propagation XXVIII, 2016. Proc. of SPIE Vol. 9739, 973904–1 to 973904-10.

[10] M.W. Wright, J.F. Morris, J.M. Kovalik, K.S. Andrews, M.J. Abrahamson, A. Biswas, Adaptive optics correction into single mode fiber for a low Earth orbiting space to ground optical communication link using the OPALS downlink, Optics Express 23 (26) (22 Dec 2015) 33705–33712.

[11] A. Biswas, J.M. Kovalik, M.W. Wright, W.T. Roberts, Optical communications telescope laboratory (OCTL) support of Space3 to ground link demonstration, in: SpaceOps Conference, American Institute of Aeronautics and Astronautics, 2014, https://doi.org/10.2514/6.2014-1710.

[12] D. Malacara, M. Servín, Z. Malacara, Interferogram Analysis for Optical Testing, CRC Press, 2010.

[13] J.D. Barchers, D.L. Fried, D.J. Fink, G.A. Tyler, W. Moretti, T.J. Brennan, R.Q. Fugate, The performance of wave-front sensors in strong scintillation, in: Proc. SPIE, vol. 4839, 2003, p. 217.

[14] http://www.laserfocusworld.com/articles/print/volume-50/issue-11/features/adaptive-optics-laser-beam-control-takes-advantage-of-advanced-wavefront-sensing.html as reported for the paper, 11/04/2014 "A self-referencing point-diffraction interferometer and a new technique for unwrapping phase help adaptive optical systems for laser beam projection over long atmospheric paths compensate well for strong turbulence" by M.J. Steinbock, J.D. Schmidt, and M.W. Hyde IV.

[15] D.L. Fried, J.L. Vaughn, Branch cuts in the phase function, Applied Optics 31 (1992) 2865–2882.

[16] D.L. Fried, Branch point problem in adaptive optics, Journal of the Optical Society of America. A 15 (1998) 2759–2768.

[17] B.L. Edwards, D. Israel, K. Wilson, J. Moores, A. Fletcher, Overview of the laser communications relay demonstration project, in: Proceedings of SpaceOps 2012, the 12th International Conference on Space Operations, Stockholm, Sweden, June 11-15, 2012, in: http://www.spaceops2012.org/proceedings/documents/id1261897-Paper-001.pdf.

[18] Laser Communications Relay Demonstration (LCRD), "Overview for Future In-Space Operations," Bernard Edwards, NASA, October 30, 2013, NASA Goddard Space Flight Center, October 2013, http://www.nasa.gov/mission_pages/tdm/lcrd/lcrd_overview.html#.VSySAZPuz_W.

[19] B. Edwards, D. Israel, A. Caroglanian, J. Spero, T. Roberts, J. Moores, A day in the life of the laser communications relay demonstration project, in: Proceedings of the 14th International Conference on Space Operations (SpaceOps 2016), Daejeon, Korea, May 16-20, 2016 paper: AIAA 2016-2590, http://arc.aiaa.org/doi/pdf/10.2514/6.2016-2590.

[20] E. Luzhansky, B. Edwards, D. Israel, D. Cornwell, J. Staren, N. Cummings, T. Roberts, R. Patschke, Overview and status of the laser communication relay demonstration, in: SPIE Photonics West 2016 Conference, 14−18 February 2016. San Francisco, CA; USA, http://nt1rs.nasa.gov/archive/nasa/casi.ntrs.nasa.gov/20160001877.pdf.

[21] https://ntrs.nasa.gov/archive/nasa/casi.ntrs.nasa.gov/20160014546.pdf.

[22] Y.I. Xiang, L.I.R. Zengji, Y.U.E. Peng, S. Tao, BER Performance Analysis for M-ary PPM over Gamma-Gamma atmospheric turbulence channels, in: Wireless Communications Networking and Mobile Computing (WiCOM), IEEE Conference, 23-25 Sept 2010. ISBN: 978-1-4244-3709.

[23] http://www.dlr.de/dlr/en/desktopdefault.aspx/tabid-10081/151_read-19914/#/gallery/24870. http://www.dlr.de/dlr/en/desktopdefault.aspx/tabid-10081/151_read-19914/.

[24] J. Poliak, D. Giggenbach, F. Moll, Terabit-throughput GEO satellite optical feeder link testbed, in: Telecommunications (ConTEL), 2015 13th International Conference on Date of Conference, 13−15 July 2015.

[25] R. Mata Calvo, D. Giggenbach, A. Le Pera, J. Poliak, R. Barrios, S. Dimitrov, Optical feeder links for very high throughput satellites − system perspectives, in: Conference: Proceedings of the Ka and Broadband Communications, Navigation and Earth Observation Conference 2015. Ka Conference 2015, 12−14 October 2015. Bologna, Italy.

[26] D. Giggenbacha, J. Poliaka, R. Mata-Calvoa, C. Fuchsa, N. Perlotb, R. Freundb, T. Richterb, Preliminary Results of Terabit-per-second Long-Range Free-Space Optical Transmission Experiment THRUST, in: Proc. of SPIE, 2015, 9647−21.

[27] A. Carrasco-Casado, H. Takenaka, D. Kolev, Y. Munemasa, H. Kunimori, K. Suzuki, T. Fuse, T. Kubo-oka, M. Akioka, Y. Koyama, M. Toyoshima, LEO-to-Ground Optical Communications Using SOTA (Small Optical TrAnsponder) −Payload Verification Results and Experiments on Space Quantum Communications, 2017. https://authors.elsevier.com/a/1VSJELWHFgB1g.

[28] J.A. Maynard, D. Begley, Airbrne LaserCommunication: past, present and future, in: Proc. Of SPIE, vol. 5892, 2005.

[29] R.R. Parenti, J.M. Roth, J.H. Shapiro, F.G. Walther, J.A. Greco, Experimental observations of channel reciprocity in single-mode free-space optical links, Optics Express 20 (19) (September 10, 2012).

[30] F. Moll, J. Horwath, A. Shrestha, M. Brechtelsbauer, C. Fuchs, L.A. Martin Navajas, A.M. Lozano, D.D. Gonzalez, Demonstration of high-rate laser communications from a fast airborne platform, IEEE Journal on Selected Areas in Communications 33 (9) (September 2015).

[31] G.G. Oritz, S. Lee, S.P. Monacos, M.W. Wright, A. Biswas, Design and development of robust ATP subsystem for the Altair UAV-to-ground lasercomm 2.5-Gbps demonstration, in: Proc. SPIE Vol. 4975, Free-Space Laser Communication Technologies XV, 2003, https://doi.org/10.1117/12.478939.

[32] G.C. Gilbreath, W.S. Rabinovich, Research in free space optical data transfer at the U. S. Naval research laboratory, in: D.G. Voeltz, J.C. Ricklin (Eds.), Proc. SPIE Vol. 5160, Free-Space Laser Communication and Active Laser Illumination III, SPIE, Bellingham, WA, 2004.

[33] A. Carrasco-Casado, R. Vergaz, J.M. Sánchez-Pena, Design and early development of a UAV terminal and a ground station for laser communications, in: SPIE. Proc., vol. 81814, September 29, 2011.

[34] Y. Gouy, E. Steck, C. Sanchez, G. Faulkner, D. O'Brien, et al., The C3PO Project: a laser communication system concept for small satellites, in: Proc. SPIE Vol. 10096, Free Space Laser Communication and Atmospheric Propagation, SPIE LASE, 2017.

[35] L. Li, R. Zhang, Z. Zhao, G. Xie, P. Liao, K. Pang, H. Song, C. Liu, Y. Ren, G. Labroille, P. Jian, D. Starodubov, B. Lynn, R. Bock, M. Tur, A.E. Willner, High-Capacity Free-Space Optical Communications between a Ground Transmitter and a Ground Receiver via a UAV Using Multiplexing of Multiple Orbital Angular-Momentum Beams, Scientific Reports, 7: 17427, December 12, 2017, https://doi.org/10.1038/s41598-017-17580-y.

[36] F. Fidler, M. Knapek, J. Horwath, W.R. Leeb, Optical communications for high-altitude platforms, IEEE Journal of Selected Topics in Quantum Electronics 16 (5) (September/October 2010).

[37] J. Horwath, M. Knapek, B. Epple, M. Brechtelsbauer, B. Wilkerson, Broadband backhaul communication for stratospheric platforms: the stratospheric optical payload experiment (STROPEX), in: Proc. SPIE, vol. 6304, 2006, 63041N-1–63041N-12.

[38] M. Knapek, J. Horwath, F. Moll, B. Epple, N. Courville, H. Bischl, D. Giggenbach, Optical high-capacity satellite downlinks via high altitude platform relays, in: Proc. SPIE Free-Space Laser Commun. VII, vol. 6709, 2007, 6709E-1–6709E-12.

[39] R. Lange, F. Heine, H. Kampfner, R. Meyer, High data rate optical inter-satellite links, Presented at the Eur. Conf. Opt. Commun. (ECOC 2009), Vienna, Austria, Paper 10.6.1, 2009.

[40] F. Fidler, Optical backhaul links between HAPs and satellites in the multi-Gigabit regime, in: Proc. Globecom Workshop Collection, 2008, pp. 1–5.

[41] S. Betti, E. Duca, I. Toselli, HAP-GEO optical links: performance analysis under weak turbulence conditions, in: Proc. Symp. Commun. Syst., Netw. Digit. Signal Process., 2008, pp. 77–81.

[42] E. Katimertzoglou, D. Vouyioukas, P. Veltsistas, P. Constantinou, Optical interplatform link scenarios for 20 km altitude, in: Proc. 16th IST Mobile Wireless Commun. Summit, 2007, pp. 1–5.

CHAPTER 7

Principles and Implementation of Secure Free-Space Optical Wireless Communications

7.1 INTRODUCTION

The interest in fast and secure free-space optical (FSO) communication link on the ground and in space has grown in the last few years, driven by the increasing demand in data throughput for applications spanning from terrestrial to small satellites for global services to even larger satellites for secure communication. It is also desirable in many applications over distances up to several km between rapidly moving platforms, such as air or ground vehicles, to establish bursty, high-speed, FSO links while minimizing the probability that a link is detected or intercepted over distances. There is also particular interest demonstrated by military market. In addition to speed, for secure communications a modern information society demands the most confidential and authenticated transfer of data between users, which include companies, financial institutions, and public administration. This is crucial for maintaining the competitiveness of industry and for individuals. Free-space optics offers new revolutionary tools to provide security in an unprecedented manner. This chapter discusses some of the recent advanced techniques used and applied to achieve the required security in communication.

In order to achieve secure communication, the first step is to develop encryption to prevent unauthorized access to transmitted information—a security need that is critical to modern-day electronic communication. Starting with conventional computationally secure encryption, information-theoretic secrecy, chaos-based and quantum cryptography offer progressively higher levels of security. Chaos key distribution (CKD) and quantum key distribution (QKD) are both powerful, depending on the extent of applications. Two distant parties generate shared secret keys over a lossy-noisy channel in the QKD, which is the most powerful adversary allowed by physics. When using the shared secret with the one-time pad cipher to subsequently encrypt data is the most powerful form of encryption today. Even applying encryption does not eliminate or mitigate the threat to users' privacy from the discovery of the very existence of the message itself, nor does it provide the means to communicate when the adversary forbids it.

Optical Wireless Communications for Broadband Global Internet Connectivity.
https://doi.org/10.1016/B978-0-12-813365-1.00007-2

Therefore, in order to not only protect the message content but also to prevent the detection of the transmission attempt it is essential to design low-probability of detection/intercept, or to truly covert the communication system. FSO communications can still suffer from optical trapping risks; usually the main lobe of the transmitting laser beam footprint is much wider than the receiver/detector size. For example, in urban terrestrial communications an eavesdropper can hide in the top of the same building as the legitimate receiver, or in satellite-to-ground laser communications the footprint can be a few kilometers, where tapping a portion of the downlink laser beam is not difficult. Perfectly secure communications using laser links still remain a challenging task. The security of conventional encryption techniques with a preshared secret key exchanged via a public key cryptosystem that is proved with algorithm means can be very weak because the computer technologies and decryption algorithms are advancing very fast. This is also true since the number of communication nodes are rapidly growing, making key distribution management extremely difficult and causing a larger overhead and latency to the system. Thus it is absolutely necessary to introduce and apply totally different concepts of encryption when dealing with FSO communications such as chaos-based and quantum-based key distribution for exchanging information at very high-speed in presence of the turbulent and free-space atmosphere.

This chapter concentrates mainly on these two categories of establishing secure optical communications. Subsequent sections will also describe some recent demonstrations and experiments for establishing FSO secure communications for terrestrial, air-to-ground, and satellite-to-ground links.

This chapter starts with a discussion of one of the most important benefits of optical wireless communication (OWC) offered in terms of secure communications compared to any other existing technology including radio frequency (RF). Improved user security minimizes information links and addresses the *weakest element* in the security chain—the *human aspect*. The chapter explains the basic principles involved in achieving various secure OWC concepts. The chapter clarifies the need and requirement of different techniques and methods for different communication channels and scenarios as follows:

- Indoor OWC: No turbulence, but attenuation/scattering loss due to multipath
- Outdoor OWC: Ground-to-ground, terrestrial; ground-to-aerial/ balloon platforms with uplink and downlink communications (atmospheric turbulence/scattering, scintillation)
- Space-to-space: Between satellites (attenuation in free-space because of geometry of such a long path, acquisition, tracking, and pointing misalignment loss) between transceivers

The chapter includes reviews of representative free-space and atmospheric quantum communications field experiments and discusses the experimental aims and progress achieved. The role of distances, protocol, laser pulse rates, encryption rates, technology approaches, technical limitations, and meteorology are discussed together with the advantage of adding adaptive optics to quantum communications. Secure OWC will be the ultimate issue in global Internet connectivity and is of the highest priorities in today's world to establish secure information exchange globally without information leaks or compromised security for safer connectivity. The chapter is organized as follows, providing different system architectures, approaches, and most recent experimental results:

- Indoor secure OWC using visible light communication (VLC) for Li-Fi technology
- Ultrashort soliton transmission for secure indoor and outdoor optical communication
- Acousto-optics chaos-based secure FSO communication links (secure through chaos encryption)
- Chaotic techniques implementation of secure atmospheric OWC and FSO communication
- Quantum Internet for global use using free-space and atmospheric quantum communications: mobile information, teleportation networks (including aerial/balloon and satellites) to teleport quantum information; ground-to-ground, ground-to-aircraft, and ground-to-satellite experiments—challenges and recent developments
- Interferometric optical communications for secure satellite-to-satellite optical communication (also applicable to future space—vehicle to space-station communication, ideal for secure server-to-server and router-to-router applications

The chapter concludes that no single secure OWC technique can solve the secure global Internet connectivity problem. It will require the integration of a number of different techniques to be able to handle indoor, outdoor, and terrestrial and space links for successful secure global Internet connectivity.

7.2 DEVELOPMENT OF SECURE OPTICAL COMMUNICATION LINKS

Secure communication is the method and technology to protect information using passwords. Encryption generally refers to a way of converting plaintext; the information needs to be protected from ciphertext, the masked information. Encryption and decryption keys are used to control the operation of encryption and decryption algorithms and therefore its security lies in the confidentiality of the key. Because of the possibility of passive eavesdropping, it is impossible to establish a secret key for communications between two parties through a classic communication channel. There are constant research and development efforts going on all over the world to improve the authenticity of how the secret key is distributed. CKDs and QKDs are providing the potential solutions in establishing secure FSO communications for ground and space applications.

7.2.1 Related Current Research Relevant to Secure Communication Through the Atmosphere: Physical Layer

7.2.1.1 Physical-Layer Security in Free-Space Optical Communication Using Orbital-Angular Momentum Multiplexing

Researchers have demonstrated [1] that from the transmission perspective, orbital-angular momentum (OAM) multiplexing technology can increase the capacity of FSO communion under different atmospheric turbulence conditions. Information theoretic security analysis based on optical taping scenarios is provided in Ref. [1]. Assuming the security is ensured at the transmitter and receiver telescopes, the goal is to establish a secure communication in the presence of an external eavesdropper (Eve) who can be anywhere between the transmitter and the receiver telescopes. When Eve's device intercepts the beam by collecting a fraction of the beam without being detected by Alice and Bob, the received signal power at Bob's end should not be significantly continuously reduced. Eve's detection efficiency depends on the size and position of Eve's intercept device. The results show a lower bound of letting Eve have a perfect detection efficiency. Consider the worst-case scenario when Eve is close to the transmitter end to get more precise information and before the beam propagates through the atmospheric turbulence. For each multiplexed beam, the capacity of the eavesdropper channel, from Alice to Eve is given by

$$C_{AE} = \log(1 + \gamma_e) \tag{7.1}$$

where $\gamma_e = P_e/N_0$ is the signal-to-noise ratio (SNR), and $P_e = r_e$. P_{TX} is the receive power at Eve's end, and r_e is the fraction of the power Eve can collect. The cross-talk between OAM modes after propagating through the atmospheric turbulence is increased. The capacity of the transmission from Alice's to Bob's channel can be similarly written as

$$C_{AB} = \log(1 + \gamma_b) \tag{7.2}$$

where $\gamma_b = P_b/N$ is the SNR. The total secrecy capacity is the part where Eve is unable to extract any information and is given by Ref. [1].

$$C_S = C_{AB} - C_{AE} \tag{7.3}$$

The probability of positive secrecy was defined in Ref. [1] as follows

$$P_S^+ = Pr(C_s > 0) \tag{7.4}$$

In [1] the performance of OAM-multiplexing FSO communication using Laguerre-Gaussian beams as carriers are analyzed in the presence of different turbulence strength levels and the aggregate secrecy capacity; that is, the summation of the secrecy capacity for all multiplexed channels as the total capacity where Eve cannot obtain any meaningful information was studied. The analysis is important to design an FSO communication system using OAM-multiplexing under different atmospheric turbulence conditions to increase the levels of security in communication. The concept developed is applicable to point-to-point terrestrial link as well as air-to-ground and satellite-to-ground FSO links for achieving secure communication.

7.2.1.2 Demonstration of the Physical-Layer Security Estimation in an Atmospheric Free-Space Optical Link Using a Testbed

Experimental data on message transmissions are presented [2] to estimate information-theoretic metrics including secrecy rate, secrecy outage probability, and expected code lengths for given secrecy criteria using a testbed consisting of one sender, one legitimate receiver, and another eavesdropper receiver. Potential FSO communication links for establishing Internet connectivity include a last-mile link from fiber backbone to the client's premises, unmanned aerial vehicles (UAVs), high altitude platforms (HAPs), and satellite laser communications. Together with the advanced concepts, technology development security requirements are also becoming more demanding for each of these links under different propagation paths (horizontal, uplink, and downlink) and atmospheric turbulence levels. For example, the eavesdropper would hide in the top of the same building as the legitimate receiver or tap a fraction of the received optical signal for a satellite-to-ground laser link where the footprint can be a scale of km. In the wireless wiretap channel described in Ref. [2] the sender (Alice) encodes a confidential message, using on-off keying modulation, into a code word variable (RV) X^n for transmission where n is the code length. At the receiver with a legitimate observer (Bob) the output via discrete time-varying quasi-static fading channel is

$$Y = H_B X + N_B \tag{7.5}$$

where H_B and N_B are the channel gain random variable and the additive white Gaussian noise (AWGN) random variable, respectively. The eavesdropper (Eve) taps a portion of the transmission signal from the output as follows

$$Z = H_E X + N_E \tag{7.6}$$

where H_E is the channel gain random variable and N_E is the AWGN random variable.

Alice also introduces some redundancy by adding l random dummy bits to the code word, which increases its length to n redundant bits. The two rates are defined as message rate $R_B = m/n$ and the randomness rate $R_E = l/n$.

The three information theoretical quantities as secure performance measures are the secrecy rates, the secrecy outage probability, and the expected code lengths for given secrecy criteria [2]. The sender (Alice) and the receiver (Bob) are therefore able to calculate the probability of the fatal information leakage vis the secrecy outage probability; the detailed analysis is reported in Ref. [2]. The channel estimation for physical layer security can be used for all practical types of FSO communication links involving different aerial and space platforms for establishing global Internet connectivity.

7.2.1.3 Security in Visible Light Communication System

VLC is a promising wireless communication technology through which baseband signals are modulated by an LED or diode laser. This is becoming a significant alternative to radio-based wireless communication such as Wi-Fi, Bluetooth, and similar devices. Because of a number of features, VLC data transmission networks (the so-called Li-Fi) provide an attractive and superior alternate to traditional wireless techniques. The directivity and high obstacle impermeability of optical signals offer a secure way to transmit data within a closed indoor environment to make it difficult to intercept from outside. The classic security of confidentiality, authenticity, and integrity still apply to address the VLC secure communication. The details of VLC communication and recent developments are already described and discussed in detail in another chapter of this book and are therefore not repeated here. The three basic classes of VLC devices are considered in this section: *infrastructure, fixed,* and *mobile*. The infrastructure includes data streaming integrated with room light; fixed includes PC, laptops, other desktop appliances like printers, projectors; and mobile includes smart phones. The four aspects of VLC communication with infrastructure, fixed, and mobile classes can be expressed as *availability, confidentiality, authenticity,* and *integrity*. Each threat level should be considered for each communication scheme to identify the areas of highest threat level. A qualitative threat characteristics is discussed in Ref. [3]: low, medium, and high, based on the communication scheme's physical characteristics. Qualitative classification of data transmission range (R), power (P), and radiation angle for communication (A) between mobile, fixed, and infrastructure devices can define some of the secure communication parameters as follows:

Jamming (J): $J = R/P$, Snooping (S): $S = P.A$, Data Modification (M): $M = J.S = R.A$

Jamming is directly proportional to the communication range and it is easier to introduce a concealed transmitting device for a longer range. Snooping is directly proportional to the transmission power and the radiation angle and therefore with a wider and more powerful transmission beam it is easier to oversee the communication link. Finally, the data modification risk is estimated as a product of the risks of jamming and snooping. In Ref, [3], qualitative estimations are described as follows: mobile-to-mobile range is considered low (~ 10 cm), fixed-to-fixed and fixed-to-mobile range is medium (up to 1 m), and all communications with infrastructure are considered high range (up to 3 m). For power, it is low for mobile devices, medium for fixed, and high when infrastructure is the sender. For mobile and fixed devices, the radiation angle is typically 30–60 degrees and is high when infrastructure ambient lighting is used.

The greatest risk of violating VLC security arises with infrastructure communication [3]. For example, for non–line-of-sight (NLOS) channel and line-of-sight (LOS) communication with infrastructure, snooping with an unauthorized receiver may be easily introduced without being recognized. One possible scheme for introducing jamming into the VLC infrastructure channel for the attacker is to use both directed and nondirected light sources from NLOS or LOS paths using optical beamforming. In some instances, given the channel capacity limitation, a signal source with sufficient transmitting power will be able to saturate the channel obscuring the data source. In conclusion, VLC infrastructure is not free from its own security issues, and VLC infrastructure is particularly prone to data-security risks.

7.2.2 Chaos-Based Secure Free-Space Optical Communication

Secure communication can be achieved using chaos, and chaotic secure communication is gaining wide acceptance for its inherent simplicity and ability to secure a communication link from the physical layer compared to other application layer–software encryption methods. Chaotic system features some special characteristic properties such as unpredictability, noise-like dynamics, and so on. This makes the chaotic secure communication one of the important chaotic application areas. Basically, it involves generating a chaotic signal in the form of current and adding it to the message at the transmitter where the whole of this current drives an optical source-laser. The same chaos signal is generated at the receiver by prior synchronization scheme so that the difference between this current and the internally generated matched chaos yields the message. A chaotic-based system uses the chaos for security and a very easy way to safeguard the message is to safeguard the message to hide it with the chaos and then send it. By generating the same chaos, the message can be completely retrieved at the receiver's end. In order to retrieve the message when the system is fully synchronized, a filter adaptive to the changes in environmental conditions and robust, designed to be the inverse channel can be implemented at the receiver's end so that the distortionless output can be obtained. For a system to be symmetrical, a fully synchronized system requires the receiver and transmitter to be identical. This means that the receiver is capable of producing the same chaos as that produced by the transmitter. It is important that the whole system is completely synchronized and introduction of a small mismatch can severely affect the correct recovery of message. Novel chaotic secure communication scheme design has also been recently investigated [4,5] so that the encrypting and decrypting of digital signals are carried out to realize the secure communication. Since no system is really secure enough, more securing schemes are constantly being developed.

7.2.2.1 Basics and Mathematical Representations of Generating Chaos

Chaos-based FSO communication is discussed in detail in Ref. [6]. Chaotic behavior can be expected from any dynamic system that shows sensitivity toward initial conditions; that is, a very high relationship with the previous values exists so that the value of the system at any point of time depends on the previous values. Small differences in initial conditions yield widely diverging outcomes for chaotic systems. Chaotic systems possess the ideal characters to be employed in crypto systems and by using chaotic methods we can prevent all kinds of intrusions. The message can only be retrieved at the receiver's end by generating the same chaos. Chaos can be generated mathematically. A recursive algorithm scan is used to calculate the values. Any X_i^{th} value depends immediately on X_{i-1}^{st} value so that the value can be recursively calculated. Mathematical equations can be used to calculate in a simpler way. Consider the following function:

$$f(x) = p * x * (1 - x) \tag{7.7}$$

This second order function can be used to generate mathematical chaos. Eq. (7.7) is bounded for the limits $0 < p < 4$. The equation can be written as:

$$x_{n-1} = p * x_n * (1 - x_n) \tag{7.8}$$

The starting value is x_0 and in this iterative form every n^{th} value depends on all other previous values. The plot of such functions is also called chaotic maps. For $0 < p < 3$, the function converges to a particular value after some number of iterations. As p is increased to just greater than 3, the curve splits into two branches. This splitting is termed bifurcation. Mathematically this tends to chaos. As the parameter p is further increased, the curves bifurcate again. With further increased value of p the bifurcation becomes faster and beyond a certain value of p known as the "point of accumulation," where periodicity gives way to complete chaos. This happens for $p > 3.57$ whereas for $p = 4$, chaos values are generated in the complete range of 0–1. This is the point we are interested in. During $3.6 < p < 4$, complete randomness and chaotic behavior is observed.

Chaotic signals generated in nonlinear electrical circuits [7–9] and lasers [10] can potentially be used as carriers for information transmission in a communication system. The advantage of a broadband information carrier is that it can enhance the robustness of communication channels to interferences with narrow-band disturbances. The broadband coding signal in a chaos-based communication is generated at the hardware level where chaotic carriers offer a certain degree of privacy in the data transmission. Thus, a new type of high data rate communication system can be designed using waveforms generated by a deterministic chaotic system to carry information in a robust manner. Chaotic communication systems are based on chaos synchronization where synchronized chaotic emitters and receiver lasers are employed to encode and decode information at the hardware level. The generated chaotic signal at the emitter hides the message, which can be recovered when using the appropriate receiver. Messages (information) are embedded within a chaotic carrier in the emitter, and recovered after transmission by a receiver that is synchronized with the emitter. A nonlinear filtering process is performed at the receiver where a message-free chaotic signal is generated locally, which is then subtracted from the encoded transmitted signal to recover the message (information). Chaotic optical communication is possible when the broadband chaotic emissions from two spatially separated emitters (lasers) are synchronized to each other. In order to satisfy the requirement for synchronization of the two lasers, the irregular time evolution of the emitter laser optical power must be perfectly reproduced at the receiver laser. Decoding the message from the chaotic carrier is based on the nonlinear phenomenon of chaos synchronization between the emitter and the receiver so that the message can be extracted by subtracting the chaotic carrier from the input (chaotic carrier + message).

7.2.2.2 Chaotic Free-Space Optical Communication Over Atmospheric Turbulent Channel

Feasibility of chaos-based communications using fiber-optic links have already been proposed and demonstrated in the past [11,12]. Chaotic communication in several optical systems are discussed where an erbium-doped fiber ring laser produces chaotic fluctuations of light intensity onto which a message consisting of a sequence of pseudorandom digital bits [11] is modulated. Chaos and message together propagate through a standard single-mode optical fiber from the transmitter to a receiver. The message from the chaos is recovered at the receiver. The fiber link is 35 km and a data rate up to 250 Mbps is achieved by the researchers. High-speed, long-distance, chaos-based communications over a commercial fiber-optic channel is presented [12] for a transmission over 120 km of optical fiber at data rates of Gbps ranges achieving bit-error rates

(BER) below 10^{-7}. The results provide a convincing proof-of-practical concept for optical chaos communications technology.

Although much research has been done for chaotic optical communications using fiber channels, FSO chaotic communication over a real-life highly random channel such as turbulence has not been as thoroughly investigated. This subsection discusses the concept for developing and designing FSO communication systems based on encoding and decoding of chaotic signals for recovering messages in the presence of atmospheric turbulence media.

7.2.2.2.1 Double-Pass Chaotic Secure Free-Space Optical Communication Link

One of the first chaotic free-space laser communications experiments over a turbulent channel is reported [13] where a chaotic self-synchronizing free-space laser communication in the presence of severe communication signal distortions caused by atmospheric turbulence is studied experimentally. A double-pass propagation link (~5 km round-trip path) using a corner cube reflector is used in their demonstration. Fig. 7.1 shows a block diagram of a double-pass chaotic FSO communication. A 10 mW semiconductor laser beam ($\lambda = 690$ nm) coupled to a single-mode fiber is transmitted toward a corner-cube retroreflector 2.5 km away. The reflected beam is received by the same telescope (used by the transmitter) and is detected by a photodetector. The scintillation index for the turbulence is about 0.8–0.9, indicating a strong turbulence regime. A chaotic laser communication transceiver consists of a laser

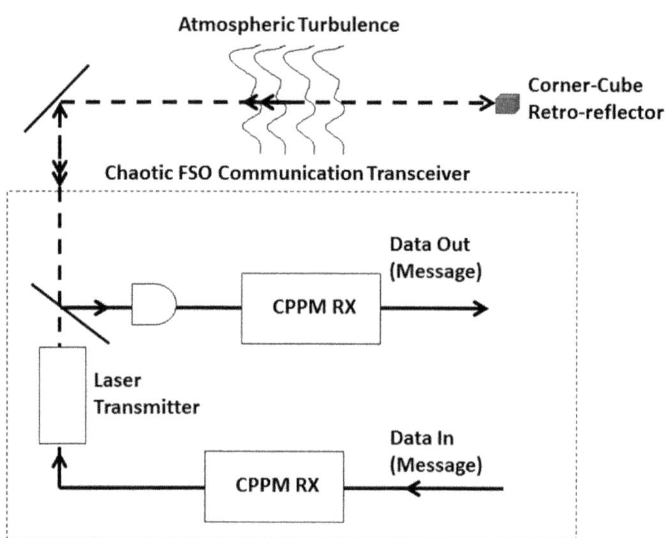

FIGURE 7.1
Block diagram of a double-pass chaotic free-space optical communication through atmospheric turbulence.

generating a chaotic sequence of short-term (~1.0 μs) on-off pulses, triggered by transistor-transistor logic pulse signals from a chaotic transceiver controller where the chaotic sequence of the time intervals corresponds to iterations of a chaotic process with the binary information signal added to the chaotic signal. This way a chaotic pulse position modulation is used. The interpulse interval fluctuates chaotically ranging from 10 to 25 μs at a ~60 kbps bit rate. A chaotic pulse position modulation receiver receives the distorted chaotic pulses detected by the photodetector and the information signal is finally recovered from chaotic iterations. This method of chaos communication is referred to as chaotic pulse position modulation (CPPM). The BER from the real-time transmission of binary pseudorandom code data is measured to be 1.92×10^{-2}. This research thus proves the concept of chaos communication in the presence of atmospheric turbulence using the self-synchronizing method.

7.2.2.2.2 One-Way Chaotic Free-Space Optical Communication Link

Fig. 7.2 shows a concept block diagram of a chaotic FSO communication system in the presence of an atmospheric turbulence channel. At the transmitter end, there is a master laser routed to chaos through a delayed optoelectronic feedback and then a message is embedded in the generated chaotic carrier. At the receiver end, a twin semiconductor laser (the slave laser) is required to synchronize to the chaotic master laser so that the message can be recovered by operating a difference between chaos and received signal and by low-pass filtering. The system described here effectively masks the message at the transmitter so that an eavesdropper is not able to recover the hidden message. The encoded message *m(t)* can be modulated either by a chaotic intensity modulation (CIM) or additive chaos modulation (ACM) method. In the first

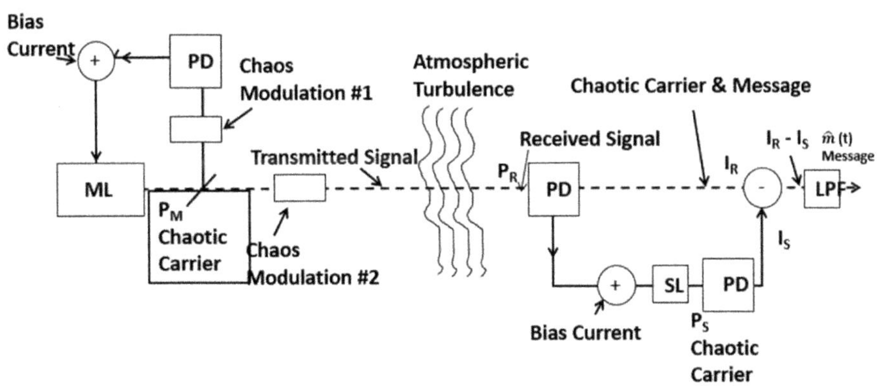

FIGURE 7.2

Block diagram of a chaotic free-space optical communication system in atmospheric turbulence. The encoded message *m(t)* can be modulated either by CIM or ACM methods.

CIM method, the message is superposed just before the transmitter output by an intensity modulation so that the transmitted power $P_T(t) = [1 + m(t)] \cdot P_M(t)$ where $P_M(t)$ is the chaotic carrier power. In the ACM method, the message can be summed inside the feedback loop to influence the chaos generation and create a symmetric scheme (the message is sent to both master and slave lasers). The signal power at the receiver P_R is converted into the photo-current I_R. A bias current is added to the slave laser and the chaotic carrier P_S is generated through the synchronization process. If perfect synchronization, the two carriers are equal, $P_S = P_M$. The message now can be recovered by the difference $I_R - I_S$. The low-pass filter eliminates the high frequency noise and chaos components with a bandwidth B_{LPF} equal to that of the pulse position modulation (PPM) pulse: $B_{LPF} = R_b \cdot M(d\log_2 M)^{-1}$ where R_b is the data (bit) rate, M is the PPM order, and d is the slot duty-cycle. Note that here the message is encoded through a PPM method.

7.2.2.3 Secure Free-Space Optical Communications in Atmospheric Turbulence Using Acousto-Optic Chaos

Secure free-space communication through atmospheric turbulence using acousto-optic (AO) chaos has been discussed [14], performing modulation of RF chaos via first-order feedback with simulation results showing the application of AO chaos in this area. Applications based on encryption with profiled optical beams and results are presented for the use of chaotic encryption for image restoration during propagation through atmospheric turbulence.

Chaos generated by an AO system with feedback to encrypt a laser beam carrying data was reported [15]. Preliminary results on using such an encryption and decryption technology in building FSO communication systems are discussed. Chaos encryption of data using external signal modulation of the diffracted light from a hybrid AO cell is reported. Numerical simulation shows that decryption of the encoded data is possible by using an identical AO system in the receiver. Encryption of optical signals using external modulation of the diffracted light in AO modulators and retrieval and deencryption of the encoded signal using parametrically synchronized chaotic demodulation with another AO cell were shown to be possible [15].

Most of the chaotic encryption and decryption reported were performed using nonlinear dynamics of external cavity feedback in semiconductor lasers. Researchers also generated the chaos using the nonlinearities in Er-doped fiber lasers. The researchers presented [15] some results to show the possibility of (1) encryption of optical signals using external modulation of the diffracted light in AO modulators and (2) retrieval and decryption of encoded signal using parametrically synchronized chaotic demodulation with another AO cell.

In [14] simulation of free-space applications of AO chaos is discussed and was found that by assuming profiled transverse beams, the performance of the hybrid AO feedback device improves considerably compared to the case of uniform-profiled beam propagation. For secure communication schemes using appropriate masking, the key parameter tolerances can be made arbitrarily small (<0.01%), which makes the encryption strategy highly robust. The hybrid AO feedback-based chaos may also be used for propagation through atmospheric turbulence with a modified von Karman-type atmospheric model. The researchers found that incorporating modulated chaos waves from hybrid AO feedback devices with random time-domain statistics into turbulence problems can also mitigate the atmospheric turbulence effects. A heterodyne scheme for encrypting and decrypting using AO chaos is shown in Fig. 7.3 where the local chaos wave is generated using a second Bragg cell with matched parameters. The local chaos multiplied with the incoming photodetected modulated chaotic signal and the product waveform is then passed through a low-pass filter (LPF). With the proper choice of the hybrid AO feedback parameters the signal $s(t)$ is completely encrypted and after transmission through an atmospheric turbulence channel can be recovered. The simulation for defining the atmospheric propagation slant path included both turbulent and some nonturbulent components, which can be modeled like a real situation of

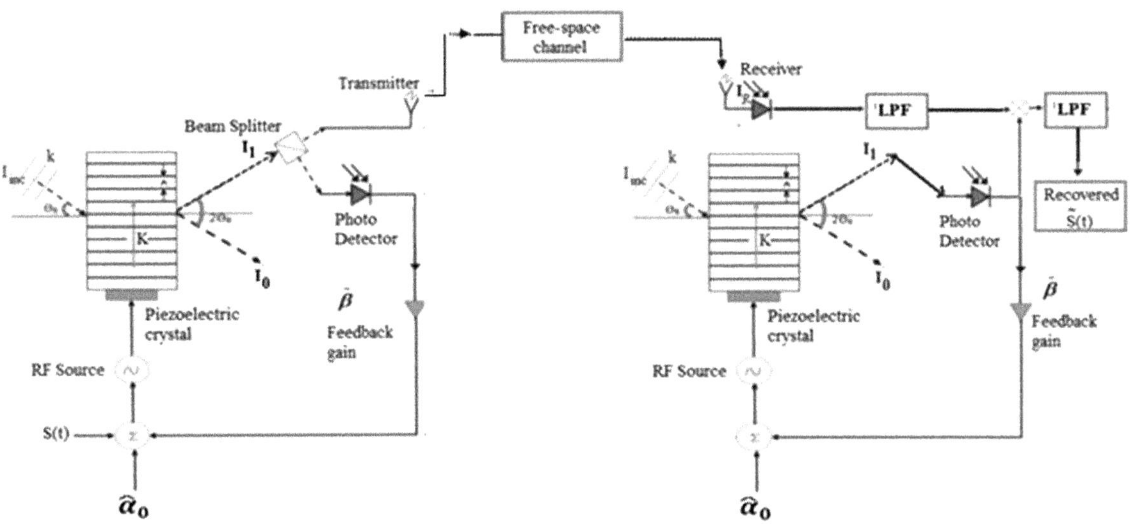

FIGURE 7.3

Free-space secure chaos-based secure communication scheme with a heterodyne scheme for encrypting and decrypting. *Figure credit to Monish R. Chatterjee, Ali A. Mohamed, Department of Electrical & Computer Engineering, University of Dayton, Dayton, OH 45469, USA and Fares A. Almehmadi, Faculty of Engineering, University of Tabuk, Tabuk 71491, Saudi Arabia. Reprinted with permission from Monish Chatterjee (similar to Fig. 10 of AO paper.) Permission from the Optical Society of America, 2018.*

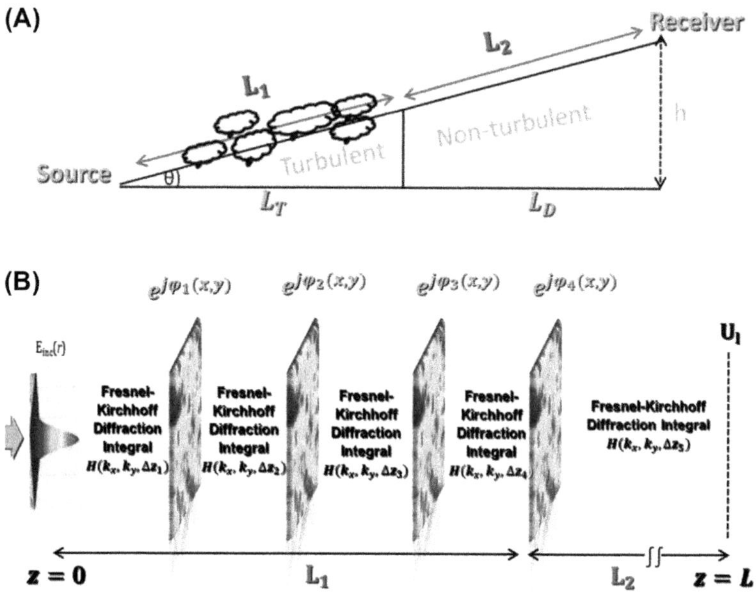

FIGURE 7.4

(A) Schematic diagram for slant path propagation under a modified von Korman turbulence model; (B) schematic illustration and physical interpretation of propagation through the modified model. *Figure credit to Monish R. Chatterjee, Ali A. Mohamed, Department of Electrical & Computer Engineering, University of Dayton, Dayton, OH 45469, USA and Fares A. Almehmadi, Faculty of Engineering, University of Tabuk, Tabuk 71491, Saudi Arabia. Reprinted with permission from Monish Chatterjee (similar to Fig. 10 of AO paper). Permission from the Optical Society of America, 2018.*

satellite-to-ground laser communication where turbulence strength is much more near the ground.

Fig. 7.4(A) shows a schematic diagram for slant path propagation under a modified von Korman turbulence model, and Fig. 7.4(B) shows a schematic illustration and physical interpretation of propagation through the modified model. The encrypted chaos is propagated through the various turbulent layers and the transmitted image signal is reconstructed using photodetection, heterodyning, and filtering process. The cross-correlation values between nonturbulent and turbulent image signals for both nonchaotic and chaotic cases are shown in Fig. 7.5. The results show the highest cross-correlation for weak turbulence and decrease monotonically from moderate to strong turbulence regime.

The results clearly demonstrate the potential use of chaos-based secure communication for many practical scenarios such as aircraft-to-ground and satellite-to-ground laser communications for high data rate communications.

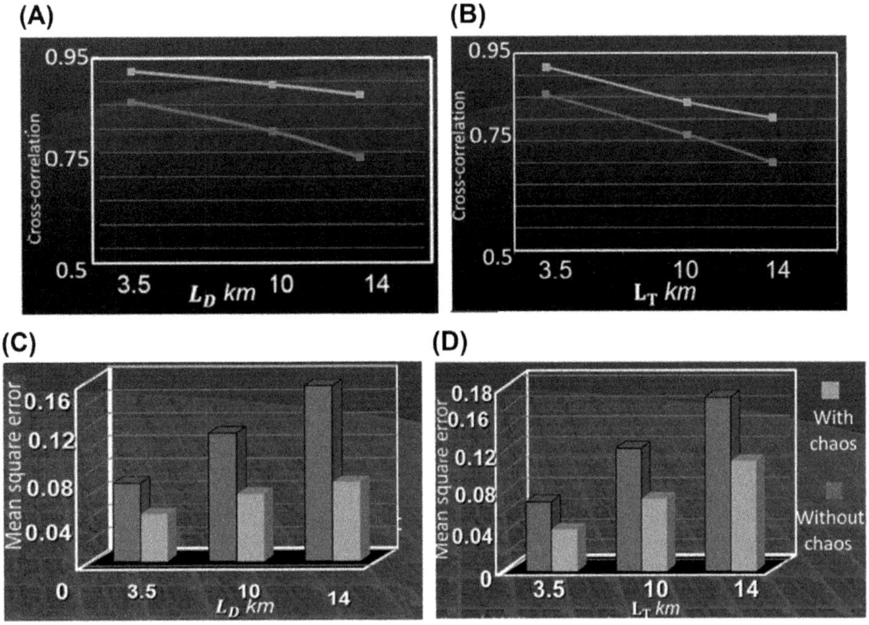

FIGURE 7.5

Chaotic versus nonchaotic transmission cross-correlation products and mean squared errors under moderate turbulence with increasing (A and C) nonturbulent distance L_D, and (B and D) turbulent distance L_T. *Figure credit to Monish R. Chatterjee, Ali A. Mohamed, Department of Electrical & Computer Engineering, University of Dayton, Dayton, OH 45469, USA and Fares A. Almehmadi, Faculty of Engineering, University of Tabuk, Tabuk 71491, Saudi Arabia. Reprinted with permission from Monish Chatterjee (similar to Fig. 10 of AO paper). Permission from the Optical Society of America, 2018.*

7.2.2.4 Feasibility of Chaotic Secure Duplex Optical Communications Through Atmospheric Turbulence Using Channel Reciprocity: Proposed Concept Technology

As the demand for broadband Internet connectivity is growing tremendously including availability of global access *anytime, anywhere* and simultaneously the need for proposing a more secure scheme is also increasing. With the development of faster, compact, modern electronic technology, the traditional chaotic secure communication is facing an unprecedented challenge. FSO technology based on satellite-to-ground, ground-to-satellite, and links between satellites are also showing tremendous growth as evidenced by vast improvements in information technology. Secure communication is still a major issue in FSO-based networks. Although schemes of chaos-based communications for fiber transmission can be used for investigating the fundamental concepts regarding controlling chaos for spatio-temporal dynamics and patterns, the problem is the FSO channel such as atmospheric turbulence and scattering medium. Atmospheric channel is basically random for treating optical propagation. If the chaos-based communication scheme is to be explored for FSO channels, signal distortions altering the chaotic waveform shape must be taken into account; otherwise received chaotic signals do not precisely represent the transmitter oscillations. Therefore, a true synchronization scheme will not be achieved, causing data/information loss and increase in BER without a thorough understanding of the interactions of optical wave with random turbulent medium.

Very few experiments and discussions of chaotic FSO communications are reported as of today. Chaotic free-space laser communications over a turbulent channel based on CPPM have been studied [13]. *One of the main problems in achieving high data rate chaos-based optical communications is the difficulty in synchronizing signals both at the transmitter and receiver ends in the presence of random turbulence.* Because of randomly dynamic characteristics of the optical propagation through atmospheric turbulence, it is extremely important to determine the limitations to worsen the quality of synchronization between transmitter and receiver because they modify the injection process on which the synchronization is based. The proposed concept technology provides an innovative solution to achieve synchronization based on *reciprocity, which provides real-time (instantaneous) turbulence state information without any external complex adaptive optics system in order to improve communication performance using chaos.* In order to perform duplex (two-way) communications with chaos, two fiber-based coupled transceivers are placed at each end of the terminals where the pairs of transmitters and receivers complete the two-way communication links using a chaos-masked message in the presence of atmospheric turbulence.

The conceptual architect for studying the feasibility of establishing chaos-based communications in an atmospheric turbulent path are valid even for communications between moving platforms such as UAVs or satellite-to-ground platforms without tracking and alignment. This can be realistically possible because of the basic physics of the nature of reciprocity, which ideally should hold for any turbulence strength or communications ranges. In order to achieve FSO secure communication is based on the initial concept reported by one of the investigators (A.K.M) [16] for remote sensing of the target-plane laser beam characteristics, which does not require any assumptions regarding the nature of refractive index inhomogeneities along the propagation path. Theoretical foundation for that paper is based on the optical reciprocity principle [17] and then derived from it conservation law for an integral characteristic that links complex amplitudes of the counterpropagating optical waves. *The overlapping integral or the interference metric of this characteristics does not change along the propagation path and thus provides direct measurements of the target-plane scintillation index* $\sigma^2{}_I$. This is the main idea where this turbulence state information can be used in chaos-based communication for achieving synchronization in either end of the duplex FSO communication system proposed here.

In Ref. [16], we consider an atmospheric sensing technique based on the so-called target-in-the-loop (TIL) wave propagation geometry. In the TIL atmospheric sensing (TILAS) technique, laser beam and turbulence characterization is achieved using a special-type of monostatic laser transceiver (TILAS transceiver) pointed to a target with a small-size retroreflector or retrotape attached to it. With the proposed TILAS sensor, laser beam and turbulence characterization can be performed along the LOS to the target of interest. The TILAS concept for remote sensing of the target-plane laser beam characteristics does not require any assumptions regarding the nature of refractive index inhomogeneities along the propagation path. The important property of this characteristic, referred to as overlapping integral or interference metric, is that its value does not change along the propagation path. The interference metric conservation law is used in the TILAS system for direct measurements of the target-plane scintillation index σ_I^2 without the need for the knowledge of C_n^2. Consider two counterpropagating monochromatic (quasimonochromatic) waves transmitted toward the target wave and scattered off the target's surface wave that propagates back to the transmitter plane. Complex amplitude of the transmitted wave is denoted by $A(\mathbf{r}, z, t)$, while the amplitude of the target-return or backscattered wave is denoted by $\psi(\mathbf{r}, z, t)$. Here $\mathbf{r} = \{x, y\}$ is a vector in the plane orthogonal to the propagation direction (optical axis oz) and t is the time.

It can be shown that in the framework of parabolic approximation of optical diffraction theory, the following integral relationship (interference metric) preserves its value along the propagation path [16,17]:

$$\int A(\mathbf{r}, z, t)\psi(\mathbf{r}, z, t)d^2\mathbf{r} = J_{\text{int}}(t), \quad 0 \leq z \leq L \tag{7.9}$$

where the integration is performed over the entire transverse plane.

We now apply this relationship for analysis of the TILAS system in Fig. 7.6. The target-return wave is obtained in this system via scattering the transmitted laser beam off a remotely located retroreflector. The complex amplitude of the wave scattered by an ideal retroreflector can be written as follows:

$$\psi(\mathbf{r}, L, t) = M_T(\mathbf{r})A(-\,\mathbf{r}, L, t), \tag{7.10}$$

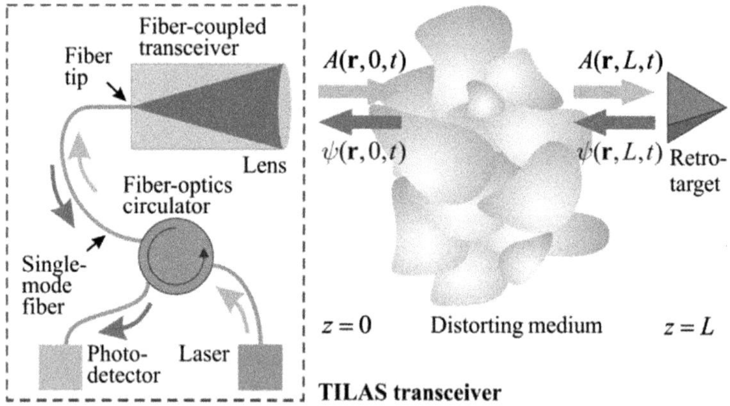

FIGURE 7.6

Target-in-the-loop (TIL) sensing geometry with the TIL atmospheric sensing (TILAS) transceiver located at the pupil plane $z = 0$ operating with a retrotarget located at the plane $z = L$. *Reprinted with permission from the Optical Society of America, OSA, 2018; M.A. Vorontsov, S.L. Lachinova, A.K. Majumdar, Target-in-the-loop remote sensing of laser beam and atmospheric turbulence characteristics, Applied Optics 55 (19) (July 1, 2016).*

where $A(\mathbf{r}, L, t)$ is the complex amplitude of the transmitted wave that enters the retroreflector and $M_T(\mathbf{r})$ is the stepwise function $M_T(\mathbf{r}) = 1$ inside the circular retroreflector aperture of diameter d_T and zero otherwise. For simplicity, we neglect power losses due to absorption in the atmosphere and inside the retroreflector. Using the scattering condition, we obtain

$$\begin{aligned} J_{\text{int}}(t) &= \int A(\mathbf{r}, 0)\psi(\mathbf{r}, 0, t)d^2\mathbf{r} \\ &= \int M_T(\mathbf{r})A(\mathbf{r}, L, t)A(-\mathbf{r}, L, t)d^2\mathbf{r}. \end{aligned} \tag{7.11}$$

Note that the complex amplitude of the transmitted beam in Eq. (7.11) is considered to be stationary: $A(\mathbf{r}, 0, t) = A(\mathbf{r}, 0)$. Sensing of the interference metric at the transceiver plane $z = 0$ can provide information regarding characteristics of the transmitted laser beam at the target plane $z = L$. This property of the interference metric is utilized in the TILAS technique discussed in the following subsection.

7.2.2.5 Interference Metric Sensing With Target-in-the-Loop Atmospheric Sensing Transceiver

As shown, the modulus of the interference metric $|J_{\text{int}}(t)|$ can be directly measured using a specially designed TILAS transceiver system presented in Fig. (7.6). The system consists of a fiber-coupled transceiver (as a collimating lens) and an optical train solely based on a single-mode fiber and fiber elements that are used for (1) delivery of a collimated Gaussian beam generated at the fiber tip to the transceiver telescope pupil and (2) delivery of the target-return optical wave to the fiber-coupled photodetector.

For the case of a single-mode fiber, the signal $P(t)$ measured by the photodetector is proportional to the optical power coupled into the fiber tip and is given by the following overlapping integral:

$$P(t) = \left| \int M_F(\mathbf{r})\psi_F(\mathbf{r},t)\, d^2\mathbf{r} \right|^2, \tag{7.12}$$

where $\psi_F(\mathbf{r},t)$ is the complex amplitude of the optical field that enters the fiber tip at the focal plane F of the collimating lens in Fig. (7.6). The normalized (dimensionless) function $M_F(\mathbf{r})$ describes the shape of principal eigenmode of the single-mode fiber and hence is proportional to the normalized complex amplitude of the optical field transmitted through the fiber tip, $A_F(\mathbf{r})$. Eq. (7.13) denotes the signal measured by the TILAS transceiver. Thus, in the introduced notations, from the earlier equations we have

$$P_{z=0}(t) = P_{z=L}(t). \tag{7.13}$$

The normalized intensity variance can then be written as

$$\sigma_I^2(0,L) = \frac{\langle P_{z=0}(t)\rangle_T}{\left\langle \sqrt{P_{z=0}(t)}\right\rangle_T^2} - 1 = \frac{T\int_0^T P_{z=0}(t)dt}{\left[\int_0^T \sqrt{P_{z=0}(t)}dt\right]^2} - 1 \tag{7.14}$$

The system does not require installation and accurate alignment of optical transmitter and receiver systems at each end of the propagation path and hence can be used to measure the impact of atmospheric effects along the LOS to a moving target with a small retroreflector or retrotape attached to it. The technique could be also applied for atmospheric turbulence profiling using as a target a small drone with retroreflector flying along the LOS of interest. Another possible application includes atmospheric sensing in maritime environments near the sea surface with a retrotarget installed on a buoy or a ship mast.

The technique can also be extended to include a simple optical feedback mechanism to adjust transmitter powers to compensate channel loss; to develop techniques for adjusting the bit rate or code rate, which can gate the data flow when the transmission drops below the threshold defined by the detection process; and also to include reliable communications between moving (floating) platforms. This knowledge of atmospheric parameters at each end can have the potential solution to develop accurate and precise synchronization of chaos-based secure laser communications at both ends of the communication transceiver systems to establish secure communications between the transmitter and receiver ends and therefore to achieve Internet connectivity between two optical nodes.

Recent works are reported describing communication systems and techniques that exploit *atmospheric reciprocity* to overcome turbulence applicable to high data rate communications [18,19]. The past research efforts include both development of theoretical foundations as well as some experimental observations of channel reciprocity. Their results are very encouraging to consider a reciprocity concept for applications in FSO communications for developing chaos-based laser communications. The two-way concepts using reciprocity at both ends of transceivers to establish duplex communications are of practical importance. The ultimate benefit will also be to improve secure communication link performance and thus to develop broadband, high data rate FSO extremely secure communication systems to decrease decoder complexity and reduce system delay.

7.2.3 N-Slit Interferometry-Based Communications

Interferometric optical communications have the potential for secure and naturally encrypted long-distance laser communications in space using the concepts of quantum entanglement [20]. For terrestrial laser communications the optical signal is distorted by the atmospheric turbulence and the scattering medium and therefore affects the communications system performance in terms of achieving high data rate. Cryptography has been extensively investigated to protect the secrecy of communications. Quantum cryptography combines the single-photon optical communication and cryptography by introducing quantum entanglement via polarization of quanta. Secure interferometric communications, based on quantum principles, utilizes the large populations of indistinguishable photons. In an atmospheric propagation path the information carrying quanta is quantum encoded to start with, to be decoherent and finally lost. However, the coupled concepts of optical communications and quantum cryptography can be valid for space-to-space optical communications where there is practically no propagation distortions or interaction with the carrier (signal carrying) quanta. In quantum cryptography, the quantum probability amplitude is described as [20] using the standard *bra* and ket formalism of the Dirac notation

$$|\psi\rangle = (|x,y\rangle - |y,x\rangle) \tag{7.15}$$

and a normalized version as follows

$$|\psi\rangle = \frac{1}{\sqrt{2}}(|x\rangle_1|y\rangle_2 - |y\rangle_1|x\rangle_2) \tag{7.16}$$

For applying to interferometric communications originating in the Dirac-based N-slit interferometric equation [21,22] is as follows:

$$|\langle x|s\rangle^2 = \sum_{j=1}^{N} \psi(r_j)^2 + 2\sum_{j=1}^{N} \psi(r_j)\left(\sum_{m=j+1}^{N} \psi(r_\mathrm{m})\cos(\Omega_m - \Omega_j)\right) \tag{7.17}$$

The equation also applies to large populations of indistinguishable photons, or highly coherent beams such as those from narrow-linewidth lasers. The concept of security in the interferometric communications can be described as follows: any attempt to extract information in the propagation from the source a to the detector x via an intrainterferometric distance $D\langle x\rangle D_{\langle x|j\rangle}$ severely distorts the propagating interferogram or the interferometric character [21,22]. This is due to the quantum wave function interaction with the distorting feature, which causes collapse or decoherence of the information wave function. The secure concept can be easily designed with an N-slit laser interferometer with a signal detection scheme. A standard narrow-linewidth laser, a multiple-prism beam expander, an N-slit grating set, and a room temperature digital detector (CCD or CMOS-type) can be assembled to develop the interferometric communication system. In [23,24] transmission of interferometric characters over distances of 35 m and 527 m through the atmosphere at or near sea level have been demonstrated. There is therefore a potential for establishing extremely secure interferometric optical communications for a number of applications in space-based communications such as satellite-to-satellite and space-vehicle-to-space-vehicle (including HAPs), which eventually might prove ideal for secure server-to-server and router-to-router applications for Internet connectivity.

7.2.4 Quantum Communication-Based Secure Communication

7.2.4.1 Basics Background of Quantum Secure Communication

Secure communications for sensitive information over the Internet is very critical, especially in the tremendous development of the Internet and applications of Internet of Things, requiring high speed connectivity. Free-space and atmospheric quantum communications will play a critical role in extending the quantum Internet to global use. Quantum information will be teleported through mobile information teleportation networks that necessarily will include satellites. Recent developments in quantum physics have the potential to add a physical layer of quantum security and increased bandwidth and speed beyond classical communications capabilities to free-space and atmospheric communications. Fundamentals of free-space quantum communications and some experiments are described and discussed in detail in Ref. [25]. Some of the highlighted free-space quantum communications field experiments have been reported to demonstrate the feasibility and practical use of free-space QKD systems, quantum transportation, and single photon exchange over extremely long distances.

Secure quantum communication is based on encoding information on photon properties, such as their polarization to distribute encryption keys between two communicating transmitters and receivers with absolute security. In this scheme, the transmitter sends a series of single photons in one or four possible polarization states: horizontal, vertical, left-circular, or right-circular while the receiver then performs a measurement that differentiates between these polarization states. Secure quantum communication depends on such polarization-based key distribution (QKD) protocols to distribute keys at very high speed and over long distances. The concept created a pathway to achieve a type of encryption such as a *one-time pad* that cannot be cracked. Atmospheric propagation effects such as turbulence introduces random noise, which also affects core quantum communication protocols such as distribution of entangled photons whose states are intertwined even at a distance. Entanglement-based QKD for transporting for a satellite-to-ground or aerial platform-to-ground communication therefore requires development of advanced payload of the equipment for the generation and detection of quantum states. The satellite or aerial platform will need to be equipped with entangled-photon sources.

For establishing secure quantum communication between two communicating parties, there are a number of scenarios possible in which the transmitter of entangled photons is located on the ground or aboard a satellite or aerial platform. These scenarios will provide different performances for different applications. Earth-based transmitter terminal will allow shared quantum entanglement between ground and satellite or aerial platform or between two satellites or aerial platforms, thus communicating between such terminals employing quantum communication protocols. Each of the photons of an entangled pair can be directed toward another Earth-based terminal or satellite or aerial platform or between two space-based separate terminals. For a space-based scenario a transmitter with an entangled photon source placed on a space-based or aerial-based platform not only allows longer link distances but also flexible network configurations. A downlink thus makes quantum cryptography possible. For two ground station applications, two of the ground stations need to establish a separate quantum key with the satellite or aerial terminal with access to both keys, transmitting a logical combination of the keys to be used by either ground station or both. Fig. 7.7 shows quantum communication architectures for space- or air-borne terminals and ground terminals. A quantum key exchange can also be performed between arbitrarily

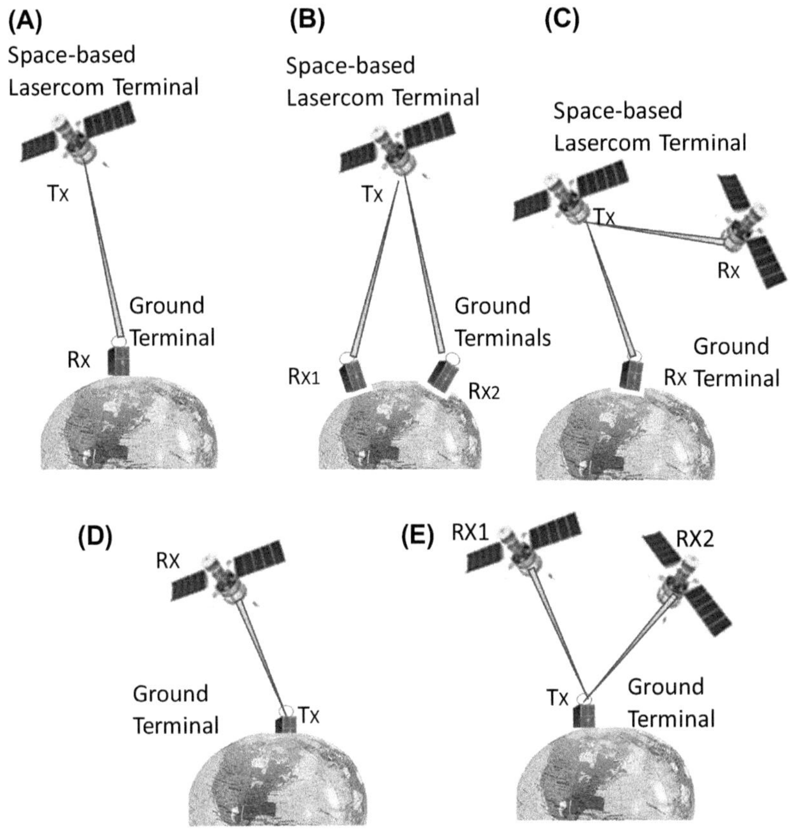

FIGURE 7.7

Quantum communications architectures for space- and air-borne terminals and ground terminal. (A) Space-based terminal transmitter to ground-terminal receiver, (B) space/air-borne terminal transmitter to two ground receiver terminals, (C) interstate terminals (transmitter and receiver) and ground terminal receiver, (D) ground-based terminal transmitter to space-based receiver, (E) ground-based terminal transmitter to space-based Lasercom receivers terminals.

located ground stations at different times, which has the potential of establishing interconnectivity for LAN or WAN at very high data rates and with the highest security possible.

Researchers have [26] proposed a different type of QKD protocol over FSO than the conventional QKD protocols. The proposed QKD protocol is based on physical characteristics of quantum mechanics, which can be implemented on standard FSO systems operating in the presence of atmospheric turbulence, and uses subcarrier intensity modulation binary phase shift keying and direct detection with a dual-threshold receiver. The protocol takes into account the atmospheric channel's characteristics such as turbulence-induced fading and receiver noises. Conventional key distribution protocols rely on computational complexities and are vulnerable to future advances on computer hardware and sophisticated algorithms. As mentioned earlier, QKD promises the principle of unconditional security by securing the key distribution based on quantum physics. Originally the first QKD protocol was proposed as the BB84 protocol. The novel free-space continuous variable QKD was described in [26]. The security parameters such as secrecy performance of the proposed system, the quantum bit error rate (QBER), ergotic secret-key rate, and final key creation rate were analyzed in [26] using the atmospheric models for log-normal and gamma-gamma distribution statistics for propagation. Their proposed system used intensity modulation binary phase shift keying and dual-threshold/direct detection with an avalanche photo detector scheme. The results proved that the QKD function can be achieved based on the pulse-based level of laser beam in the standard optical system and will reduce the deployment cost of future secured networks already using standard telecommunication devices. The analysis can be applied to both horizontal and slant path communication links including satellite and aerial platforms for establishing high speed secure Internet connectivity.

Research and development are being conducted worldwide to enable QKD with satellites as well as on-going satellite QKD initiatives [27]. Satellite QKD is now being used to enable global coverage. With the development of quantum random number generator, it can be incorporated with the advanced QKD technology to be extremely useful for cryptographic scenarios relevant to both ground and space-based terminals and nodes for an Internet network. As a part of quantum Internet, the distribution of entanglement to perform will be an important building block in a global network of distributed quantum computers. A quantum Internet will also require an interface to material systems that can act as quantum memory or processing units for global Internet connectivity at very high speeds, which will also be secure. Worldwide communication privacy can be maintained at the same time powerful computers are being developed and advanced so that a global quantum Internet can be achieved. Future ground-to-aircraft and ground-to-space quantum experiments will provide additional progress toward achieving highly secured worldwide communications and thus Internet connectivity. Quantum communication with satellites can therefore be possible because laser communications in space-based platforms are already technologically matured [28]. QKD between ground station and moving air-borne or satellite-borne platforms are therefore achievable using already established optical links. Because the uplink and downlink propagation characteristics are different due to the presence of atmospheric turbulence and scattering medium, compensation for wavefront distortions in both transmission directions as well as considering the asymmetry in atmospheric optical link between moving-receiver and moving-sender architectures will be required. Fig. 7.8 shows the conceptual architectures for future airborne and space missions with QKD.

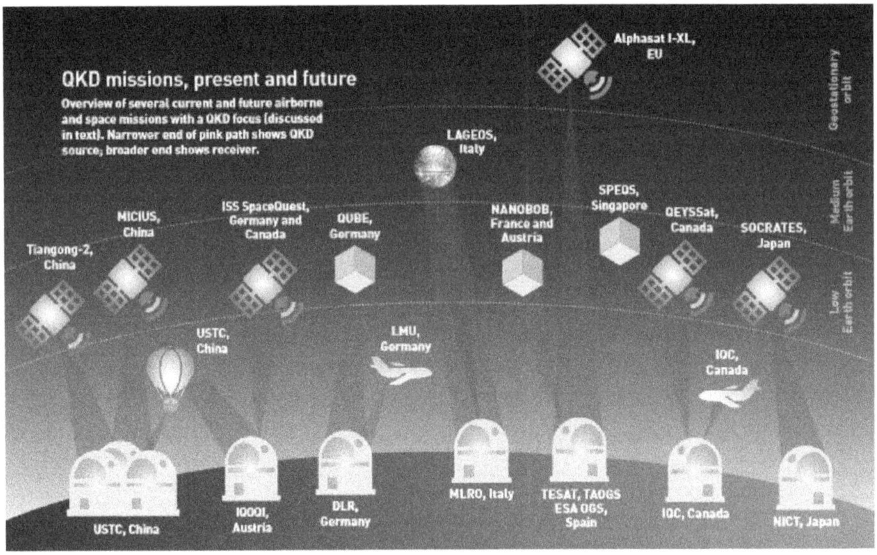

FIGURE 7.8

Airborne and space missions for future QKD integrated with the systems. *Reprinted with permission from The Optical Society of America, OSA, 2018; I. Khan, B. Heim, A. Neuzner, C. Marquardt, Satellite-based QKD, Optics & Photonics News, February 2018.*

7.2.5 Free-Space and Atmospheric Quantum Communications Demonstration Experiments

The tremendous success and progress of establishing a broadband lasercom data link between a satellite and an optical ground station (OGS) leads the way for the dream of constructing unhackable *quantum Internet.* Recent experiments demonstrate satellites to set distance records for beaming photons usable in quantum crypto in presence of atmospheric turbulence. The quantum Internet is possible because of the capabilities of quantum physics laws and growing needs and requirements of future speed, bandwidth, and cybersecurity. Free-space and atmospheric quantum communication will make it possible to develop quantum Internet for global use. An excellent recently published paper reports the fundamental foundations of quantum communications as applied to free-space and atmosphere and a review of related experiments for atmospheric quantum communication [25].

Entangled Photon Quantum Communications: Quantum physics makes possible a phenomenon known as *entanglement,* which is essentially described as, in theory, two or more particles such as photons that get linked or "entangled" can influence each other simultaneously independent of how far apart they are. One of the uses of entangled photons for quantum communications is to utilize the quantum features of photons to enable quantum information over long distances in free-space by entangling remote quantum memories. The newly developed technologies such as quantum cryptography and QKD send and receive information with the highest possible security. Fundamentals of free-space and atmospheric quantum communications are described and explained in the excellent paper by Parenti et al. [19]. Quantum cryptography, a theoretically unhackable means of information exchange, can, in principle, be established. The challenge therefore is demonstrating the feasibility of entanglement in presence of atmosphere and for a long distance such as between a satellite and a ground optical communication terminal.

A number of successful free-space and atmospheric quantum communications demonstration experiments have been performed. This section will describe the recent results of space-based (satellite) quantum communications and QKD by various researchers from all over the world, which includes various scenarios for the low earth orbit (LEO) and geostationary equatorial orbit (GEO) satellites and microsatellites technologies.

7.2.5.1 Germany Demonstration Experiment of Quantum Signals

Recent demonstration [29] in Germany clearly paves the way to a future possible global QKD network for secure communication advancing toward a quantum communication satellite. The work reports the results of quantum-limited measurements on optical signals from a geostationary satellite-to-ground terminal. The object of the space-to-ground experiment was to validate whether the quantum coherence properties are preserved after long-distance propagation of 38,600 km, through a large part of Earth's gravitational potential and through all the atmospheric turbulence layers. Atmospheric turbulence causes random intensity of an optical signal due to the random refractive index statistics of the atmosphere, which affects the optical propagation most severely close to the ground. The optical propagation characterization of quantum features will not only help in designing the global quantum communication satellite-based network, but also to develop extremely secure metropolitan area quantum communication networks on the ground. The experiment is published [29] with detailed experiment setup, data analysis, and measurement results, and will not be repeated here. The essential link setup and the results are summarized as follows.

The laser communication terminal consists of an Nd-Yag laser transmitting at a wavelength of 1064 nm continuous wave mode flying on the GEO satellite Alphasat I-XL. The laser output signal is modulated by an amplitude modulator and a phase modulator at a frequency of 2.8 GHz, which further imprints an alphabet of binary phase modulated coherent states $|\alpha\rangle$ and $|-\alpha\rangle$. The transmitting laser beam is pointed toward the OGS, which contains AOs equipped with a quantum signal acquisition system with homodyne detection scheme. The weak quantum signal mixed with a local oscillator reference beam and then detected filtering any stray daylight signal, which is noise in this situation. The total signal output loss at the receiver was measured to be 69 dB; most of the loss is due to the fraction of the actual footprint signal measured by the receiver aperture because of such a long-distance propagation. Turbulence contributed to a loss of about 1 dB in the link budget. The amplitude and variance of the received quantum states using a double Gaussian fit were determined with the assumption that variances of the two states $|+\alpha>$ and $|-\alpha>$ were equal. The measured signals were normalized to a measured reference signal of the quantum vacuum state (i.e., without input communication signal). The excess noise measured contains excess noise picked up during atmospheric propagation plus any noise in the ground station before detection and is very important as an equivalent of QBER in discrete-variable QKD in determining the BER. Considering the ground station losses, the researchers reported an upper bound on excess noise to be of 0.8 ± 2.4. For a lower-altitude LEO communication system at an orbit of 500 km, this feasibility is much greater because of much less channel loss. Quadrature phase shift keying modulation format can also be effective if modulation formats need to improve to support quantum channels with higher losses. The researchers concluded that the detected signal states are almost quantum uncertainty limited coherent states and the satellite quantum communication is feasible and thus can provide extremely powerful and effective technology for defense against any future cyberattacks.

Related research efforts on satellite and ground terminals for the high-rate laser communication component for enabling technologies for the QKD link are reported by the researchers from DLR and Ludwig Maximilians University Munich (LMU) [30]. They described the concept and hardware of laser communication terminals for low Earth-orbiting satellites. The design will also develop to reduce the QKD source size by more than an order of magnitude to simplify its integration into future FSO communication links with CubeSats. The paper has the detailed design and development information about the QKD system, FSO system, and satellite and ground terminals. The QKD developments include miniaturization of fully integrated QKD-sender employing a Vertical-cavity surface-emitting laser (VCSEL) array with pitch 250 μm, wire-grid polarizer followed by a femto-second laser written waveguide array for overlapping signals from the four diodes [30]. The increase of coupling efficiency, miniaturization of the QKD transmitter and receiver, filtering techniques to enable daylight operation, and more sensitive single photon receivers are clearly important accomplishments toward the goal of global space Internet.

7.2.5.2 Communication and Entanglement Experiments in Austria (Research Group of Prof. Anton Zeilinger)

A number of experiments in communications through atmospheric turbulence and in entanglement in Austria by the research group of Prof. Anton Zeilinger is reported. Some of their experiments in atmospheric transmission and communication are described next.

The researchers described the transmission of OAM modes through 3 km of strong turbulence over the city of Vienna by employing an incoherent detection scheme that relies on unambiguous intensity patterns of the different spatial modes [31]. The characteristic mode patterns displayed on a screen at the receiver was used to identify the characteristic mode patterns. By this scheme 16 different OAM mode superpositions were distinguished with only ~1.7% error rate and were used to encode and transmit small greyscale images. The results showed that the relative phase of the superposition modes was not affected by atmospheric turbulence. This experiment clearly demonstrated the potential feasibility of performing long-distance quantum experiments with the OAM of photons. Fig. 7.9 shows a sketch of the experimental setup for the 3 km free-space experiment in Vienna.

In [32] twisted photon entanglement through atmospheric turbulence across Vienna was demonstrated. Photons with a twisted phase front can carry a discrete unbounded amount of OAM. The researchers demonstrated the possibility of distributing quantum entanglement encoded in OAM over a turbulent atmospheric link of 3 km over Vienna. Quantum optical experiments involving entanglement have the potential use of the result that the spatial phase structure of single photons can be preserved for a long-distance link. By using the

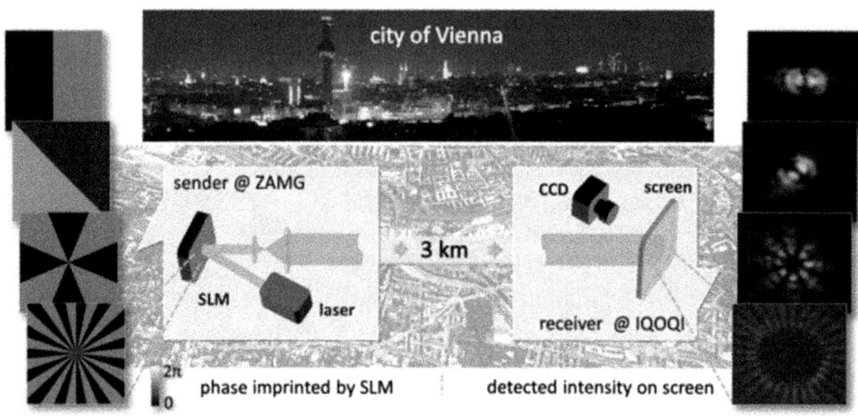

FIGURE 7.9
Sketch of the experimental setup for 3 km free-space experiment in Vienna. **Top**: Picture of an alignment laser between two terminals; **Left**: Different phase holograms modulating the laser beam at 532 nm with a spatial light modulator showing the superpositions of orbital-angular momentum modes; **Right**: At the receiver the transmitted modes recorded with a CCD camera. *Courtesy of and reprinted with permission from Prof. Anton Zeilinger and Dr. Mario Krenn, University of Vienna, Austria, March 2018; M. Krenn, R. Fickler, M. Fink, J. Handsteiner, M. Malik, T. Sheidl, R. Ursin, A. Zeilinger, Communication with spatially modulated light through turbulent air across Vienna, New Journal of Physics 16 (2014) 113028.*

first two higher-order structures the scientists in Vienna were able to show at least four additional orthogonal channels that can permit the long-distance quantum communication. Entanglement encoded in OAM was identified after long-distance transmission. Their quantum link allowed up to 11 orthogonal channels of OAM.

Fig. 7.10 shows the experimental setup for the twisted photon entanglement through turbulent air across Vienna for 3 km transmission path.

The experiment clearly demonstrated and showed the potential for quantum communication between widely separated parties, which is the basis of quantum repeaters as nodes in a global quantum network.

One experiment involved transmission of orbital angular momentum modes of light over a distance of 143 km between two Canary Islands [33]. The researchers used superpositions of these modes to demonstrate the transmission quality and at the receiver an artificial neural network was used to distinguish between the different twisted light superpositions. With this technique, they were able to identify different mode superpositions with an accuracy of more than 80% up to the third mode order and decode the transmitted message

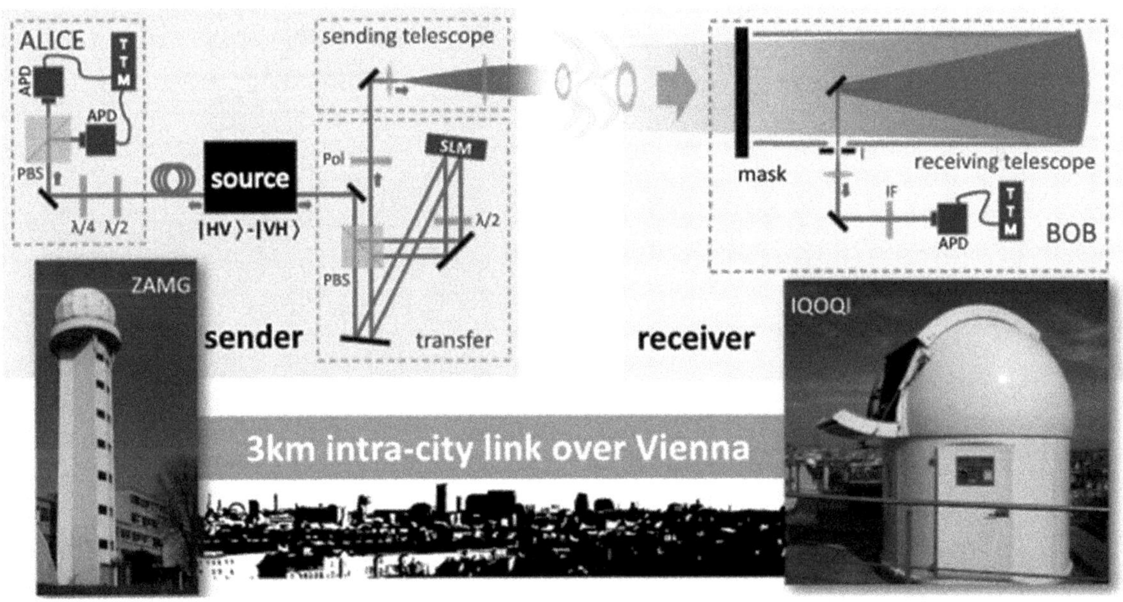

FIGURE 7.10

The experimental setup for the twisted photon entanglement through turbulent air across Vienna for 3 km transmission path. **Left**: Sender with a high-fidelity polarization entanglement source; **Right**: Shows the receiver portion of the setup. *Courtesy of and reprinted with permission from Prof. Anton Zeilinger and Dr. Mario Krenn, University of Vienna, Austria, March 2018; M. Krenn, J. Handsteiner, M. Fink, R. Ficker, A. Zeilinger, Twisted photon entanglement through turbulent air across Vienna, Proceedings of the National Academy of Sciences 12 (46) (November 17, 2015) 14197–14201.*

with an error rate of 8.33%. The demonstration indicates the feasibility of long-distance distribution of quantum entanglement in the future.

The paper [34] presented a free-space entanglement-swapping experiment between the Canary Islands of LA Palma and Tenerife and showed the feasibility of the swapping protocol in a long-distance scenario by consecutive generation of the two required photon pairs and spike-like separation of the relevant measurement events. This experiment was important because the independence of nodes must be highly demanded. Efficient implementation of entanglement of purification based on these results clearly show the potential for an efficient quantum repeater over a realistic high-loss and turbulent quantum channel. Future space and ground-based worldwide quantum Internet and distributed quantum computation as well as efficient quantum repeater for global quantum-communication are all feasible (Fig. 7.11).

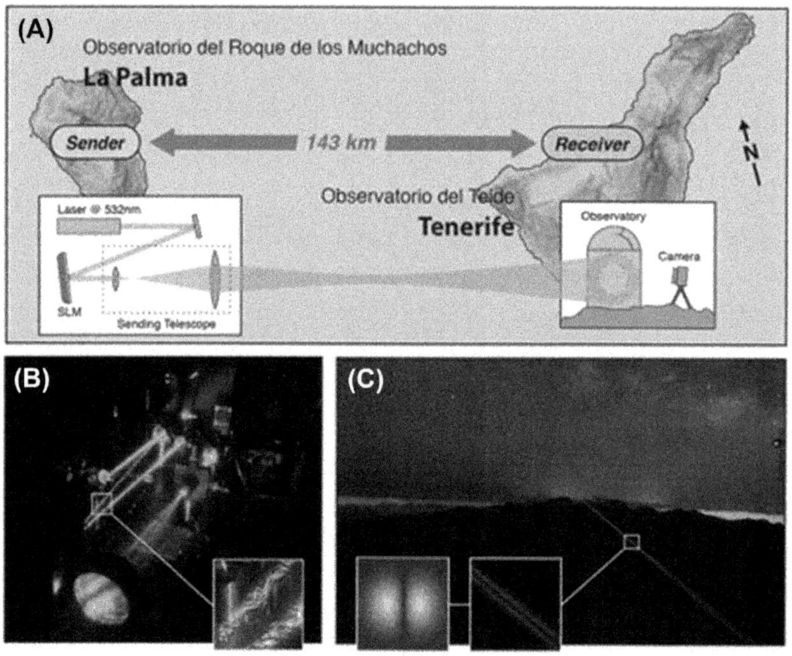

FIGURE 7.11

Sketch of the experimental setup for the twisted light transmission over 143 km: (A) sender has 60 mW laser at 532 nm modulated by SLM, the modes are magnified sent over 143 km to the receiver; (B) photo of the sender during strong turbulence showing small vortices and eddies formed by the water vapor; (C) long-time exposure photo showing an orbital-angular momentum superposition, also showing the double-lobed modal structure. *Courtesy of and reprinted with permission from Prof. Anton Zeilinger and Dr. Mario Krenn, University of Vienna, Austria, March 2018; M. Krenn, J. Handsteiner, M. Fink, R. Fickler, R. Ursin, M. Malik, and A. Zeilinger, Twisted light transmission over 143 km, Proceedings of the National Academy of Sciences (November 15, 2016) 201612023.*

Fig. 7.12 shows a free-space entanglement-swapping experiment over 143 km between the Canary Islands and Tenerife.

7.2.5.3 Long Distance Over 1000 km Quantum Communication and Entanglement Experimental Demonstration

A recent report on the possibility of sending signals for long distance quantum communication is described in [35]. Recent results from the researchers provide proof of performing quantum communication for a long distance exceeding 1000 km in space. Liao et al. [36] and Ren et al. [37] demonstrated satellite-based QKD or quantum teleportation use a satellite in a low Earth orbit at 500 km where the quantum information has been encoded in the hardware module in the polarization of photons that have a near-infrared wavelength. Fig. 7.13 shows the sketches of the experimental setups for the downlink and uplink scenarios. For Laio and colleagues' experiment setup the transmitter is

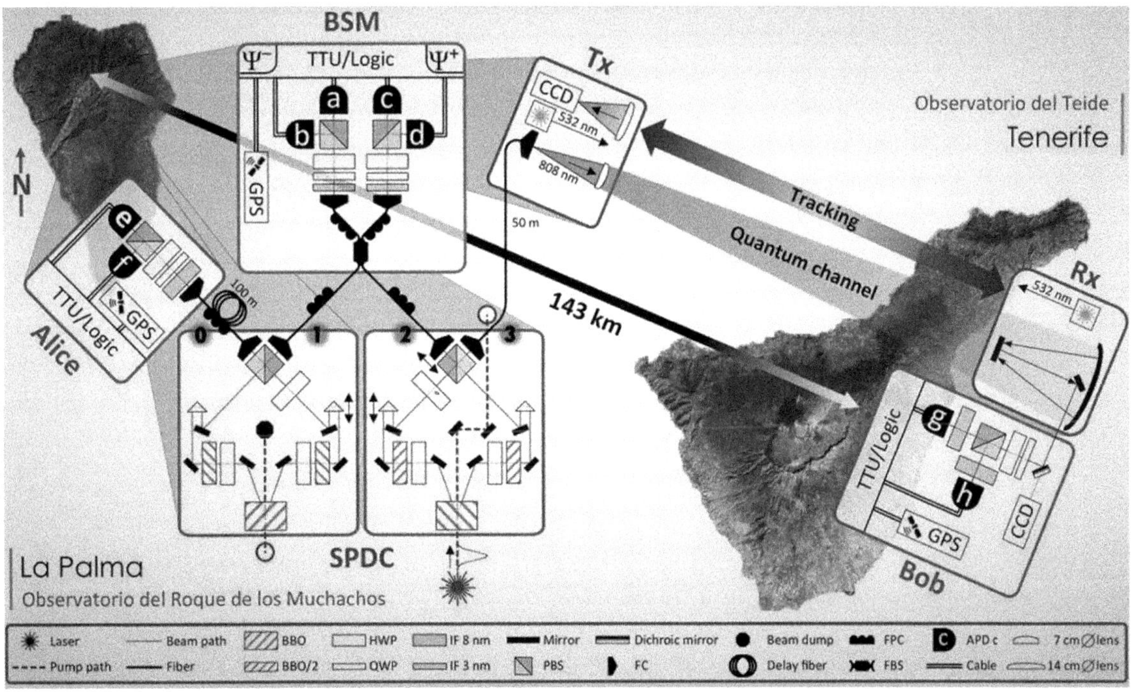

FIGURE 7.12

The free-space entanglement-swapping experiment over 143 km between the Canary Islands of La Palma and Tenerife. Alice, the SPDC sources, and the BSM module were located at La Palma, and Bob was located on Tenerife. Alice and Bob spectrally filtered their photons; the receiver on Tenerife captured photons where Bob performed his polarization-dependent measurement. *Courtesy of and reprinted with permission from Prof. Anton Zeilinger and Dr. Mario Krenn, University of Vienna, Austria, March 2018; T. Herbst, T. Scheidl, M. Fink, J. Handsteiner, B. Wittmann, R. Ursin, A. Zeilinger, Teleportation of entanglement over 143 km, Proceedings of the National Academy of Sciences 112 (46) (November 17, 2015) 14202–14205.*

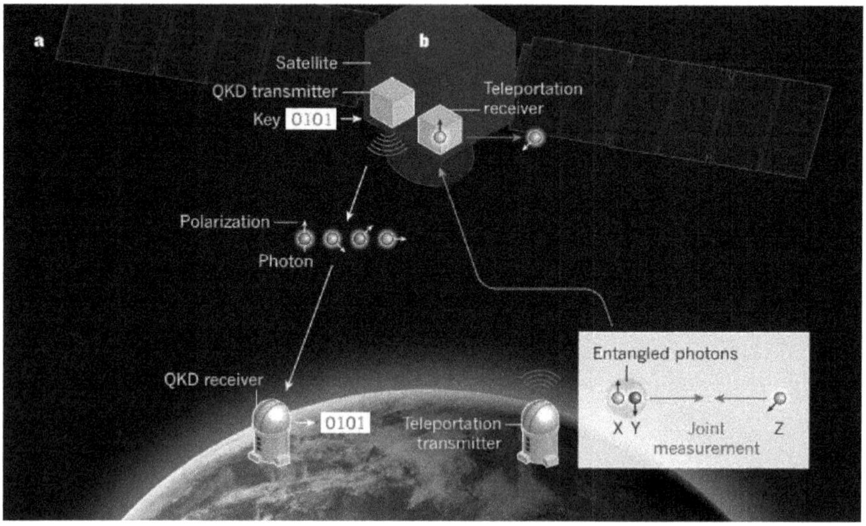

FIGURE 7.13

Long distance over 1000 km quantum communication. *Reprinted with permission from publisher Springer Nature, March 2018; E. Diamanti, Quantum signals could soon span the globe, Nature 549 (September 2017) (7).*

located on the satellite and the receiver on a ground station with a separation distance of up to 1200 km (Fig. 7.13A) in which case downlink light propagates through atmospheric turbulence close to the ground. Ren and colleagues' quantum teleportation experiment uses an uplink propagation from the ground station to a receiver on the satellite across a distance up to 1400 km (Fig. 7.13B). Ren et al. produced a pair of entangled photons; one of the pair was to the satellite, which performed a joint measurement of the photon on Earth and a third photon whose polarization state was to be teleported. The successful demonstration results is therefore a potential of performing quantum communication over long distances exceeding 1000 km, which is not possible with terrestrial links.

7.2.5.4 China Demonstration Experiment of Quantum Communication From Low Earth Orbit Satellite

Chinese researchers, at about the same time as the European researchers, reported the record for transmitting a pair of entangled particles using a beam splitter mounted on a satellite, which split the laser signal from the satellite into two distinct beams. After being filtered through a crystal onboard, the beams produced a pair of entangled photons. In the Chinese demonstration, these photons traveled through up to 1240 km of space to two different ground stations in China. The quantum communication demonstration experiment used the Chinese satellite named *Micius* (SOM) at an altitude of ~500 km.

For the entanglement distribution experiment the satellite cooperated with three OGSs located in Delingha in Qinghai, Nanshan in Urumqi, Xinjiang, and Gaomeigu Observatory in Lijiang, Yunnan. The distance between the two OGSs Delingha and Lijiang (Nanshan) is 1203 km. The distances between the orbiting satellite and the ground stations vary from 500 km to 2000 km. Bidirectional distribution of entangled photon pairs through satellite-to-ground two-downlink channels was successfully demonstrated [38]. The experiment reported establishing the entanglement between two single photons separated over 1203 km with an average two-photon count rate of 1.1 Hz and state fidelity of 0.869 ± 0.085. The experimental details are already published [38] and are not repeated here. The essential results and the conclusions from the demonstration are summarized next.

A CW laser diode at 405 nm wavelength was used to pump a periodically poled $KTiOPO_4$ (PPKTP) crystal. The pump laser split by a polarizing beam splitter passes through a nonlinear crystal in clockwise and anticlockwise directions simultaneously to produce down-converted photon pairs at ~850 nm wavelength in polarization-entangled states. The source emits 5.9 million entangled photon pairs per second, which propagate through two ground stations at 500–2000 km apart. The atmospheric effects such as attenuation and turbulence-induced distortions become severe close to the ground, for about 10 km from the ground. This results in a limiting factor in the results of entanglement distribution of the photons. The researchers claim that the effective link efficiency at 1200 km in their work is over 12 orders of magnitude higher than the direct bidirectional transmission of the two photons through the best commercial telecommunication fibers with a loss of 0.16 dB/km. The results of the demonstration lead a way to develop future global quantum Internet.

7.2.5.5 Experimental Satellite Quantum Communications: Italy Experiment Physical Review Lett 24 July 2015

Instead of a laser transmitter installed on a GEO satellite, the demonstration in Italy used corner cube retroreflector (CCR) as quantum transmitters in orbit and the Matera Laser Ranging Observatory of the Italian Space Agency as a quantum receiver. The quantum transmitter was simulated by CCRs of an orbiting satellite and the discrimination of different polarization states was done by the ground receiver. Qubit stream is realized from a 100 MHz train of pulses sent to the satellite from the ground and retroreflected by the CCRs back to the single photon level from the satellite. A QKD source in space was thus simulated with retroreflectors equipped with a modulator to make a very compact payload in space. Bright laser ranging pules at a repetition rate of 10 Hz helped to accomplish synchronization. Four LEO satellites ranging from 1000 to 2600 km distance, namely

Jason-2, Larets, Starlette, and Stella, were selected with such polarization preserving CCRs. The setup and the results were published [39] by the researchers in 2015. The observed QBER was to a level suitable for several quantum information protocols with an average QBER = 4.6% for a total link duration of 85 s. The estimated mean photon number per pulse leaving the satellites was estimated to be of the order of one. A two-way protocol can also be established as a practical QKD system since both beams share the same path so that the polarization transformation induced by the telescope movement can be compensated for the downlink. This simple trusted device can be turned into QC stations around the world.

7.2.5.6 NICT Experimental Demonstration of Satellite-to-Ground Quantum-Limited Communication Using Microsatellite

NICT successfully demonstrated the first quantum communication experiment from space using the world's smallest and lightest quantum communication transmitter (SOTA) onboard the microsatellite SOCRATES [40–42]. The information was received from the satellite in a single-photon regime in an OGS in Koganei city. SOTA has the dimension of 17.8 cm × 11.4 cm × 26.8 cm and weighs 6 kg and transmits a laser signal to the ground at 10 million bits per second from an altitude of 600 km at a speed of 7 km/s. Some basic experiments on space QKD are also included in the ongoing projects. Ranges of QKD links can be significantly increased with satellites to allow intercontinental key exchanges in the future. The signal photons transmitted from the SOTA are mostly lost before reaching the receiver because of the divergence of the laser beam and the limited aperture to collect the photons. Furthermore, many photons are absorbed or scattered because of the atmospheric propagation effects and therefore the received signal can be very weak, carrying an average of fewer than 0.1 photons per pulse, which can be detected by a quantum receiver. Quantum cryptography however can detect the presence of an eavesdropper even with less than one photon per pulse to deliver the secret keys securely. NICT has been able to synchronize the signals between SOCRATES and the OGS using microsatellites. This demonstration clearly has the potential for the global long haul and a truly secure satellite communication network using small low-cost satellites.

References

[1] X. Sun1, I.B. Djordjevic1, Physical-layer security in orbital angular momentum multiplexing free-space optical communications, IEEE Photonics Journal 8 (1) (February 2016).

[2] H. Endo, M. Fujiwara, M. Kitamura, T. Ito, M. Toyoshima, Y. Takayama, H. Takenaka, R. Shimizu, N. Laurenti, G. Vallone, P. Villoresi, T. Aoki, M. Sasaki1, Free-space optical channel estimation for physical layer security, Optics Express 24 (8) (April 18, 2016).

[3] G. Blinowski, Security issues in visible light communication systems, ScienceDirect (2015) 234–239. IFAC-Papers On Line 48-4.

[4] N.F. Rulkov, A.R. Volkovskii, M.M. Sushchik, L.S. Tsimring, L. Illing, Digital communication using self-synchronizing chaotic pulse position modulation, in: L.E. Larson, L.S. Tsimring, J.M. Liu (Eds.), Digital Communications Using Chaos and Nonlinear Dynamics, Institute for Nonlinear Science. Springer, New York, NY, 2006.

[5] B. Wang, S.M. Zhong, X.C. Dong, On the novel chaotic secure communication scheme design, Communications in Nonlinear Science and Numerical Simulation 39 (2016) 108–117. Elsevier.

[6] A.K. Majumdar, Advanced Free Space Optics (FSO): A Systems Approach, Springer, New York, 2015.

[7] K.M. Cuomo, A.V. Oppenheim, Physical Review Letters 71 (1993) 65.

[8] L. Kocarev, et al., International Journal of Bifurcation and Chaos in Applied Sciences and Engineering 2 (1992) 709.

[9] T.L. Carroll, L.M. Pecora, IEEE Transactions on Circuits and Systems 40 (1993) 646.

[10] P. Colet, R. Roy, Optics Letters 19 (1994) 2056.

[11] G.D. Vanwiggeren, R. Roy, Chaotic communication using time-delayed optical systems, International Journal of Bifurcation and Chaos 9 (11) (1999) 2129–2156.

[12] A. Argyris, D. Syvridis, L. Larger, V. Annovazzi-Lodi, P. Colet, I. Fischer, J. Garcia-Ojalvo, C.R. Mirasso, L. Pesquera, K. Alan Shore, Chaos-based communications at high bit rates using commercial fibre-optic links, Nature Publishing Group 438 (17) (November 2005) 343–346.

[13] N.F. Rulkov, M.A. Vorontsov, L. Illing, Chaotic free-space laser communication over a turbulent channel, Physical Review Letters 89 (27) (December 30, 2002), 277905-1–27705-4.

[14] M.R. Chatterjee, A. Mohamed, F.S. Almehmadi, Secure free-space communication, turbulence mitigation, and other applications using acousto-optic chaos, Applied Optics 57 (10) (April 1, 2018).

[15] M.R. Chatterjee, M. AlSaedi, Examination of chaotic signal encryption, synchronization and retrieval using hybrid acousto-optic feedback, in: Proc. of OSA FiO/LS/META/OF&T, Paper No. FWC3, 2008.

[16] M.A. Vorontsov, S.L. Lachinova, A.K. Majumdar, Target-in-the-loop remote sensing of laser beam and atmospheric turbulence characteristics, Applied Optics 55 (19) (July 1, 2016).

[17] J.H. Shapiro, Reciprocity of the turbulent atmosphere, Journal of the Optical Society of America 61 (1971) 492–495.

[18] A.L. Puryear, J.H. Shapiro, R.R. Parenti, Reciprocity-enhanced optical communication through atmospheric turbulence- Part II: communication architectures and performance, Journal of Optical Communications and Networking 5 (8) (2013).

[19] R.R. Parenti, J.M. Roth, J.H. Shapiro, F.G. Walther, J.A. Greco, Experimental observations of channel reciprocity in single-mode free-space optical links, Optics Express 20 (19) (September 10, 2012).

[20] F.J. Duarte, T.S. Taylor, Interferometry: Quantum Entanglement Physics Secures Space-to-Space Interferometric Communications, Article in LaserFocusWorld, 04/08/2015. https://www.laserfocusworld.com/articles/print/volume-51/issue-04/features/interferometry-quantum-entanglement-physics-secures-space-to-space-interferometric-communications.html.

[21] F.J. Duarte, Optics Communication 205 (2002) 313–319.

[22] F.J. Duarte, Optics Communication 103 (1993) 8–14.

[23] F.J. Duarte, T.S. Taylor, et al., The N-slit interferometer: an extended configuration, Journal of Optics 12 (015705) (2010).

[24] F.J. Duarte, T.S. Taylor, et al., N-slit interferometer for secure free-space optical communications: 527 m intra interferometric path length, Journal of Optics 13 (035710) (2011).

[25] R.E. Meyers, K.S. Deacon, A.D. Tunick, Free-space and atmospheric quantum communications (Chapter 10), in: A.K. Majumdar (Ed.), Advanced Free Space Optics (FSO): A Systems Approach, Springer, New York, 2015.

[26] P.V. Trinh, T.V. Pham, N.T. Dang, H.V. Nguyen, S.X. Ng, A.T. Pham, Design andSecurity analysis of quantm key distribution protocol over free-space optics using dual-threshold direct-detection receiver, IEEE Access 6 (2018). Digital Object Identifier 10.1109/ACCESS.2018.2800291.

[27] R. Bedington, J./M. Arrazola, A. Ling, published in npj Quantum Information, Progress in Satellite Quantum Key Distribution, vol. 3, 2017, p. 30, https://doi.org/10.1038/s41534-017-0031-5.

[28] I. Khan, B. Heim, A. Neuzner, C. Marquardt, Satellite-based QKD, Optics & Photonics News, February 2018.

[29] K. Günthner, I. Khan, D. Elser, B. Stiller, Ö. Bayraktar, C.R. Müller, K. Saucke, D. Tröndle, F. Heine, S. Seel, P. Greulich, H. Zech, B. Gütlich, S. Philipp-May, C. Marquardt, G. Leuchs, Quantum-limited measurements of optical signals from a geostationary satellite, Optica 4 (6) (June 2017), https://doi.org/10.1364/OPTICA.4.000611.

[30] F. Moll, H. Weinfurter, M. Rau, C. Schmidt, G. Melén, T. Vogl, S. Nauerth, C. Fuchs, Aerospace laser communications technology as enabler for worldwide quantum key distribution, Proceedings of SPIE 9900 (2016), 99000K-99001K to 99000K-99009K.

[31] M. Krenn, R. Fickler, M. Fink, J. Handsteiner, M. Malik, T. Sheidl, R. Ursin, A. Zeilinger, Communication with spatially modulated light through turbulent air across Vienna, New Journal of Physics 16 (2014) 113028.

[32] M. Krenn, J. Handsteiner, M. Fink, R. Ficker, A. Zeilinger, Twisted photon entanglement through turbulent air across Vienna, Proceedings of the National Academy of Sciences 12 (46) (November 17, 2015) 14197–14201.

[33] M. Krenn, J. Handsteiner, M. Fink, R. Fickler, R. Ursin, M. Malik, A. Zeilinger, Twisted light transmission over 143 km, Proceedings of the National Academy of Sciences (November 15, 2016), 201612023.

[34] T. Herbst, T. Scheidl, M. Fink, J. Handsteiner, B. Wittmann, R. Ursin, A. Zeilinger, Teleportation of entanglement over 143 km, Proceedings of the National Academy of Sciences 112 (46) (November 17, 2015) 14202–14205.

[35] E. Diamanti, Quantum signals could soon span the globe, Nature 549 (7) (September 2017).

[36] S.-K. Liao, et al., Nature 549 (2017) 43–47.

[37] J.-G. Ren, et al., Nature 549 (2017) 70–73.

[38] J. Yin, Y. Cao, Y.-H. Li, S.-K. Liao, L. Zhang, J.-G. Ren, W.-Q. Cai, W.-Y. Liu, B. Li, H. Dai, G.-B. Li, Q.-M. Lu, Y.-H. Gong, Y. Xu, S.-L. Li, F.-Z. Li, Y.-Y. Yin, Z.-Q. Jiang, M. Li, J.-J. Jia, G. Ren, D. He, Y.-L. Zhou, X.-X. Zhang, N. Wang, X. Chang, Z.-C. Zhu, N.-L. Liu, Y.-A. Chen, C.-Y. Lu, R. Shu, C.-Z. Peng, J.-Y. Wang, J.-W. Pan, Satellite-based entanglement distribution over 1200 kilometers, Science 356 (6343) (June 16 , 2017) 1140–1144.

[39] G. Vallone, D. Bacco, D. Dequal, S. Gaiarin, V. Luceri, G. Bianco, P. Villoresi, Experimental satellite quantum communications, Physical Review Letters (July 24, 2015) week ending.

[40] H. Takenaka, A. Carrasco-Casado, M. Fujiwara, M. Kitamura, S. Sasaki, M. Toyoshima, Satellite-to-ground quantum-limited communication using a 50-kg-class microsatellite, Nature Photonics 11 (2017) 502–508, https://doi.org/10.1038/nphoton.2017.107.

[41] https://phys.org/news/2017-07-world-space-quantum-microsatellite.html.

[42] http://www.nict.go.jp/en/press/2017/07/11-1.html.

Further Reading

[1] A.K. Majumdar, J.C. Ricklin, Free-Space Laser Communications: Principles and Advances, Springer, New York, 2008.

CHAPTER 8

Free-Space Optical Communications: Role and Integration With the Internet of Things

8.1 INTRODUCTION

This chapter starts with a brief description of the concept of *Internet-of-Things* (IoT) wherein machines and everyday objects are connected via the Internet. Optical wireless communication (OWC) offering broadband capability is a perfect candidate to control and monitor remotely the devices within IoT. By the end of 2020, IoT is predicted to include more than 75 billion networked devices globally. Connecting mobile items wirelessly to the Internet via Wi-Fi, Bluetooth, or other proprietary wireless communications protocols has become one of the important drivers behind the huge growth of IoT. The chapter discusses the role of OWC/free-space optics (FSO) in the goal of IoT to enable things to be connected anytime, anywhere, with anything and anyone using any path/network. Consumers and enterprise applications include and involve appliances and services in smart houses, the electric vehicle, integrated security, energy-saving, automation, telecommunication, computers/portable devices, and entertainment into a single system with a shared user interface. Very high data rates, a dense number of users, and a huge number of devices are new challenges that can only be met by OWC/FSO technology. The chapter points out the importance of essential, reliable, bidirectional signaling for collecting and routing data between devices where IoT data streams come into play. The chapter describes and provides potential architecture of OWC/FSO integration as an integral part of IoT and machine-to-machine (M2M) communication to meet the need for fast, high-capacity networks and to achieve global IoT. The chapter finally provides a potential future OWC feasibility for a worldwide IoT market in order to develop hybrid terrestrial/satellite access solutions into the IoT for seamless and ubiquitous communications across the globe.

8.2 BASICS OF INTERNET OF THINGS

IoT enables the networking of "things," allowing different devices and items to sense and then communicate with each other and their surrounding environment. A sensor and a communicating device are therefore embedded using transceivers to send and receive information. In addition, since many standard bodies are developing the fifth generation (5G) around 1 Gb/s, they should

Optical Wireless Communications for Broadband Global Internet Connectivity.
https://doi.org/10.1016/B978-0-12-813365-1.00008-4

adapt to handle IoT. It is predicted that around 50 billion things will connect to IoT and Internet of Everything (IoE). This is where increasing customers' numbers impacting the spectrum bands already congested will find a potential solution with FSO links. IoT is one of the most exciting emerging topics to make a significant impact on technical, social, and economic issues and basically addresses how all these *things* in the world should connect to the *Internet*. IoT combines all the possible everyday consumer products goods, cars and trucks, utility products, sensors, and almost every object with Internet connectivity. Thus the concept of IoT extends Internet connectivity beyond traditional devices to a diverse range of devices so that they all communicate and interact with the external environment, all via Internet.

IoT is the network of interrelated objects ("things") to collect and exchange data with one another through embedded electronics, software, and sensors, all over the Internet. This IoT approach can change the way we do business, live, and play. There will be potential for 100 billion connected IoT devices, which can have a projected global economic impact of more than $11 trillion by the year 2025. However, there are many technical challenges and concerns for making IoT successful. Some challenges include security of Internet-connected devices, surveillance and privacy concerns with growing technical and development challenges, and resolving legal issues. There has been tremendous success of mobile computing using laptops, smart phones, and tablets because of the widespread and convenient W-Fi and cellular network access to the Internet and cloud-powered applications. Recent developments of low-powered, tiny devices such as mobile and wearable devices like small watches or health monitors, rely on low-power wireless communications like Bluetooth, and therefore require an application layer gateway to connect to the Internet. Connecting a new IoT device requires a new smart phone app or a new laptop dongle to provide the Internet gateway. Therefore, in order to accommodate low-powered devices, the Internet gateway for IoT needs to be addressed and solved. This will require a new networking architecture for low-powered devices to be part of the global network of smart phones.

IoT can achieve a number of key areas including wearables, connected cars, connected buildings and homes, connected cities, and industrial Internets. Industrial Internets can include different markets such as transportation, oil/gas industry, healthcare hospitals and clinics, just to name a few that affect everyday lives in our society. Web-enabled devices are being developed to provide common platforms on which they can communicate and develop new future applications. This will also require increased share for Wi-Fi, sensors, and micro-controllers, and various software applications for managing communications between devices. These requirements will modify the direction of current technology from the fixed to the mobile Internet anywhere, anytime.

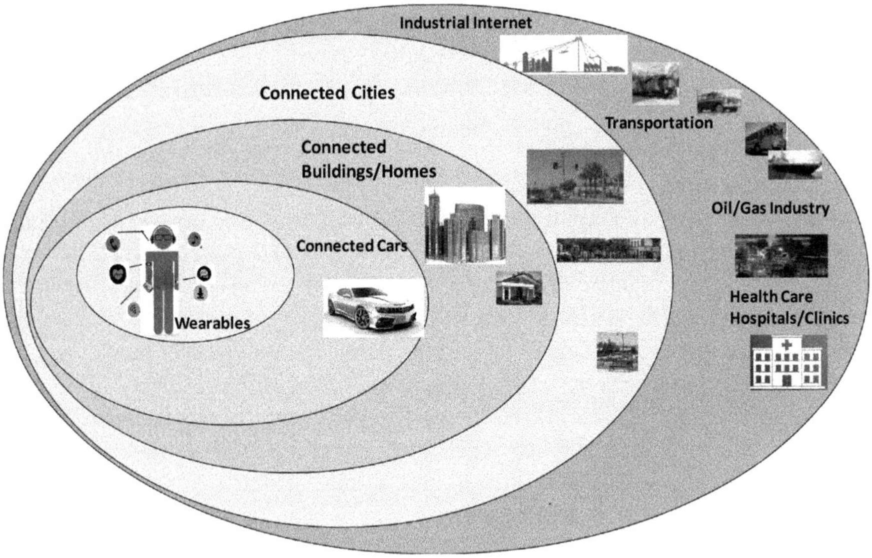

FIGURE 8.1
Some general aspects of IoT applications affecting everyday life.

The impact of potential communication technology will occur in three areas, namely, Wi-Fi, cellular, and cloud (aka Fog) computing. Fig. 8.1 shows some general areas of everyday life today where IoT can be adopted to connect devices such as everyday consumer objects and industrial equipment onto the network to enable information gathering and via software manage to provide new services and benefits for modern life. The most profound impact from the IoT sector across three areas of communications technology are Wi-Fi, cellular, and the Fog. New software needs to be developed for IoT that is different from the software on our desktops, laptops, tablets, or smart phones to create new platforms in order to smoothly manage communication with the connected devices and sensors as well as for storing and securing data for analyzing and visualizing the data. In addition, it is also required for developing consumer software.

8.2.1 Key Concepts of Internet of Things

8.2.1.1 Internet of Things Common Uses and Applications

IoT systems enhance data collection and automation, and remotely control capability and flexibility through smart devices and enabling technology. IoT has potential in many applications that affect personal lives, workplace productivity, and consumption, which include wearables, cars, homes, cities, and industrials. Applications include broad areas such as government and safety, home and office, health and medicine, just to name a few. IoT provides a common platform for web-enabled devices to communicate for different uses. By connecting the devices like everyday consumer objects and industrial equipment onto the network, the *IoT* connects things to communicate with each other.

8.2.1.2 Internet of Things Key Features

Some of the features of IoT that specifically integrate small devices are *artificial intelligence*, to make virtually anything smart with the power of data collection using artificial intelligence algorithms and network; *connectivity*, creating small networks between its system devices; *sensors*, defining individual and different types of instruments, making them into an active system for real-world integration; and *active engagement*, making active content, product, or service engagement. *Hardware* for IoT includes sensors (for defining individual instruments) and wearable electronic devices worn on the head, neck, arms, or feet (e.g., smart watches and smart glasses). IoT also requires software for data collection and integration, and real-time analytics. Standard protocols such as radio frequency identification (RFID) and near-field communication, Bluetooth, and Wi-Fi-Direct needed to support specific networking functionality in an IoT. Some of the major issues for developing IoT include significant challenges such as security, privacy, and compliance with regulations, which need to be addressed while IoT technology is being developed. IoT *software* addresses the various areas of networking using platforms and embedded systems and also performs data collection and device integration, and analyzes information to accomplish automation-related tasks.

8.2.1.3 Secure Internet of Things: Microsoft's Solution

To connect everything to the Internet and build IoT with millions of online devices with no maintenance, security is a big problem. Microsoft recently proposed a solution to secure IoT based on custom Linux, custom chips, Azure [1]. The solution, named Azure Sphere, has three components: (1) a new class of microcontrollers, which is the necessary foundation, have seven critical hardware features with encryption keys protected by hardware and incorporate an application processor, a real-time processor, flash storage, and memory with network connectivity; (2) a new operating system—Azure Sphere OS—providing a secure platform that scales down smaller systems than Windows can reach; and (3) Azure Sphere Security Service, which is a cloud service to detect security issues (recognizing failures and errors on devices) that can be used to mediate secure communications between devices and to the cloud [1].

8.2.1.4 Internet of Things Communication Models

It is important to understand how IoT devices connect with each other and communicate between the devices and the environment, or simply networking of smart things. General configurations of IoT communication models may

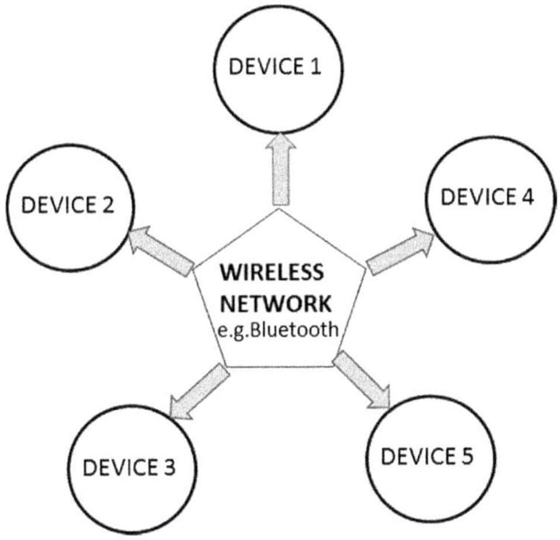

FIGURE 8.2
Device-to-device communication model for a number of devices in a wireless network.

include communications between devices, as well as between cloud and gateway. Device-to-device can be directly connected and can communicate with each other over IP networks or the Internet and can use protocols like Bluetooth. Some of the examples in this category of communication are applications of home automation systems like light bulbs and switches, which require low data rate. In the device-to-cloud communication model, devices connect directly to an Internet cloud service with existing Wi-Fi connections and the IP network to exchange data and control message traffic. The cloud enables the user to obtain remote access to their devices, for example, via a smart phone or Web interface. For the device-to-cloud model, the IoT device connects to the cloud via a local gateway device, which can provide security and other functionality. Fig. 8.2 depicts a device-to-device communication model for a number of devices, and Fig. 8.3 depicts device-to-cloud and device-to-gateway communication modes.

The communication architectures are important to the end users and can open the possibilities of using information in innovative ways.

8.2.1.5 Internet of Things Architecture

It is important to understand how all these *Things* should be connected with the *Internet*. An excellent survey of and fundamentals behind IoT architectures was presented recently [2,3], discussing the IoT-oriented architectures that are capable of improving the related tools, technology, and methodology to facilitate developer's requirements, an architecture that leverages the increasingly ubiquitous presence of Bluetooth to connect IoT peripherals to the Internet [2,3]. The main difference of today's IoT demand from the conventional

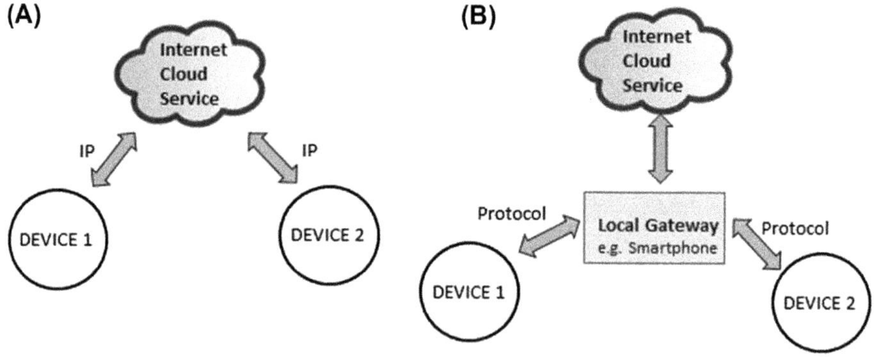

FIGURE 8.3
(A) Device-to-cloud and (B) device-to-gateway communication models to connect and communicate between a number of devices.

requirement is that the tiny, embedded, low-power, wireless, mobile, and wearable devices do not have the same level of ubiquitous and universal access to the Internet as other devices like laptops, tablets, and smart phones. These devices require an *application layer gateway*—a system capable of translating data from the low-power link to the Internet, which is obviously not provided by the current implementations. Connecting a new IoT device requires a new smart phone app, a new laptop dongle, or a new base station device [2,3]. This simply means that a new networking architecture for a low-power wireless device is required. The proposed architecture [2–4] consists of two data transmission mechanisms: one is via IPv6, which uses the smart phone as a temporary IPv6 router and treats the peripheral as an IP-connected end host, and the other data transmission mechanism is via proxy, which uses the smart phone to forward the peripheral's Bluetooth low-energy profile to the cloud. The researchers pointed out that the proposed architecture could expedite the growth of a global, highly connected, robust *IoT*.

8.2.1.6 Some Design Considerations for Internet of Things Systems

Some design considerations for a practical IoT network include the following:

Sensors: Processing and networking capabilities for designing the sensors are important, and will also depend on the power consumption for the sensors. Typically, sub-mW boards are used to handle a few hundred mW of power.

IoT Nodes: The choice of the networking technology used is important, and depends on the power required to transmit and receive messages in the sensor. This will also depend on the distance between the sender and the receiver and the propagation characteristics like signal distortion and ambient noise. This will lead to the choice of a given wireless networking protocol. For example, communication inside a small building needs a different protocol than communication in a smart city.

Middleware: The choice will depend on the ability to interoperate with all kinds of communication protocols and devices and compatibility with certain devices and protocols. Communication protocols need to be compatible with the firewalls in the organizations involved; for example, a protocol based on hypertext transfer protocol. Also, it is important to choose between transmission control protocol (TCP) and user datagram protocol as well as options to store sensor data streams and such.

Application Layer: The application developers need to consider creating the application layer, which can include data mining, data processing, and visualization application programming interfaces.

8.3 5G WIRELESS COMMUNICATIONS FOR FUTURE INTERNET OF THINGS: THE NEED FOR FREE-SPACE OPTICAL/VISIBLE LIGHT COMMUNICATION TECHNOLOGY

5G wireless communication is expected to offer high data rates and improved quality of service, and therefore IoT needs to be implemented in 5G architecture. FSO communication is the key technology that can offer high-capacity communication performance and therefore is the best choice for wireless communications for satisfying the fullest capabilities of 5G expectations offering 1 Gb/s or more speeds. The performance and coverage area of IoT will also be improved with high-speed FSO communication when 5G features are implemented in IoT. The requirements of IoT such as sensing data, processing, and connectivity can all be satisfied for IoT devices' signaling, security, omnipresent direction, power consumption, and bandwidth [5]. Techniques like device-to-device and M2M communications are useful to understand how IoT works as discussed in [5]. 5G, the next generation for mobile system technology using high data broadcast within the latest mobile operating system, provides high-resolution bidirectional connectivity. Visible light communication (VLC) technology can offer very high data rate communication using off-the-shelf LEDs as transmitters and avalanche photo detectors as receivers. How VLC and specifically Li-Fi (equivalent of Wi-Fi using light and VLC communication) can be integrated for IoT will be discussed later. The details of VLC and Li-Fi are discussed in another chapter and therefore are not repeated here.

8.3.1 Integrating High Bandwidth Fiber With 5G and Internet of Things

The tremendous growth in connections and network traffic require building infrastructure that can fulfill the upcoming demands on capacity and latency, both of which are essentials in high data rate wireless communications. Sensors embedded in devices and infrastructure will demand 5G wireless communication systems to integrate effectively. 5G mobile networks will be needed to support IoT to connect potentially more devices with high bandwidth. 5G wireless networks will use fiber for backhaul and fronthaul to connect cell sites to mobile control backbone. To fulfill the requirements of both 5G and IoT, an optical network as the backbone for the transport of those networks will be required to satisfy the high bandwidth and very low latency demands. The main emphasis in this section is that microdata centers are integrated into 5G towers by fiber, which offers the high capacity data transfer benefits. Network accesses closer to the individual users, along city streets, in buildings, and everywhere in between can be possible using VLC technology, especially developing Li-Fi, which will serve as the conventional Wi-Fi except operating at visible wavelength. The flexibility and high bandwidth offered by VLC and Li-Fi are discussed in detail in another chapter. Furthermore, 5G across macrosites, in-building, and within small cells and integration with IoT are all possible using VLC and Li-Fi technologies.

8.4 FREE-SPACE OPTICS INTEGRATION FOR INTERNET OF THINGS

8.4.1 Internet of Things and Photonics/Free-Space Optical

IoT will benefit the bottom line of photonics products manufacturers and laser companies worldwide such as the optical fiber communication business, RFID and barcode reader technology suppliers, and imaging and sensor companies. Because of the large bandwidth requirement FSO links will play a major role in all aspects of developing IoT architectures whether in home or outdoor environments or even for global applications accessible by the Internet.

8.4.2 Need for Optical Links

The increase of connected devices for IoT is predicted by CISCO to be about 50 billion by the year 2020. This simply will lead to real congestion in wireless traffic among devices, causing interference. In order to handle this wireless traffic to accommodate the increasing wireless bandwidth demand, FSO links can provide a potential solution. Besides offering huge bandwidth, an FSO link is license-free, physically secure, has no electromagnetic interference, and can provide spatial diversity to prevent multipath fading. In earlier chapters of this book, the design and implementations of optical connectivity for indoor, terrestrial, and space- and air-borne platforms are described and discussed in detail. The concept of global connectivity with optical wireless communications was discussed in earlier chapters. Recent ideas of creating a "smart city" is a broad concept in an urban development vision to integrate many information and communication technologies and IoT solutions. The smart city, with all the connectivity of many "things," will improve the quality of life using combined 5G and IoT technology. The future 5G networks with the concept of smart cities with lots of things working in synergy are becoming a reality and practically achievable. OWC, both FSO and VLC, can provide a viable solution to make the 5G concept successful. Ubiquitous data links are realizable with optical links with their opportunity to adopt a system in many flexible ways for the users for many IoT applications and also for IoE in the future.

8.4.3 Visible Light Communication and Li-Fi for Internet of Things

The number of IoT devices is predicted to be over 28 billion in 2020 and beyond. Because of the shortage of the conventional radio-frequency spectrum, VLC can be a better alternative to meet the tremendous demand for IoT. VLC integrated with IoT has the potential for a wide range of applications; for example, indoor, smart city, industry, hospital, and transportation. VLC applications for device-to-device communication require reliable uplink along with high-speed downlink, which uses LEDs and photodetectors integrated into the device. Some of the recently developed VLC work in darkness and therefore are much more energy efficient than the conventional VLC, and can be extremely suitable for IoT applications. A recently proposed system design to overcome the limitations of traditional VLC combines VLC in dark and an orthogonal frequency division multiplexing method for data transmission and mapping the serial data using quadrature amplitude modulation to accomplish a high data rate of 100 Gb/s or more as well as taking advantage of lower energy consumption [6] suitable for IoT. In some other applications IoT networks may only require very low data rate communication, for example, for transmitting sensing or identity information. For implementation of a VLC link on existing computer communication some of the standards and interfaces such as universal receiver/transmitter (UART) are used. A VLC-over-UART system has been discussed for these types of applications [7]. The details of VLC are described and discussed in an earlier chapter.

Li-Fi is a new technology that allows mobile devices and other connected objects to connect to each other by using LED lights, which transmit data by modulating the light signals. Light signals are received and converted back into data by a *dongle* connected to the device. Every one of the billions of light bulbs in use today in the world equipped with Li-Fi technology can be thought of as a wireless hotspot delivering connectivity at very high speeds. A Li-Fi—enabled light bulb can transmit data at speeds of 224 Gb/s (demonstrated in laboratory tests). The details of Li-Fi and the working principles were discussed in an earlier chapter and are not repeated here. Li-Fi is therefore a technology breakthrough with potential for a new future of global communications. With proper security to secure IoT infrastructures, the required level of reliability and privacy to the Li-Fi devices can be achieved. This concept will therefore provide a powerful solution for connecting global IoT implementing Li-Fi technology. Note that IoT is part of the Internet, which comprises billions of intelligent communicating things or Internet-connected objects with sensing, and data capabilities. Embedded sensors in each Internet-connected object will capture enormous amounts of data that need to be transmitting very quickly, which requires high bandwidth. Li-Fi can therefore provide the potential solution. Li-Fi can be used to connect with the Internet-connected objects to communicate quickly only in the local area. There is a need for a flow control protocol that synchronizes Li-Fi and Internet for very fast transmission and reception of information that requires design of a new protocol for this purpose. A Li-Fi node has strong communication and networking capabilities of optical wireless physical and above TCP/IP layers, which can connect things and everything anytime and everywhere in a secure manner. Any light source anywhere can be turned into an operational Li-Fi node that can communicate with the rest of the Li-Fi communication network and Internet infrastructure. A light node can be integrated with a controller area network bus in many objects. In short, Li-Fi communication and networking technology has the true potential element for IoT and ubiquitous communications.

Li-Fi consists of an array of FSO transmitters that modulate the Li-Fi—enabled LEDs placed directly under the ceilings of offices, airport terminals, entertainment centers, healthcare hospitals and clinics, and different buildings or homes equipped with security sensors. Fig. 8.4 shows an artist's concept of

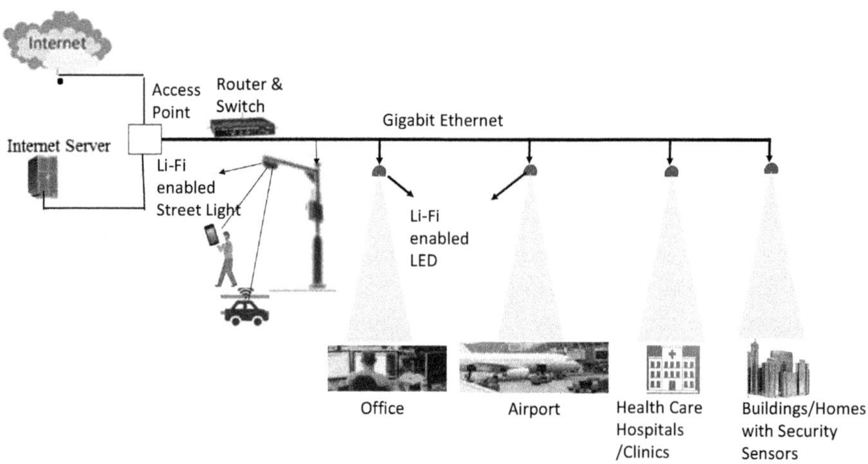

FIGURE 8.4
An artist's concept for the configuration of Li-Fi—enabled LEDs for Internet of Things for various places in everyday life.

configuration of the Li-Fi–enabled transmitter array and the typical coverage areas of various buildings and terminals to be connected with the IoT devices assigned in each of these buildings. Also shown in the figure are the possibilities of mobile connectivity (with smart phone or from a car) with any of these IoT devices, which also can be achieved using streetlights converted to Li-Fi–enabled LEDs. To transmit data, each FSO transmitter creates the LED light cones about one square meter directly below in which the data can be received. Digital bits 1 and 0 are transmitted by switching the LEDs on and off at the modulating rate. The LED transmitters are connected to a >100 Gb/s Ethernet network controlled by the access point. Each Li-Fi receiver is equipped with a silicon pin photodiode, which converts light intensity into electrical currents that can be interpreted as the digital bits 0 and 1. The received data is decoded to get the intended message. For data transmission and reception, the readers are referred to an earlier chapter for details.

Expanding usage of IoT technology in extensive industry and everyday life applications is developing continuously. This fuels the IoT-enabled Li-Fi technology market with a positive effect. Also, the IoT-enabled Li-Fi technology market depends on the extensive accessibility through its cloud sending.

Terrestrial node connectivity, including moving platforms and people carrying iPhones, will need optical links for high bandwidth for connecting many hundreds and thousands of devices belonging to IoT.

8.5 SATELLITES AND INTERNET OF THINGS

For IoT to deliver all of its promises, the network carrying all that communication has to extend everywhere without any interruption. IoT will have to depend on the satellites—the invisible infrastructure extending beyond the reach of the ground-based network and delivering truly ubiquitous coverage for IoT.

8.5.1 Importance of Satellite Communication for Internet of Things

Satellite technology has played a vital role in developing global connectivity. Many consumer-centric devices depend on mobile networks that are terrestrial-based; many other applications are constantly being developed, which require much greater reach and reliability for the global space using satellite technology. IoT relies heavily on the Internet definitely benefits from this satellite communication. Some of the immediate benefits of satellites for IoT include (1) connecting to and monitoring remote assets across various geographical areas extending the mobile network coverage, (2) improving mobile connectivity beyond smart phones such as connecting with remote sensors and mobile vehicles, (3) improving reliability and speed of IoT device services beyond land-based networks requiring broadband speeds to support bandwidth-intensive applications in real-time, and (4) facilitating M2M communication and connection for IoT. These benefits clearly show the continued growth of the connection between IoT and satellite communication technology. By 2023 it is predicted that there will be more than 5.8 million M2M and IoT connections.

8.5.2 Role of Satellite Connectivity in an Internet of Things World

The role of the satellite is the driving the vision for a truly connected world. In this uber-connected world full of connected devices that can track almost everything man and machine can do as billions of devices talk to each other is being made possible using combined satellite and IoT technology. This will still require more development of satellite technologies and hardware and software tools specifically for satellite networks of the future. More development of software to manage the next generation of satellite networks and innovative, compact, fast components for IoT applications will also be necessary. At the same time, new commercial applications using satellite-enabled data connecting commercial end users using IoT devices will also be needed. Some of the in-orbit low earth orbit (LEO) constellations are some kind of IoT systems and LEO and geostationary equatorial orbit (GEO) systems will also be able to serve massive uncoordinated sensors. Because of the continued improvements in satellite connection services and the development of associated technologies, in the maritime sector, for example, they could allow operators to increase transport efficiency with an impact of $4.5 and $9.3 billion in 2025 (https://www.ecnmag.com/blog/2016/03/why-internet-things-iot-wont-survive-without-satellite), which will also increase the efficiency of their sensor systems. The increased connectivity around the world including remote locations, and continued miniaturization of computers and smart devices will have a direct effect on the IoT to enter into every aspect of our lives. It will also be necessary to integrate satellite into local area network components to address end-user demands for small homes and environments. Underlying communication infrastructure in smart cities, driving cars, e-health, telemedicine, and medical surgery in remote areas will all be represented by IoT.

Implementation of a global anywhere-Internet in LEO satellites to create a world hub for IoT is being considered. The global connectivity using laser-equipped satellites such as LEO are discussed in an earlier chapter. A constellation of satellites with LEO satellites around the earth will allow global connectivity. Thus increasing the number of LEO satellites in orbit will increase network capacity and enlarge the IoT network of devices and users. Developments are being made for direct or backhaul connectivity to deploy new LEO or highly elliptical orbit constellations that can be optimized for the IoT market. There are plans to deploy 64 CubeSat-class spacecraft by 2021 to support IoT applications. All these satellite communication systems can be combined with terrestrial IoT devices to develop hybrid terrestrial–satellite IoT systems. IoT is therefore expected to continue driving up market demand for the integration of satellite into the overall communications mix. Satellite technology has a pivotal role in driving IoT growth and unlocking the promise of interconnected devices anywhere, anytime.

8.5.3 Issues for Low Earth Orbit Satellite Constellation for Internet of Things

LEO satellite has the advantage of low propagation delay (latency issue), small propagation loss, and wider global coverage than GEO satellite. However, existing protocols will be required to enhance the compatibility of the LEO satellite constellation-based IoT with terrestrial IoT systems. In order to minimize any LEO constellation-based IoT interference sharing the same spectrum band (such as same optical wavelength as in FSO communications) with terrestrial IoT networks will be very practical for LEO constellation-based IoT to design a true satellite–terrestrial hybrid system. The compatibility between two systems should also be considered in order to generate an integrated IoT network. IoT depends on a TCP/IP protocol suite where IP provides routing information and TCP guarantees reliable data transmission on top of IP. TCP/IP protocol therefore should be implemented in the network design to ensure the compatibility between LEO constellation-based IoT and terrestrial network as well as to distinguish it from M2Mcommunications [8]. The design of an LEO constellation-based IoT system should also include an efficient routing scheme and a united higher layer design.

8.6 DISCUSSION

In summary, three broad options to handle the tremendous additional IoT data worldwide of about 44 trillion gigabytes are local, cloud, or hybrid architectures. The ideal solutions may consist of three levels of IoT system architectures:

- Home, indoor: Uses a combination of both Li-Fi and FSO for some uplink or downlink. This way the huge bandwidth required for all the devices and sensors can be solved for anywhere and anytime. Examples include wired devices, smart phones, and mobile devices that can use both Li-Fi and FSO simultaneously. VLC can be used easily in streets using Li-Fi. But the devices need to be integrated with optical devices.
- Smart city: Can still use Li-Fi and FSO, but needs optical cells on the roof or some optical towers.
- Last-mile problem solution used for connecting to Internet providers where there are no high bandwidth fibers available.
- Space- or air-based solutions using space- or air-based platforms and FSO to connect with IoT either on the ground or in space. Satellite laser communication links can be used for establishing global IoT.

The concept of constellation of satellites for IoT integrating with FOS communication links has the potential for global IoT. However, this will require satellites to be equipped with laser communication units (transceivers), and an advanced optical network architecture for IoT will need to be developed to deal with a constellation of satellite handover and global connectivity issues. Terrestrial node connectivity, including moving (mobile) platforms and people carrying iPhones, will need high bandwidth optical links for connecting many hundreds and thousands of devices belonging to IoT.

References

[1] https://arstechnica.com/gadgets/2018/04/microsofts-bid-to-secure-the-internet-of-things-custom-linux-custom-chips-azure/.

[2] The Internet of Things: An Overview, Internet Society, October 2015. www.Internetsociety.org.

[3] P.P. Ray, A survey on Internet of Things architectures, Journal of King Saud University-Computer and Information Sciences (2016), https://doi.org/10.1016/j.jksuci.2016.10.003.

[4] T. Zachariah, N. Klugman, B. Campbell, J. Adkins, N. Jackson, P. Dutta, The Internet of Things Has a Gateway Problem, HotMobile'15, Santa Fe, New Mexico, USA, February 12–13, 2015, https://doi.org/10.1145/2699343.2699344. ACM978-1-4503-3391-7/15/02.

[5] M. Cedrickk, 5G wireless communications-the ideal candidate for future internet of things, International Journal of Advanced Technology in Engineering and Science 4 (8) (August 2016).

[6] K. Kadam, M.R. Dhage, Visible light communication for IoT, in: Applied and Theoretical Computing and Communication Technology (ICATcCT), 2016 2nd International Conference on, IEEE Publication, 2016, https://doi.org/10.1109/ICATCCT.2016.7912007.

[7] C.-W. Chen, W.-C. Wang, J.-T. Wu, Y.-T. Chen, Visible light communications for the implementation of internet-of-things, Optical Engineering 55 (6) (June 2016) 060501.

[8] Z. Qu, G. Zhang, H. Cao, J. Xie, LEO satellite constellation for internet of things, IEEE Access 5 (2017), https://doi.org/10.1109/ACCESS.2017.2735988.

Further Reading

[1] IoT, 5G and Significance of Fiber, Networks Asia Special Projects Team, January 8, 2018. Monday, https://www.networksasia.net/article/iot-5g-and-significance-fiber.1515370319.

CHAPTER 9

Potential Applications for Broadband Global Internet Connectivity

9.1 INTRODUCTION

The Internet has changed our everyday life and is constantly creating new ways to connect with friends and family everywhere, and to do business effectively at a much higher speed than before. The public web with the World Wide Web and the Internet is still growing. The future of the Internet and applications over the next 10 years are therefore important to understand because they all will depend on global Internet connectivity. The Internet will become more and more important and embedded in everyday life [1–3]. This communication tool gives us the power and flexibility to restructure practically our entire existence of everyday life. Over the past 20 years, a global computer-based communication network has fundamentally changed our social, cultural, and political landscape, affecting our lives in many ways.

With organized and powerful search engines, the Internet is becoming more and more intelligent. In the future, the Internet will eventually have all our lives recorded in some way and saved in some form of memory. We will be able to use smaller and smaller devices to access all this information, and probably will make a transition from "wearables" to be part of our bodies, without carrying any external physical devices from wherever and anytime. Eventually a true global culture will be built. Global IP traffic is predicted to increase nearly threefold over the next 5 years at a compound annual growth rate of 24% from 2016 to 2021 and globally, mobile data traffic will grow nearly tenfold from 2016 to 2021. Business IP traffic is also predicted to grow at the compound annual growth rate of 21% for the similar 5-year period.

> As the world becomes increasingly digital ... affordable broadband must be within reach of people, businesses, and governments in all corners of the world —UN Secretary Ban Ki-Moon, in an address to the Broadband Commission for Digital Development.

The future of the Internet may be described as follows:

- More live view of a physical, real-world environment augmented by computer-generated perceptual information and wearable devices

Optical Wireless Communications for Broadband Global Internet Connectivity.
https://doi.org/10.1016/B978-0-12-813365-1.00009-6

will be implemented, which will monitor in real-time, for example, personal health information.

- With the spread of the Internet global connectivity will be increased, affecting societies worldwide.
- Online education via Internet will become more flexible and less expensive worldwide where the students can log on to take classes from anywhere and learn from worldwide experts in various subjects.
- The Internet of Things (IoT) is creating one more level of convenience to remotely turn on/off Internet-enabled devices at home and also to be an active part of the smart city.

Broadband global Internet connectivity is discussed in an earlier chapter of the book. Some of the relevant developments include Facebook's development of a huge, solar-powered aircraft to help achieve Internet connectivity in remote areas. The unmanned aircraft called Aquila, is made of carbon fiber will be responsible for beaming Internet signals back to rural areas on Earth presently unable to access the Web. The aircraft will hover between altitudes at around 60,000 to 90,000 ft and can stay up in the sky for 3 months, able to provide the Internet to people on Earth in a 50-km radius. These solar-powered Internet drones will also be able to stream data between drones using laser technology at multigigabit speeds.

A cloud-based collaboration platform can provide online data storage and advanced modeling tools for sharing information quickly and easily, which can be accessed from a browser from a laptop of mobile devices like smart phones.

The Internet is now really a part of our social fabric and is essential to how we connect, communicate, share, and collaborate. Digital technologies, information and communication technologies, and broadband global connectivity when combined offer various opportunities to improve people's lives in worldwide societies by exchanging and sharing information. Real-time accessing of information helps transform various sectors ranging from healthcare and education, to business and many others. Global Internet connectivity will make it possible to collect and analyze larger and more complex kinds of data.

9.2 POTENTIAL APPLICATIONS FOR GLOBAL INTERNET CONNECTIVITY

Information communication technologies and Internet access working together have opened many applications of today's society to improve people's lives in various sectors as mentioned earlier. Some of the potential applications of broadband global connectivity are discussed.

9.2.1 Social Connections, Friends and Family, Finding/Locating Friends by Internet

Networking is less intimidating when people start with people they know or have something in common with such as family, friends, coaches or advisors, professors or coworkers, and others. Most likely people can be connected with friends and family via social media using the Internet to share their profiles, photos, and interactions. However, the Internet must be global, open, and secure and the key to the future is to put humanity at the center of the online world [1].

9.2.2 Quality Education for All

Education is a human right and a force of transformation providing individuals the knowledge to expand their choices including health and productivity. Broadband high capacity technologies can help with knowledge dissemination and can widen access to information for advancing quality and effective learning. Online education via Internet will become more flexible and less expensive worldwide where the students can log on to take the classes from anywhere and learn from worldwide experts in various subjects. Broadband Internet connectivity will be helpful in planning, curricula, teacher training, and schools, not added onto regular curricula. Information and communication technologies will be able to provide innovative digital services possible for education systems in low-income countries at very low cost. This will dramatically expand access to knowledge and other unprecedented opportunities for everyone and everywhere worldwide, independent of their income categories. All that is required is easy access to devices for the students, online resources, and Internet connections in their schools, local communities, or their homes, which the global Internet can provide. Lifelong learning, including higher education, is essential in a modern economy can thus be promoted. Advanced information communication technologies can improve learning quality and diversify education and reach rural communities, men and women, boys and girls, out-of-school youth, and people with disabilities. Internet access with tablets, laptops, smart phones, and other devices will also enable students to gain free access to high-quality skills development opportunities and university courses. Students will have access to qualified teachers with free-of-cost or very low-cost high-quality study programs delivered by quality teachers in urban areas through videoconferencing services.

9.2.3 Improving Food Production, Management, and Productivity

Internet access and information communications technology together will provide useful tools to improve nutrition by enhancing agricultural systems for rural people to access services and information for development. The Internet technologies along with digital literacy skills to use them will help people create their solutions according to their economic, climatic, and local circumstances. Farmers will be able to use the Internet to diversify their methods and improve the productivity of their farms; for example, using high resolution satellite imagery and Earth observation to map land use, water resources, water scarcity, crop yields, and environmental degradation, improves production [4].

9.2.4 Business and Economic Growth

The Internet offers opportunities for economic growth in developing countries by providing access to information so that people can be connected to businesses everywhere to explore opening up new markets. Global Internet connectivity access increases the productivity of labor and capital by improving human resource qualifications and specializations. Information flows lead to innovation and more financial capital access to encourage entrepreneurship. Information communications technologies using Internet access can be used to support many of the business activities for both internal business systems to business communications to electronic commerce. Electronic business, or "e-business," can improve business efficiency by saving time and streamlining business processes. New opportunities can be developed with many collaborators, which ultimately may lead to increase in revenue. Using Internet-based technologies and various tools necessary to support the technology, some of the advantages are internal business systems (e.g., intranet and extranet, human resources, and customer relationship management) and electronic commerce (e.g., business-to-business, business-to-customer, Internet marketing). All these resources should improve business productivity with technology with the business website to reach potential customers and interested investors worldwide. Internet financing for investing and financing operation structure can be developed for expanding the business.

Banking Industry Undoubtedly the globally spread Internet has enhanced banking business because of the vast market developed on the Internet and online purchasing. E-business is creating a new business environment to improve the business process, and e-banking will also contribute to future developing countries. Most banks have potential opportunities arising from the new marketplace. The issue of service marketing in general and banking services in particular is one of the most important and modern directions in all societies where banks need to provide an excellent service to customers. The banking industry can be generally characterized by developing business models, customer satisfaction, technology and operation, regulation and supervision, and other related items. Electronic banking to carry out efficient business transactions of a bank need to use electronic devices such as computer systems, global system for mobile communication phones, automated teller machines, Internet facilities, optical card recognition, smart cards, and so on. Customers should be able to electronically transfer funds across the globe without any problem or delay as compared to the traditional method. Internet access to connect bank networks is therefore essential. In e-banking, the opportunities can be local, global, and intermediate. There is no doubt that banks have been equipped with information communication technology and Internet access.

The Effect of the Internet on Businesses Internet has truly created new communication forms and other business methods that companies use when processing financial and business information. One example is the e-business mentioned earlier, which represents the use of Internet and business technology in a company's operations.

The Impact of Information Communication Technologies and Internet Access on Effective Financial Management Given the fact that Internet access and information communication technologies have enhanced the business world, proper implantation and maintenance of information and communication technology need to be considered. Effective financial management is therefore very important using advanced technology and Internet communication networks. All departments of financial organizations such as management, marketing, credit administration, human resources, operational management and strategic management, procurement, and accounting and financial management need to be considered [5]. Financial management deals with raising financial resources and their effective utilization to accomplish the goals of the organization. Information communication technologies and Internet access have helped organizations in financial management to be successful and perform various operations related to financial transactions efficiently. Global Internet connectivity provides further opportunity to reach remote locations for organizing and managing businesses successfully.

9.2.5 Global Internet Connectivity: Space-Based and Upper-Atmosphere Solutions for Sustainable Development

In addition to connecting individuals to the Internet, space-based platforms can offer diverse applications in achieving United Nation's 17 Sustainable Development Goals [6]. The effects will be across various aspects of society from industry, health, and education to e-government and various developments of future smart cities and societies. Some of the applications where specifically *space-based platforms* can be used for establishing Internet are the following [6]:

1. *Education*. Space–based platforms with the potential advantage of establishing robust broadband communications providing global Internet access can connect to remote national educational programs and international resources to ensure equality of educational opportunity anywhere worldwide.

2. *Health.* Healthcare and well-being services with Internet connectivity can be provided for healthcare workers to receive remote training in detecting deceases and preventative care, and for clinics and medical providers to be connected with each other to collect data and share information. The patients in remote areas across the country can also have in–home monitoring systems, saving costs.
3. *Banking.* To be a part of growing digital economy, remote locations should have access to banking apps so that many of the banking transactions can be performed by the customers for local businesses. Space-based communication technologies can provide this opportunity.
4. *Oil/gas exploration.* It is also possible to explore new reserves and extraction of gas and minerals with deployable, portable, and mobile Internet connectivity. At the same time remote monitoring of operations are possible.
5. *Monitoring municipal services.* Sensors for monitors can be used for services like electricity, water, waste management, transportation, and lighting for future smart cities to manage resources from a distance in a very flexible way.
6. *Agriculture.* Internet connectivity can help in remote monitoring of fields and hydroponic installations so that the framers can plan for efficient cropping and timely yields with remote sensing tools, models, and software.
7. *Environmental management.* Remote sensing and data collections to track environmental and climate changes can help predict global environment. The Internet-based information communications technology will play a vital role to help preserve ocean health, the seas, and marine ecosystem and resources.
8. *Emergency response to disaster relief.* Global Internet connectivity can help with real-time tracking of any natural disasters such as hurricanes, earthquakes, and fires.
9. *Search and rescue.* Global Internet connectivity can track and locate lost hikers, lost aircraft after an accident, remote vessel assistance/diagnostics, and can provide rescue by connecting with appropriate emergency services.

9.2.6 Improving Health and Well-Being

Access to the Internet provides better medical information for both patients and health practitioners. The Internet therefore enables health improvements by providing information, prevention, diagnosis, treatment, and monitoring.

Internet access and broadband communications technology can provide a foundation for future health systems and can include connecting clinics and remote communities and training healthcare providers, thus improving the quality of health care through digital solution [7]. The Internet-based concepts can also improve electronic medical records, data management, and monitoring for real-time analysis and communications, thus improving the quality of care and efficiency of health systems by helping patients take care of more responsibilities in the management of their own health.

The New Paradigm in Medicine Today we are facing a new paradigm in the management of the physician–patient relationship because of a number of reasons such as advances in technology, lack of medical care and shortage of medical staff in remote areas, demands of healthcare by communities, and so on. Telehealth and telemedicine are becoming a part of future healthcare and well-being. Telehealth is the use of information and communication technology to support health in remote locations; for example, long-distance medical care, health education for patients and professionals, and prevention. Telemedicine uses information technology to provide medical care almost in real-time such as consultation with the doctor via Skype or a remote telesurgery, or telecardiology. Both telehealth and telemedicine services involve videoconferences and telerobotic systems with real-time transmissions between health providers and patients as well as remote patient monitoring where special electronic devices transmit patient health information to healthcare centers. Telemedicine is even providing patients with remote dial-in diagnostic capability. Furthermore, telemedicine can be a key lifeline after surgery; for example, postsurgical patients can benefit from virtual appointments with a smart phone. Telemedicine can be categorized into remote monitoring, store-and-forward and interactive services. Telemedicine and robotics working together is truly paving the way to the globalization of surgery. Internet access with email, websites, and clouds provide the basic tools for accomplishing these objectives.

Remote Surgery Patients are now able to utilize communication networks to contact a clinician or Skype a doctor using real-time video for a diagnosis, thus obtaining healthcare from a distance can provide access to the best care even in remote locations. As an example, a physician pushes a joystick on a console in an urban advanced research laboratory while a robotic arm responds on a patient lying on an operating table a few thousand miles away in a remote area: advanced wireless connections and information technology are reliable enough for doctors to perform this type of remote surgery. Furthermore, the next generation of wireless connectivity using 5G communication networks will enhance remote surgery. Telesurgery has been made possible as a result of extraordinary advances in the fields of telecommunications and robotics and supported by Internet access. The use of telemedicine for remote surgery now includes telementoring, teleproctoring, teleconferencing, and teleconsulting with patients. Telesurgery and telemedicine provide patients with full surgical care such as training, preoperative, operative, and postoperative follow-up. Current high-speed communication networks make telesurgery possible even from a thousand miles away or in an inaccessible area. *Global Internet access, especially with all-optical communications technology concepts described in this book, will make remote surgery entirely possible in the future.* In spite of tremendous progress being made, there are still some humanistic, legal, or ethical issues to be addressed regarding telesurgery.

Virtual Reality in Health Care Advanced healthcare solutions begin to incorporate virtual reality (VR) to improve results and overhaul the patient experience and to use VR to simulate surgery, treat phobias, and simplify the diagnosis process. Some of the areas where VR has potential uses are [8] (1) VR enhances robotic surgery, where a human surgeon performs actions executed by a robotic device that can then perform the actual operation; (2) VR can be used to accelerate recovery rate for strokes patients along with a neurorehabilitation app to encourage early motor rehabilitation in stroke; (3) VR health care for pain management program; and (4) VR can connect older people with the outside world using a headset-based platform with the ability to track movement data to catch dementia in its early stage. Internet access provides potential data collection to help healthcare professionals determine treatment plans and preventive care.

9.3 ROLE OF INTERNET IN BITCOIN

Recently an innovative payment network and a new kind of money was introduced called *bitcoin*, which is actually a cryptocurrency and worldwide payment system [9]. Bitcoin is open-source and its design is public—nobody owns or controls bitcoin and everyone can take part. Bitcoin is a decentralized digital currency where the transactions take place between users directly without an intermediary and are verified by network nodes through cryptography and finally recorded in a public distributed ledger called *blockchain* [9]. In 2017 there were 2.9–5.8 million unique users using a cryptocurrency wallet using bitcoin. The issuing of bitcoins is carried out collectively by the network. The transactions in bitcoin are taken as input and run through a *hashing* algorithm, using SHA-256, which gives an output of a fixed length. Using readily available software applications the transactions of the form *payer A sends B bitcoins to payee C* are broadcast to the network. After validating transactions network nodes add them to their copy of the ledger and then broadcast these ledger additions to other nodes [9].

It is important to understand the role of the Internet relevant to bitcoin transactions. Bitcoin is transferred initiation from the sender of the transaction through the Internet. When signing the transaction, the sender does not need to be connected to the Internet but he or she is required to be connected to the Internet to declare the transaction to the network to record it [10]. The transaction therefore needs to be signed to be recorded on the blockchain and consensus should accept it. However, the receiver does not need to remain connected to the Internet as the transaction happens. After the transaction is recorded the wallet is free to sync with the blockchain and display the funds received [10]. Therefore, the receiver does not need an active Internet connection to eventually receive the money but needs the connection only if the receiver wishes to show in his or her wallet to verify and show the received funds; thus an Internet connection is needed to broadcast the transitions after having bitcoin. Without the Internet connection, there is no connection to the blockchain to validate the transition. Also, blockchain ideas in bitcoin can be part of IoT, which has the potential of resulting data speed and the tremendous amount of data collected by the sensors installed in various smart objects, enabling information to be turned into data [11].

9.4 CONCLUDING REMARKS ABOUT INTERNET ACCESS

Although Internet access provides us with so many opportunities that open the door for information for connecting to friends/families, education, health, business, and everything else, we must also understand the responsibilities coming from using the Internet and the personal rights involved in the whole process. With Internet access, we are able to take advantage of participating in various aspects of societies. This also means that those who are excluded suffer economic, educational, and social exclusion [12]. Internet provision is most expensive in remote areas and it is still difficult or very expensive for remote users to obtain Internet access. It is difficult to deliver universal provision in remote areas for everybody. A digital divide is created with those who are unconnected or have inferior service and those who are well-connected and can take advantage of the digital economy and everything the Internet has to offer. Access to the benefits of modern society such as freedom, education, and health are closely related to fundamental rights [12]. Many rights such as rights to health or education can lead to rights to hospitals or a university. In short, the rights to access information and education must be established to avoid a digitally divided society between those who have Internet connectivity access everywhere and those who do not. Fundamental rights to store, process, and communicate information are therefore necessary to reestablish an inclusive digital society [12].

References

[1] Internet Society Global Internet Report, Paths to Our Digital Future, 2017. https://www.internetsociety.org/.

[2] Updated, The Zettabyte Era: Trends and Analysis, June 7, 2017. Document ID: 1465272001812119.

[3] S. Rushing, Analyzing "Global Access to the Internet for All," Projects, Seminars FI/IITM WS 15/16, Network Architectures and Services, July 2016.

[4] The State of Broadband 2017: Broadband Catalyzing Sustainable Development, September 2017.

[5] M.M. Kirmani, F.A. Wani, S.M. Saif, Immpact of ICT on effective financial management, International Journal of Information Science and System 4 (1) (2015) 1–14.

[6] Broadband Commission for sustainable development, Working Group Technologies in Space and the Upper-atmosphere- Identifying the Potential of New Communications Technologies for Sustainable Development, September 2017.

[7] C.M. Davidson, M.J. Santorelli, The Impact of Broadband on Telemedicine, a Report to the U.S. Chamber of Commerce, April 2009.

[8] INSIGHTS- HEALTHCARE, Virtual Reality Has a Bright Future in Healthcare, April 14, 2017.

[9] Wikipedia Website: https://en.m.wikipedia.org/wiki/Bitcoin.

[10] https://www.quora.com/Does-the-receiver-of-bitcoin-need-an-active-internet-connection-to-eventually-receive-the-money?utm_medium=organic&utm_source=google_rich_qa&utm_campaign=google_rich_qa.

[11] T. Croce, From bitcoin to the Internet of Things: the role of the block chain, Annali della Facoltà Giuridica dell'Università di Camerino (2017) 2281–3063, n. 6 – ISSN Online.

[12] Towards an Affordable Internet Access for Everyone, The quest for enabling universal service commitment, Dagstuhl Reports 4 (11) (2014) 78–137.

CHAPTER 10

Current Recent Research and Future Directions

10.1 INTRODUCTION

This book has attempted to develop and analyze the unlimited possibilities in advancing the *all-optical* concept technology for creating global Internet connectivity access for all. The technology is based on optical communications: free-space optical (FSO) and fiber-optics–based communications both take advantage of high bandwidth, high-speed data rates and high channel capacity, which are essentials to satisfy the demands of today's requirements for Internet access. A complete system with all-optical technology for establishing global Internet connectivity is still a future potential; however, based on a number of experimental demonstrations worldwide using optical communications are paving the way to finally accomplish this goal. There are many individual segments of these successful demonstrations such as indoor, terrestrial, ground-to-satellite, and intersatellites optical communications, all of which play important roles in establishing this overall all-optical global Internet connectivity. Although FSO communications have been studied for years, FSO networking is still in its infant stage when we apply it to handle global networking. Many global players are working on solutions to make the idea of global access to the Internet for all a reality, but not necessarily using our approach of all-optical technology. Some of the global players include Google and their project Loon, Facebook with intrenet.org, and Microsoft using TV whitespaces. This book provides some technology solutions where we can complement many proposed and demonstrated solutions of global players with the optical technology developed in this book. This is where this book differs, with its emphasis entirely on optical technology to accomplish this goal. Based on demonstrations worldwide by many researchers, the future of global access to the Internet for all looks very promising. Hopefully someday soon it will not be an idea anymore, but reality.

The area of all-optic technology for global Internet access is extremely interdisciplinary and requires knowledge of physics; electrical, optical, and communication engineering; satellite and information technology and communication software; and related subjects. Many research challenges and open FSO networks require research efforts for practical and realistic

Optical Wireless Communications for Broadband Global Internet Connectivity.
https://doi.org/10.1016/B978-0-12-813365-1.00010-2

solutions. This chapter discusses some of the recent research efforts and points out some future directions. The research areas relevant to the goal of this book are organized in three general areas: (1) advanced electrooptic or optoelectronic components and devices needed as part of the optical network to transmit, route, or receive the information; (2) laser or optical communication subsystems/systems including indoor, terrestrial, or satellite-based; (3) innovative optical communications architectures, concepts, and networks; (4) information communication software-based new solutions; and (5) totally new concepts useful for global Internet access, for example, quantum communications and entanglement for secure communications and quantum computation.

10.2 CURRENT ADVANCED RESEARCH RELEVANT TO ALL-OPTICAL INTERNET CONNECTIVITY

Future optical communications to address the new trends and transmission capacity is critical because of continuous demands of high-speed data rate for global Internet access. A whole new generation of ultrafast devices is needed in order to avoid electronic bottlenecks. A new class of *all-optical* has therefore been proposed and is being developed. For example, some of the initial technologies involved are micromachining, creating low-cost microelectromechanical modulator devices, and bubble technology for all-optical switching leading to the development of the next generation of large-scale switching fabrics. This section discusses some of the recent innovative research that is being developed to establish the goal of all-optical high-speed global Internet connectivity.

10.2.1 Fast Luminescent Detector With Omnidirectional Sensitivity for Free-Space Optical Communication

Researchers have reported fast detector using fluorescent materials, increasing the active area of a photodiode by orders of magnitude and its field-of-view.

They demonstrated a detector with an active area of 126 cm^2 achieving a data rate of 2.1 Gbit/s [1]. This new material-based for communication system applications has the advantage of omnidirectional sensitivity and strongly improves the field-of-view over optical systems that conserve étendue. Use of a large and fast *luminescent detector* is also promising to suppress the phase fluctuations and deep fades caused by atmospheric turbulence for FSO communication systems at multi-Gbit/s data rates suitable for high-speed Internet access.

10.2.2 Future All-Optical Networks Using Nonlinear Silicon Photonic Signal Processing Devices

Novel materials for all-optical processing nonlinear devices are promising in many applications of optical communication-based Internet connectivity.

Silicon-based nonlinear devices for all optical nonlinear processing of complex telecommunication signals are discussed using silicon-germanium and amorphous silicon-based waveguides [2]. Novel materials such as silicon-rich silicon nitride and tantalum pentoxide are also discussed. They have demonstrated wavelength conversion of complex modulation format signals, phase sensitive regeneration in integrated devices, and supercontinuum generation in cm-long waveguides. Next-generation terabit all-optical networks are promising from their demonstration of four wave mixing wavelength conversion applied on complex signals differential phase shift keying (DPSK), quadrature phase shift keying (QPSK), 16-quadrature amplitude modulation (QAM), and 64-QAM enhancing the telecom signal spectral efficiency. Silicon-rich silicon nitride and tantalum pentoxide were shown to be promising optics platforms.

10.2.3 Recent Technological Advancement in All-Optical Networking

Optical networking is really the backbone of the modern Internet and their applications in the cloud will be diverse and limited to bandwidth. Optical networking speeds are becoming continuously faster and approaching all-optical networks for global Internet access. Research efforts and advancement in optical networking include coherent detection with digital signal processing, software-defined transceiver and flexible wavelength-division multiplexing (WDM) Network, all-optical switching and high-speed photonic components. The technologies for the recent advances involve advancing in high-speed optical detection by the photodetector and improved and more efficient modulation schemes such as DPSK, differential QPSK, polarization-division multiplexing, and QAM, as mentioned earlier. Optical transport network also needs to meet the requirement of new network technology eventually to develop *all-optical switching*. Some new technology involves integrating 1x*N* wavelength selection switching into one *N*x*N* wavelength cross-connect units, enhancing energy efficiency and the performance of optical signals. Optical switching will play an important role in metronetworks and in other related Internet access services in response to the upcoming 4K/8K video, 5G, virtual reality (VR), and other high-bandwidth and low-latency applications.

10.2.4 High-Speed Photonics Components

Optical component performances are essential for establishing Internet connectivity using optical communication systems. Implementations of the various functions the optical components perform require different materials in order to develop these components. Photonic integration technology such as silicon photonics technology provides high production efficiency and high integration. Future developments of all-optical networks will involve silicon photonic technology using graphene materials to achieve various optical switching, optical routing, optical logic, optical storage, and optical signal processing functions for the next generation. Furthermore, a new type of photonic component using photonic crystals and artificial materials will find applications in the near future. 5G mobile network services, which will drive the future optical industry development and architectural transformations, will also benefit from the high-speed development of future photonics components. Some of the services of future *all-optical networks* will also include the traditional metro and backbone enterprise services evolved into cloud line access and cloud backbone services, ultrabandwidth video services (HD 4K and 8K video, e.g., for the Olympic games as well as telemedicine applications for remote surgery), and 5G mobile network services, including the Internet of Things (IoT) and Internet of Vehicles.

10.2.5 Plasmonic Device for 100 Gbit/s Optical Links

Research work in developing a microscale modulator using plasmonically active gold components is reported [3], which demonstrates fast operation for ultrabroadband signals. The broadband plasmonic modulators operating greater than 100 Gbit/s limit of photonic devices for a single carrier was presented at the Optical Fiber Communications Conference (OFC) held March 19–23, 2018 in Los Angeles, California. Plasmons are created when mapping photons to a metal surface and converting them to a particular kind of electron oscillation. The research is a collaboration effort between scientists from Switzerland, Germany, and the United States and paves a new way to meet today's high demands for the Internet traffic over optical networks for optical communication. The research relies on plasmonics rather than photonics; advantages of these modulators are their compact size and significantly wide bandwidth. This work is definitely a major step toward establishing global Internet connectivity.

10.2.6 Graphene-Based Technologies for High-Speed Optical Communications Systems

Recent works in developing faster optical detection involve a new, waveguide-integrated photodetector design for a record high bandwidth for ultrafast, high data rate graphene devices [4]. High performance optical communication systems for ultrafast and compact optoelectronic devices are enhanced by graphene-based technologies enabling data rates of up to 100 Gbit/s. High-speed optical links are the backbone of the Internet, and the faster photodetector integrated with other compact optoelectronic devices will be a giant step toward the success of all-optical Internet connectivity. For example, the next generation of communications systems will need to handle an extremely large number of individual connected devices belonging to IoT for efficient and seamless data transfer between all connections. High data transfer using high-speed photodetectors will be required for the Internet as well as next-generation mobile communications such as 5G. Because these ultrafast graphene-based photodetectors are very compact, they can be used and integrated with other optoelectronic devices including *switches* in functionally dense, integrated chips [4], and have the potential to be integrated on one CMOS chip. On-chip photodetectors of different graphene-enabled technologies will also help developing technology onto optoelectronic chips for real-world applications in communication networks.

10.2.7 High Capacity, Ultrabroadband On-Chip Twisted Light Emitter for Optical Communication Systems

Researchers have demonstrated an ultrabroadband multiplexed orbital angular momentum (OAM) emitter for OAM communication with a data rate of 1.2 Tbit/s assisted by 30-channel optical frequency combs [5]. The research has potential applications in extremely high-speed, high-capacity optical communications systems including FSO communications through the atmosphere. The light emitter can be an important part of optical communication systems by providing an additional degree of freedom for designing wavelength/frequency division multiplexing. In Ref. [5] the authors described the multiplexed OAM emitter that utilizes a novel joint path-resonance phase control concept, with a micron-sized radius and nanometer-sized features. The wavelength considered was the entire band from 1450 to 1650 nm. The approach is a major development toward establishing high-speed all-optical Internet connectivity.

10.2.8 Optical Phase-Sensitive Amplifiers

One of the projects at the Chalmers University of Technology, Gothenburg, Sweden involves research in optical phase-sensitive amplifiers (PSAs) in order to explore fundamentally new concepts and technology platforms. This research in optical amplifiers will have a number of applications in optical communications, which constitute the backbone of the Internet. PSAs are truly unique optical amplifiers that can achieve a quantum-limited noise figure of 0 dB (i.e., noiseless amplifier) [6]. Other projects involve devices enabled by PSAs including ultralow noise lasers and broadband amplifiers, new optical amplification approaches for spatial division multiplexed systems, and few-photon nonlinear optics and quantum communications [6].

10.2.9 Market Research for 400 G Transceivers

A recent market research report describes the 400 G optical transceivers market industry and their investment in high-quality technology and processes to develop leading edge broadband network capability [7]. The report discusses the market driving forces of the 400 G optical receivers, noting that high-speed serial transceivers form the backbone of networks, communications, servers, and many other systems depending on high-speed transceivers. A 400 G optical transceiver is a single, packaged device containing both transmitter and receiver. This is also timely because the global adoption of the Internet is increasing, which supports online commerce, streaming video, social networking, cloud services, and also software as a service.

10.2.10 Compact Optical Transceiver by Hybrid Multichip Integration

A compact optical transceiver formed by hybrid multichip integration is described in a recent US Patent [8] based on silicon photonics using three-dimensional (3D) hybrid multichip stacking without wire bonds and method. The demands for increased bandwidth for optical communication and decreasing package footprint of an optical transceiver have been the subject of this patent and are useful in high-speed Internet.

10.2.11 All-Optical Routing and Switching for 3D Photonic Circuitry

Experimental demonstration of 2D all-optical and switching schemes in femtosecond laser-written waveguide arrays is described in which structures can be combined in complex nonplanar photonic networks where the discrete solitons direct light signals along specific paths [9]. This research has the potential for building all-optical devices for rapid routing and processing of data of optical pulses and is necessary for Internet connectivity. The future systems based on this technique can support 3D network topologies for photonic circuitry. Another research report discusses a novel, highly scalable architecture using the scheme of wavelength routing capability of arrayed waveguide gratings with a potential possibility of expanding the switch port to exceed 1000×1000 port counts for data center application [10]. Researchers also reported demonstrating an integrated wavelength router based on the microring resonators using 2×2 multiinterference crossing on silicon-on-insulator technology [11]. The router has a footprint of $0.37 \times 0.26\ mm^2$, and the cross-talk of channels smaller than -10 dB with a total loss of one channel to be about -4 dB. The routers are the building blocks of photonic networks and provide an optical link between input and output ports by routing signals and hence to design efficient Internet connectivity. Another four-port mode-selective silicon optical router for on-chip optical interconnect is also proposed and demonstrated by other researchers [12]. They emphasize that the router consumes no power to establish the optical links. Their prototype demonstrated the insertion losses for all optical links to be lower than 8.0 dB and the largest cross-talk values lower than -18.7 and -22 dB for broadcasting and port-to-port routing modes at the wavelength range of 1525 and 1565 nm [12].

10.2.12 Compact Multitransceiver Free-Space Optical Communication Structures

Multitransceiver spherical FSO node to hand-off multiple data flows between optoelectronic transceivers is described [13]. Optical antennas used in FSO spherical structures have the advantages of achieving angular diversity via spherical surface, spatial re-use via directional optical transmitters, and are multielement integrated with multiple transceivers such as LED and photodetector pairs. A simple handshaking protocol can be utilized to electronically steer the line-of-sight alignment onto the correct transceiver. The technology has potential to be used for FSO mobile ad-hoc networks to be useful for mobile inner-city settings for achieving Internet connectivity.

10.2.13 On-Chip Light Sources for Integrated Silicon Photonics

A recent article discusses advanced silicon photonics components for the latest platform enhancements in designing more compact, high performance and power-efficient telecom networks and data center applications [14]. Some of these advances will be useful to meet the demands for huge increase in data transfer and the number of networked devices because of exploding of data traffic to reach the zettabyte range. Optical transceivers that can combine higher performance with smaller form factors will be required. High-level chip integration will therefore be needed to accomplish this. The review emphasizes that leveraging the fabrication infrastructure and processes of CMOS industry are possibilities to integrate a number of optical functions including passives, modulators, and photodetectors at a high data rate on a single chip. When combined with other schemes using WDM and the advanced modulation formats there is the potential for making silicon photonics technology scalable for a Tb/s data rate. Furthermore, with 3D integration, low-cost and small-factor transceiver modules can be designed to build complex photonic circuits for next-generation metro and access networks, data centers, and optical networks-on-chip connecting different processors. Research works in these areas will make global Internet connectivity a reality.

10.2.14 Quantum Internet: Hybrid Quantum Information Network

Quantum information network is becoming a disruptive technology to provide radically new capabilities for information processing and communication [15] that will have tremendous effects on the future of most *secure* Internet connectivity and access. Researchers published their work to have achieved an elementary hybrid quantum network link. They demonstrated photonic quantum communication between two very distinct quantum nodes placed in different laboratories [16] using a single photon as information carrier. The researchers used the telecommunication wavelength 1552 nm to demonstrate this network. The capabilities and transmitting quantum bits by means of single photons between them is an important step in the development of hybrid quantum networks. Note that the classic information network, the World Wide Web, uses the bits for the information to flow through the network, processed and modulated by electronic circuits and chips, and transmitted by light pulses to move the information through the network. Quantum information networks or *quantum Internet* process and store quantum information through quantum bits, or "qubits," in a quantum network. There is thus a potential advantage of perfectly secure data transmission, which is one of the most important features required for Internet connectivity and access.

10.2.15 Optical Chips-Integrated Smart Phones and Similar Photonics Areas

Recent technological developments to integrate optical devices to be integrated with smart phones will establish all-optical Internet connectivity and access for mobile platforms and in remote locations. Optical-based innovative solutions will help smart phones optimize cellular communications performance to leading-edge connectivity solutions that maximize optical wireless data transfer. Researchers are developing optical circuits made of chalcogenide glass for ultrafast telecommunication networks to transfer information at the speed of light [17,18]. Integrating these glass optical circuits into silicon chips will have more advanced high-speed communications systems in the future. University of Sydney, the Australian National University, and RMIT University created a compact version. The new device is smaller than a thumbnail with a size of 0.1 mm × 4 mm. Mass manufacturable optical circuits with enhanced functionalities by combining nonlinear glasses with silicon-based material was reported in Ref. [17]. This opens up a new possibility of combining this material with the current industry platform for photonic integration, silicon, and to integrate a novel nonlinear glass into an industrially scalable CMOS-compatible platform in order to develop efficient ultracompact optical circuits. In the future, this concept will miniaturize photonics functionalities from devices such as a laptops to smart phones and even smaller so that all-optical global Internet connectivity and access can be established in the future, *anywhere, anytime*. Another related technology development is reported to have developed smart phones with an integrated 36° camera to build VR smart phones with a 16-megapixel 360 camera able to shoot 4K videos [19]. Although it will take more time to develop, but the possibilities to eventually integrate with smart phones for collecting optical information and data seem to be real and useful for Internet applications.

Other areas in the future include wearable devices in medical IoT, which will benefit self-health monitoring and preventive medicine with this developed technology [20]. Other similar areas with potential applications of future Internet connectivity include progress in optical waveguide technologies to enable integration of high-density compact photonics, which will help achieve IoT in the near future to bring changes in society [21,22]. There will be a strong need for optical devices in order to achieve future high-speed optical communication network for IoT and new developments in optical waveguide technologies will be essential to establish the needed Internet connectivity and access.

10.3 OTHER RELEVANT RESEARCH FOR ACCOMPLISHING FUTURE ALL-OPTICAL GLOBAL INTERNET CONNECTIVITY AND ACCESS

This book covers the fundamentals of global Internet connectivity and access, and various technologies involved to accomplish this. The potential for all-optical technology emphasized for achieving global high-speed Internet connectivity worldwide, *anytime, anywhere* is becoming a reality because of two recent major communication technology demonstrations: (1) successful demonstration of free-space laser communications between a low earth orbit (LEO) satellite and ground, and (2) intersatellite laser communications. These two demonstrations have clearly paved a way to design and develop future global Internet connectivity everywhere including remote locations. Simultaneously other exciting optical link architectures are being designed and developed that will definitely support establishing global Internet connectivity links worldwide. Some of these are constellation of small and micro-LEO satellites for communication links everywhere and anytime, and high-altitude platforms and balloons for establishing communication in remote locations. Separately, successful demonstrations of free-space quantum communications between two distant locations clearly are moving in the right direction for developing most secure Internet connectivity and access. The concept is based on *quantum entanglement* of quantum physics and is the most unbreakable secure link possible today.

Other related research efforts include:

- Integrated optical-wireless access networks [16]: Energy-saving solution for a network architecture consisting of Backbone/Core Network, Metro/Regional Network, and Access Network
- Quantum random number generator using entangled photons [23]: Method developed by NIST for generating numbers guaranteed to be true random by quantum mechanism
- World-records access at 1.72 terabits in FSO communications by DLR researchers for future high-speed Internet possibility (discussed in

another chapter). A more recent world record achieved by DLR and ADVA in FSO data transmission includes demonstration of 13.16 terabits per second [24]. This clearly sets a new record for laser communications for satellites.

- Li-Fi technology to provide high-speed data communication using visible light spectrum and to replace Wi-Fi for Internet access: Discussed in detail in another chapter
- Quantum communication to establish entanglement over thousands of kilometers and provide the backbone for large-scale, long-distance quantum key distribution: For secure Internet connectivity (discussed in detail in another chapter)
- Project Loon: To provide Internet access to rural and remote areas using high altitude balloons (discussed in detail in another chapter)
- Facebook's solar-powered Internet drone and laser technology: Unmanned aircraft (Aquila) to help to blanket the planet with Internet connectivity (discussed in another chapter)
- Quantum router for future quantum Internet: To use a quantum control signal to determine the route of a quantum data signal, experimental demonstration [25], and a new quantum circuit [26] for communication networks to transmit quantum information over long distances

References

[1] T. Peyronel, K.J. Quirk, S.C. Wang, T.G. Tiecke, Luminescent detector for free-space optical communication, Optica 3 (7) (July 2016).

[2] C. Lacava, M.A. Ettabib, P. Petropoulos, Nonlinear silicon photonic signal processing devices for future optical networks, Applied Sciences 7 (2017) 103. https://doi.org/10.3390/app7010103.

[3] The Golden Link — Plasmonic Device Offers Broadband Modulation to Optical Links at 100 Gbit/s.

[4] S. Schuler, D. Schall, D. Neumaier, L. Dobusch, O. Bethge, B. Schwarz, M. Krall, T. Mueller, Graphene flagship/news/high performance graphene photodetectors set peed record, Nano Letters 16 (December 13, 2016) 7107.

[5] Z. Xie, T. Lei, F. Li, H. Qiu, Z. Zhang, H. Wang, C. Min, L. Du, Z. Li ans, X. Yuan, Ultra-broadband on-chip twisted light emitter for optical communications, Light: Science and Applications 7 (2018) 18001. https://doi.org/10.1038/lsa.2018.1, published online 20 April 2018.

[6] Chalmers University project, Noiseless Phase-sensitive Optical Amplifiers and Their Applications, Friday 15 Sep 2017. Published: Fri 29 Jan 2016.

[7] Global Information, Inc, Market Research Report: 4oo G Optical Transceivers Market Shares, Strategies, and Forecasts, WinterGreen Research, Inc, July 14, 2017.

[8] Patent number: 9921379, Date of Patent: March 20, 2018, Patent Publication Number: 20170261708, Assignee: INPHI Corporation (Santa Clara, CA) Inventors: Liang Ding (Singapore), Radhakrishnan L. Nagarajan (Santa Clara, CA), Roberto Coccioli(Westlake Village, CA).

[9] R. Keil, M. Heinrich, F. Dreisow, T. Pertsch, A. Tunnermann, S. Nolte, D.N. Christodoulides, A. Szameit, All-optical routing and switching for three-dimensional photonic circuitry, Scientific Reports 1 (September 15, 2011) 94, https://doi.org/10.1038/srep.00094.

[10] K. Sato, H. Hasegawa, T. Niwa, T. Watanabe, A large-scale wavelength routing optical switch for data center networks, IEEE Communications Magazine (September 2013).

[11] G. Fan, R. Orobtchouk, B. Han, Y. Li, H. Li, 8 x 8 wavelength router of optical network on chip, Optics Express 25 (20) (October 2, 2017) 23677.

[12] H. Jia, T.Z.X. Fu, J. Ding, L. Zhang, L. Yang, Four-port mode-selective silicon optical router for on-chip optical interconnect, Optics Express 26 (8) (April 16, 2018) 9740.

[13] A. Sevincer, M. Bilgi, M. Yuksel, N. Pala, Prototyping multi-transceiver free-space optical communication structures, in: Conference: Communications (ICC), 2010 IEEE International Conference, June 2010. https://doi.org/10.1109/ICC.2010.5501774.

[14] On-chip light sources for ultimately integrated silicon photonics, PIC Magazine (4) (March 2017).

[15] N. Maring, P. Farrera, K. Kutluer, M. Mazzera, G. Heinze, H.de Riedmatten, Photonic quantum state transfer between a cold atomic gas and a crystal, Nature 551 (7681) (2017) 485. https://doi.org/10.1038/nature24468.

[16] Glenda Zafir Gonzalez Diaz, Energy saving solutions for integrated optical-wireless access networks, in: Networking and Internet Architecture [cs.NI]. Institut National des Télécommunications, 2015 (English).

[17] ARC Centre of Excellence for Ultrahigh bandwidth Devices for Optical Systems (CUDOS), Taking technology to the next level: researchers develop new platform making next-generation electronic devices more advanced, ScienceDaily (July 26, 2017). www.sciencedaily.com/releases/2017/07/170726103010.htm.

[18] B. Morrison, A. Casas-Bedoya, G. Ren, K. Vu, Y. Liu, A. Zarifi, T.G. Nguyen, D.Y. Choi, D. Marpaung, S.J. Madden, A. Mitchell, B.J. Eggleton, Compact Brillouin devices through hybrid integration on silicon, Optica 4 (8) (2017) 847. https://doi.org/10.1364/OPTICA.4.000847.

[19] Meet the Small 360 Camera Module That Will Fit Into Phones, Engadget 06.29.17. https://www.engadget.com/2017/06/29/ht-optical-360-camera-module-smartphones/.

[20] M. Haghi, K. Thurow, R. Stoll, Wearable devices in medical internet of things: scientific research and CommerciallyAvailable devices, Healthcare Informatics Research 23 (1) (January 2017) 4–15, published online 2017 Jan 31.

[21] H. Nishi, T. Tsuchizawa, M. Ono, M. Notomi, T. Yamamoto, S. Matsuo, Deep-subwavelength Si-core plasmonic waveguide monolithically integrated with Si photonic waveguide, in: Proc. of the Conference on Lasers and Electro-optics, FF1B.6, San Jose, CA, USA, June 2016.

[22] K. Watanabe, Y. Kurat, T. Hiraki and H. Nishi, Recent progress in optical waveguide technologies enabling integration of high-density compact photonics, NTT Technical Review. https://www.ntt-review.jp/archive/ntttechnical.php?contents=ntr201701fa2_s.html.

[23] Website: https://www.yahoo.com/news/quantum-random-number-generator-never-201700053.htm.

[24] DLR News, May 10, 2018. http://www.dlr.de/dlr/en/desktopdefault.aspx/tabid-10081/151_read-27323/#/gallery/30516.

[25] X.-Y. Chang, Y.-X. Wang, C. Zu, K. Liu, L.-M. Duan, Experimental Demonstration of an Entanglement-based Quantum Router, August 2013 arXiv:1207.7265 [quant-ph).

[26] B.K. Behera, T. Reza, A. Gupta, P.K. Panigrahi, Designing Quantum Router in IBM Quantum Computer, March 17, 2018 arXiv:1803.06530v1 [quant-ph].

Further Reading

[1] ICFO-The Institute of Photonic Sciences, Quantum internet goes hybrid, ScienceDaily 22 (November 2017). www.sciencedaily.com/releases/2017/11/171122131435.htm.

CHAPTER 11

Conclusions and Discussion

All-optical concepts to establish global Internet connectivity and access at high speed are described in this book, covering all aspects of fundamentals of free-space optical communications and high bandwidth fiber-optics backbones to design optical links suitable for various scenarios in a number of applications. The Internet is now really part of our social fabric and is essential to how we connect, communicate, share, and collaborate. Digital technologies, information and communication technologies, and broadband global connectivity, when combined, offer various opportunities to improve people's lives in worldwide societies by exchanging and sharing information. Basics of achieving worldwide broadband Internet connectivity independent of terrestrial limitations are explained in detail with technology development in mind. As the world becomes increasingly digital, the concepts developed in a number of chapters provide potential technology solutions for affordable broadband within reach of people worldwide. Real-time accessing of information is helping to transform various sectors ranging from healthcare, education, business, and many others. Global Internet connectivity will make it possible to collect and analyze larger and more complex kinds of data. The fundamentals of all-optical Internet connectivity and the associated communication technologies described in this book with an emphasis on developing innovative architectures and design of optical communication link systems are based on the concepts and demonstrations of some of the following areas: demonstration of low earth orbit satellite-to-ground laser communications; optical antenna; all-optical small-cell solution; constellation of satellites, microsatellites, and CubeSat for establishing Internet access to remote locations via drones, balloons, and high-altitude platforms; Internet of Things to integrate with 5G technology; Li-Fi—based hotspots to replace Wi-Fi; chip-based compact optical devices such as optical routers and optical switching to be integrated with optical network systems for Internet access by smart phone; and quantum communications using quantum entanglement for achieving the most secure quantum Internet; just to name a few.

Other current advanced research relevant to establishing all-optical connectivity for future optical communications includes the following: *fast Luminescent detector, nonlinear silicon photonics, high-speed photonics components, plasmonic*

Optical Wireless Communications for Broadband Global Internet Connectivity.
https://doi.org/10.1016/B978-0-12-813365-1.00011-4

devices, graphene-based technologies, and on-chip-twisted light emitter. The concepts of the technology proposed in this book are the first step toward accomplishing the ultimate goal of providing Internet access for all. The next step in our development is to apply the concepts and the proposed optical link schemes to build true all-optical global Internet connectivity systems to provide Internet access for all *anytime, anywhere*.

Index

Note: 'Page numbers followed by "f" indicate figures and "t" indicate tables.'

G

H

M

N

O

P

AT/1015189/19-12-18

CPI Antony Rowe
Chippenham, UK
2018-10-25 10:55